A MARITIME REGIME FOR NORTH-EAST ASIA

A MARITIME REGIME
FOR NORTH-EAST
ASIA

Mark J. Valencia

HONG KONG
OXFORD UNIVERSITY PRESS
OXFORD NEW YORK
1996

Oxford University Press

Oxford New York
Athens Auckland Bangkok Bogota Bombay
Buenos Aires Calcutta Cape Town Dar es Salaam Delhi
Florence Hong Kong Istanbul Karachi
Kuala Lumpur Madras Madrid Melbourne
Mexico City Nairobi Paris Singapore
Taipei Tokyo Toronto

and associated companies in
Berlin Ibadan

Oxford is a trade mark of Oxford University Press

First published 1996

This impression (lowest digit)
1 3 5 7 9 10 8 6 4 2

Published in the United States
by Oxford University Press, New York

© Oxford University Press 1996

All rights reserved. No part of this publication may be reproduced,
stored in a retrieval system, or transmitted, in any form or by any means
without the prior permission in writing of Oxford University Press (China) Ltd.
Within Hong Kong, exceptions are allowed in respect of any fair dealing for the
purpose of research or private study, or criticism or review, as permitted
under the Copyright Ordinance currently in force. Enquiries concerning
reproduction outside these terms and in other countries should be sent to,
Oxford University Press (China) Ltd at the address below

This book is sold subject to the condition that it shall not, by way
of trade or otherwise, be lent, re-sold, hired out or otherwise circulated
without the publisher's prior consent in any form of binding or cover
other than that in which it is published and without a similar condition
including this condition being imposed on the subsequent purchaser

British Library Cataloguing in Publication Data
available

Library of Congress Cataloging-in-Publication Data
Valencia, Mark J.
A maritime regime for north-east Asia/Mark J. Valencia.
p. cm.
Includes bibliographical references and index.
ISBN 0-19-587595-8 (hb: alk. paper)
1. Maritime law—North Pacific Ocean. 2. Marine resources—North
Pacific Ocean. 3. Marine pollution—North Pacific Ocean. 4. Asian
cooperation. I. Title.
Jx4411.V35 1996
341.4'5'095—dc20 95-51770 CIP

Printed in Hong Kong
Published by Oxford University Press (China) Ltd
18/F Warwick House, Taikoo Place, 979 King's Road, Quarry, Hong Kong

This small contribution to regional peace is dedicated to the people of the region—and to my family: Shabariah, Aishah, and Bennett.

Acknowledgements

I wish to acknowledge the support of the East–West Center for the research that culminated in this work. I am particularly indebted to Center Vice-President Bruce Koppel and to International Politics and Economics Program Director Charles Morrison for their encouragement and confidence in me, and for providing me with the intellectual and bureaucratic 'space' to work in peace. I also acknowledge the invaluable assistance of Zha Daojiong in the analysis of the national interests of China/Taiwan, and of Noel Ludwig in the analysis of the national interests of Japan and Russia. Many people read portions of the initial draft and offered useful suggestions for its improvement, including Tsuneo Akaha, Douglas Ancona, Shiro Chikuni, Moritaka Hayashi, Vladimir Ivanov, Igor Kolossovsky, Ma Zhongsi, Song Yann-huei, and Kunio Yonezawa. Ann Takayesu typed the manuscript in her usual highly efficient manner.

Mark J. Valencia
Honolulu, Hawaii
December 1995

Table of Contents

1. **Introduction** — 1
   - The Political Context — 1
   - Organization of the Book — 13
2. **Regime Theory and Maritime Regime Building** — 17
   - General Principles, Definitions, Context, and Demand for Regimes — 17
   - Origin of Regimes — 19
   - The Demand for Regimes — 24
   - Designing Institutions — 26
   - Resource and Environment Regimes — 28
   - Criteria for Evaluation of Regimes — 35
   - Regime Formation and Change — 37
   - Marine Policy Regimes — 39
   - Approaches to Regional Marine Policy Cooperation — 41
   - Objectives of Maritime Regimes — 42
   - The Global Maritime Scene — 46
   - Marine Regionalism: Lessons Learnt — 47
   - A Model Maritime Regime — 58
3. **Marine Regionalism in North-East Asia: The Context** — 68
   - The Natural Environmental Setting — 68
   - The Political Context — 80
   - Jurisdictional Claims and Disputes — 83
   - Integrative and Disintegrative Forces — 86
   - Research Questions — 92
4. **National Interests in Maritime Regime Building** — 96
   - China and Taiwan — 96
   - Japan — 105
   - North Korea — 121
   - South Korea — 132
   - Russia — 144
5. **Pollution and Environmental Protection** — 175
   - Environmental Problems and Issues — 175
   - Institutional Issues — 178
   - The Nuclear Waste-Dumping Controversy — 183
   - Existing Regimes and their Inadequacies — 186
   - Current Initiatives — 197

	Recent Developments	207
	Problems and Inadequacies of Existing Regimes	219
	Transnational Environmental Protection Approaches	223
	The Ideal Marine Environmental Protection Regime	229
6.	**Fisheries**	244
	Introduction	244
	The Problems and Issues Defined	248
	Existing International Regimes	252
	The Changing Situation and the Inadequacies of the Existing Regimes	262
	Alternative Regimes	270
	The Example of the Yellow Sea	277
	Suggestions and Next Steps	280
	Elements of a Model Regime	292
7.	**Conclusions**	301

Appendix I: An Ocean Zoning System for the Yellow Sea	311
Appendix II: Specific Measures for Protection of Fishing Resources in the Yellow Sea	315
Glossary	321
Index	324

List of Tables

Table 2.1	Possible Advantages and Disadvantages of Economic and Political Cooperation in the Marine Arena by North-East Asian Nations	45
Table 4.1	North Korean Environmental Treaty Commitments or International Participation, mid-1992	128
Table 4.2	Governance of Korean Marine Policy, 1993	136
Table 5.1	IMO Pollution Convention Signatures	187
Table 5.2	Water Quality Standards in North-East Asia	192
Table 5.3	Effluent Standards in North-East Asia	195
Table 5.4	Current North-East Asian Regional Initiatives in Environmental Management	199
Table 6.1	Regulations Inside the Joint Areas Set by the Japan-South Korea Fisheries Treaty	256
Table 6.2	Incidents of South Korean Illegal Fishing in Japanese Waters	257
Table 6.3	History of the Japan–South Korea Agreement	258
Table 6.4	Incidents in the East China Sea, 1992–1993	268
Table II.1	The Proposed Fisheries Protection Zones	317

List of Figures

Figure 2.1	Maritime North-East Asia	50
Figure 3.1	The Sea of Japan Basin	69
Figure 3.2	Surface Circulation in the Sea of Japan: Summer	70
Figure 3.3	Surface Circulation in the Sea of Japan: Winter	71
Figure 3.4	The Yellow Sea Basin	75
Figure 3.5	Surface Circulation in the Yellow Sea	77
Figure 3.6	Jurisdictional Claims in North-East Asian Seas	84
Figure 5.1	Projected Trajectory of Oil Spilled in South Korean Waters in Winter	224
Figure 5.2	Projected Trajectory of Oil Spilled in Disputed Waters in Summer	225
Figure 5.3	Proposed Ocean Zoning in the Yellow Sea	228
Figure 5.4	Fish Migrations: Demersal Species	231
Figure 6.1	Japan–Republic of Korea Fisheries Regulation Zones	254
Figure 6.2	Japan–China Fishery Regulation Zones	259
Figure 6.3	FAO Reporting Area for the North-West Pacific	283
Figure 6.4	Proposed Joint Fishing Zone for North and South Korea	291

1 Introduction

This book is concerned with multilateral maritime regime building in North-East Asia. Maritime issues are of course embedded in the broader milieu of international relations in the region. Thus the first section sketches this broader political context with a focus on incipient regionalism, and in particular on the potentially critical role maritime issues and regimes may play in regional confidence building.

The Political Context

North-East Asia[1] has long been a security complex[2] involving Russia, China, Japan, and the United States. Bilateral and regional issues have global implications and vice versa. During the Cold War, international relations in the region were heavily influenced by the Soviet–United States dynamic and thus were almost indistinguishable from the global system.[3] However, with the end of the Cold War, the layering of conflict involving outside powers is peeling away to reveal an emerging Asia-Pacific system. The future central dynamic of this system is likely to depend on relations among the North-East Asian countries themselves. In the past century these relations have been characterized by cyclical patterns of amity and enmity, frequent tension between the strongest powers in the region, and resultant attempts to forge alliances with the lesser powers. Moreover, North-East Asian countries have generally operated on the basis of 'worst case' scenarios—the large powers were motivated by a fear of being isolated by the other powers, and the lesser powers most feared a hegemonic alliance of the two strongest powers.[4] Intra-regional conflict has tended to be along the boundaries between the major powers—the Korean Peninsula, Taiwan, Manchuria, Mongolia, and Sakhalin/the Kuriles. And the seas between these powers were a dangerous frontier as well.

We are currently witnessing a transformation of the political system in the region. As survival has ceased to be the prime concern of powerful states, their quest for relative gains has become less driven and consistent. Indeed some argue that their behaviour can best be understood in the context of international institutions that both constrain states and make their actions more predictable.[5] Most North-East Asian governments are now more motivated towards maximizing wealth than controlling territory, and their increasing economic interdependence makes outright conflict too costly. With the development of political multipolarity and the abandonment of Stalinist economic models, economic relationships have begun to develop a more 'natural' pattern. These economic relations have tended to concentrate in those boundary areas where the economies of adjacent regions obviously complement each other, and comprise 'natural economic territories' (NETs)[6]: southern China, the Yellow Sea Rim, the Tumen River area, and the Sea of Japan Rim.

The dominance of the United States in the region throughout the Cold War obviated the need for multilateral co-operation in North-East Asia. Now the trans-Pacific economic axis, which was so prominent in the Cold War era, is gradually being modified by more multidirectional intra-Asian relationships.[7] This multidirectional pattern implies a more diversified set of co-operative and conflictual economic relations in much of the North Pacific, creating a need for rules, codes of conduct, and harmonization of domestic practices which, in turn, affect international transactions and, thus, regional institutions. As economic relations in the region increase, co-operation among the region's countries can progress from policy consultation, or discussion of common or mutual policy problems, to policy harmonization, thus reducing and eliminating contradictory and conflicting national policies. Already, economic interaction across ideological and political boundaries is creating a 'soft' regionalism in North-East Asia, one lacking organizational structure but which is accepted and even encouraged by governments.[8] This cross-border economic exchange and the emergence of NETs raises questions of jurisdiction and control and impinges on national sovereignty. Nationalism is being rekindled and is competing with internationalism and regionalism in the formulation and implementation of national policies. The interplay between localism, nationalism and internationalism will be a major theme of political relations in North-East Asia for several decades to come.

Within this overall dynamic, some traditional concerns are likely to persist.[9] China and the two Koreas worry about an independent, militarily resurgent Japan. Japan and South Korea worry about a

China that can combine economic development and political will to attain genuine military power and the ability to project it. And all North-East Asian states wonder if, when, and how Russia will re-emerge as a major power and what that will mean for them.

The nature of Sino-Japanese relations stands as another major question for the future of the region. The nexus between China and Japan has the potential to become the central axis, not only of North-East Asia, but of the larger North Pacific or Asia-Pacific regions as well. Strategic co-operation—rather than conflict—between these two powerful countries is critical to the future of Asia.[10] They need to work out a new relationship with each other more or less by themselves, i.e., in a regional environment in which ideology and other powers are not major factors. Thus the relationship between them seems likely to be a product of the redefinition of their own respective roles in the region and the world at large.

One positive scenario would be a China and Japan that have assumed positions of influence and responsibility in global affairs, with their economic systems interacting in a mutually beneficial manner, and their political and cultural values causing no offence to each other or to the rest of the world. In this positive scenario, China and Japan would also develop effective modes of governing their own countries, and co-operating with each other on bilateral and regional issues, such as ocean management.

In a negative scenario, China and Japan remain insecure in their places in the world and in their relations with each other. Nationalism would grow in both countries as a reaction to increased economic interaction and to outside pressures to change their systems. Nationalism in China may also be promoted as a glue to hold its society together, while in Japan it may stem from a longer-term, deeper economic malaise. As nationalism grows, so would military budgets and capabilities. Doubts about the future of the US presence in the region would fuel rivalry as to which country, China or Japan, would fill the vacuum. Historical and transborder issues would be politicized, with maritime frontiers in particular becoming dangerous 'no man's lands,' and territorial and maritime disputes, such as the Senkakus/Diaoyutais, or overlapping continental shelf claims, re-emerging as 'hot spots.'

These scenarios have been painted in rather stark fashion, and the future of Sino-Japanese relations is likely to be characterized by elements both of co-operation and of tension or conflict. The contrast, however, illustrates that this relationship can have strong positive or negative effects on the overall political environment and the development of regional regimes in North-East Asia and the North Pacific.

The Korean Peninsula, as it has been for so much of this century, remains a potential locus of rivalry among the North Pacific's large powers. But because of the normalization of relations of both Russia and China with South Korea, North Korea no longer constitutes an impediment to most forms of regional political and economic relations. And, before the nuclear issue complicated matters, Pyongyang itself had extended tentative feelers towards greater regional economic co-operation, including participation in the Tumen River Area Development Project and its meetings, in international economic fora, and in environmental protection efforts, especially in the marine sphere.[11] Although the nuclear issue has put on hold North Korea's participation in broader regional co-operation efforts, the resolution of the issue to the satisfaction of all concerned will likely pave the way for including North Korea in such co-operative efforts.

Regardless of the final outcome of these changes, confrontation is turning to dialogue and tensions are relaxing:

- Japan and Russia agreed in 1991 to institute a regular security dialogue and negotiated an incidents at sea agreement (INCSEA) which was signed in October 1993.
- In 1992, Japan and South Korea agreed to exchange military observers as a confidence-building measure. In June 1994, Japanese Defense Agency officials announced that plans for naval exchanges between the Japanese and South Korean navies were underway.
- There have also been unofficial reports of an INCSEA negotiation between Japan and South Korea.
- In 1992, Russia and South Korea exchanged a Memorandum of Understanding seeking to 'substantially promote exchanges and co-operation between respective defence ministries.' Russian ships visited South Korea in August 1993 and Russian observers attended the 1992 Team Spirit US-ROK exercise.
- Following dialogue between the Russian and Chinese chiefs of staff in 1993, Russian warships visited Qingdao in China in August 1993, and in May 1994, Chinese warships visited Vladivostok in Russia.

- Since 1989 Sino-Russian border negotiations have focused on reducing troop numbers and demarcating borders; in 1992, both states renounced the use of force against each other.
- In 1994, Japan and China established a regular dialogue on bilateral and regional security issues. In June 1993, there were reports that Japan and China would 'study ways to establish a joint military force for participation in United Nations peacekeeping activities.'
- The US Pacific Command and Russian Far East command, in 1993, established a program 'of frequent military-to-military contacts at all levels.' The US Pacific Fleet visited Vladivostok in 1990. Earlier regional CBMs between the two countries include the 1989 Bering Straits Commission, and with Japan, the 1988 Regional Air Safety Agreement.
- In June 1994, Japan for the first time took part in the annual multi-nation RIMPAC exercise on the same side as South Korea. In the same month the Japanese Defense Agency announced that plans for joint training between the Japanese and South Korean navies were being implemented.
- In September 1994, Russia and Japan held their first-ever joint search and rescue naval exercise.
- In May 1993, the South Korean goverment suggested that the creation of 'a Korean peninsula-centred version of a "mini-CSCE" for North-East Asia should be considered in tandem with an expanded ASEANPMC security dialogue.'
- Regular dialogue between a number of the Asia Pacific navies was instituted in 1988 by the Royal Australian Navy with the establishment of the Western Pacific Naval Symposium (WPNS). Participants at the WPNS Workshop in Sydney in 1992 agreed to jointly develop a 'Maritime Information Exchange Directory'.

There is one official-level annual dialogue forum involving North-East Asia which focuses on security issues. It involves policy planning staff from the United States, Canada, Japan, Korea, and Australia. And numerous non-official and Track Two meetings on security are now taking place in the region. Almost all seek to encourage security dialogue at the official level. Among the established dialogue fora are: Canada's North Pacific Cooperative Security Dialogue; the North-East Asian Cooperation Dialogue run by the University of California, San Diego's IGCC institute; the Trilateral Forum involving, Japan, Russia and the United States;

the Stanford University workshops involving Japan, Russia, and the United States and possibly in the future, China, and the CSIS/Pacific Forum meetings.[12]

The time is right to propose and implement concrete steps that will contribute to this process. I argue that transnational ocean resource issues and conflicts can be turned into opportunities to build confidence, dampen frontier tension, and improve relations in this region that is so critical to world peace and prosperity.

Incipient Regionalism

North-East Asia and the North Pacific are almost unique for their lack of regional institutions. Bilateralism dominates both political and economic relations. Buzan notes that Asia is remarkable for 'its combination of several quite highly industrialized societies, with a regional international society so impoverished in its development that it compares poorly with even Africa and the Middle East.'[13] This impoverishment reflects the conflicts among the governments in the region, particularly, between the divided countries, which create enormous obstacles to the establishment of any regional institution.

Although solutions to divided states and regional problems are primarily the responsibility of the parties immediately concerned, they cannot be resolved solely by those parties since external states are also involved, directly or indirectly.[14] Thus solutions to North-East Asian regional issues must be sought through a series of concentric arcs: the immediate parties, the vitally interested external nations, and the regional or international organizations that can exercise influence or provide assistance.

The end of the Cold War has not necessarily made the prospects any brighter for a stable regional order. Instead, the two major powers in the region—China and Japan—are suspected by some of the other powers within it, of aspiring to impose a post-American hegemonic stability in the region. But both presently lack the 'comprehensive power' capacity and status to do so.[15] The question, then, is whether North-East Asian countries can embrace an institutional mechanism for regional stability. Since there is no recent example in the region of such a regime, I will turn to neorealist theory for intellectual guidance.

Neorealism as a theory gained credence in the 1970s—particularly in America—as an intellectual response to the realization that the dominant nation-state, the United States, could no longer afford unilaterally to manage world order. This realization was the stimulus for proposals for multilateral approaches to international energy, financial, and environmental protection regimes, through which the United States could retain strong influence. Indeed, neorealism has been criticized as charting new routes for the advancement of American hegemony.[16] Nonetheless, political trends portend a continual relative decline of American hegemony in North-East Asia.

This decline of a Pax Americana is relevant to the study of maritime issues in North-East Asia. During the Cold War, North-East Asia achieved some stability as a product of American hegemony.[17] With the end of the Cold War, that hegemonic stability is increasingly being called into question. At the same time, the dynamics of North-East Asian regional political-economic competition calls for more concerted regional efforts to deal with the management of critical natural resources, particularly those shared by all the North-East Asian nations, such as the marine environment and its living resources. And the absence of an external hegemony in maritime management[18] satisfies an important condition for the application of multilateralism in a regional context. Relevant here is the Clinton administration's new security strategy for the East Asia-Pacific region which emphasizes multilateralism, inclusiveness, regional co-operation, co-operative security and a separate subregional security dialogue for North-East Asia.[19]

There are several other factors favouring multilateralism. As the global system becomes more interdependent, multilateral institutions have increased in number to deal with the growing number of problems which states cannot resolve unilaterally or bilaterally. Multilateral fora provide small and medium powers with opportunities for coalition-building among themselves to balance the influence of larger powers within the collectivity. Multilateral institutions are thus ideal fora for the practice of 'middle power' or 'niche' diplomacy in setting regional or global agendas. Regular convening of multilateral fora helps create a 'habit of co-operation' among participants. And in multilateral relationships, issue-linkage can create opportunities for creative problem solving not available in a bilateral relationship. Nevertheless, multilateral security co-operation may be extremely difficult to achieve in North-East Asia. Thus an *a la carte* approach focusing on specific issues is probably more appropriate.[20] Indeed, present political trends provide an unequalled opportunity to think boldly, and to be innovative about solutions to problems of international relations in general, and about maritime regime building in North-East Asia in particular.

Regionalism is also becoming more attractive to developing states because of their growing political maturity, and the perceived potential of regionalism to promote their economic development and to mitigate their disadvantaged position in the international system.[21] Some argue that multilateral norms and institutions can make significant contributions towards stabilizing the peaceful transformation of the international system, and that they are likely to become increasingly important in the management of change at the regional level.[22] UN Secretary-General Boutros Boutros-Ghali, for example, argues that regional arrangements can render great service by contributing to a deeper sense of participation, consensus, and democratization in international affairs.[23] However, there will be no simple organizational solution to the security, economic, and political opportunities and challenges of the vast and varied Asia-Pacific region. Multiple organizations will form; there will be overlapping layers to economic co-operation and security collaboration; and subregional or *ad hoc* arrangements will proliferate.[24]

Already apparent is a gradual development of a thin net of regional institutions covering the region in the economic, environmental, and to a lesser degree, the political arenas, but within a broader Asia-Pacific framework. The world, and the Asia-Pacific region in particular, are definitely witnessing a renewed interest in multilateralism.[25] Economically, the principal broad-gauged quasi-governmental institution is the Pacific Economic Co-operation Council (PECC), which grew out of a 1980 conference in Canberra, Australia. The intergovernmental Asia-Pacific Economic Co-operation Forum (APEC) followed in 1989 and consists of annual ministerial meetings and 10 working groups. The heads of government of the APEC countries met in a 'leadership conference' alongside the APEC ministerial meeting in Seattle in November 1993 and again in 1995 in Osaka.[26] Moreover APEC has spawned calls from within it for closer triangular co-operation between South Korea, China, and Japan, particularly on environment and resource issues.[27] PECC includes among its membership all the

North Pacific economies (except Mongolia and North Korea), but the APEC does not include Russia although it has applied to join.[28]

In the security arena, the outstanding recent event has been the decision of the Association of South-East Asian Nations (ASEAN), taken at the ASEAN Post-Ministerial Conference held between 23–24 July 1993, to sponsor an 18-member ASEAN Regional Forum (ARF) to discuss Asia-Pacific region-wide security issues.[29] The ARF includes all the great powers of the North Pacific, but not Mongolia, North Korea, or Taiwan. The forum first met in Bangkok in July 1994. At this meeting South Korea tabled a proposal for a North-East Asia Security Dialogue.[30] Also, an informal or Track-Two approach has since emerged—the Council for Security Cooperation in the Asia Pacific (CSCAP). This is a multilateral, non-governmental organization dedicated to promoting security dialogue in the Asia-Pacific region. The Council supports, among other things, multilateral approaches to regional conflicts and the creation of multilateral mechanisms to promote dialogue on effective governance. In this context, a working group on maritime security has been formed.[31]

Indeed, the development of *both* formal and informal channels of communication has been a characteristic of Asia-Pacific co-operation, as well as an essential step in the process of institutionalization.[32] In this sense, the ASEAN approach to political understanding and co-operation may be quite relevant to North-East Asia.[33] Through ASEAN, habits of consultation and even a nascent feeling of regional identity were built up, increasing sensitivities within the governmental elites to each other's interests and developing some norms on conducting their relations with each other. The new Forum, supplemented by CSCAP as well as more frequent meetings of permanent officials, might build confidence in North-East Asia for similar but separate arrangements.

The growing acceptance of the notion of comprehensive security[34] is a positive development for maritime regime building. Comprehensive security implies that security should, and can, be achieved through a web of interdependence, including co-operation in economic development and scientific research and a general enhancement of human interactions. In this perspective, military might alone does not define security nor generate long term peace. Indeed, the failure to comply with basic standards of good neighbourliness, for example, preventing transnational pollution or notifying neighbours when it occurs, or carrying out transboundary environmental clean up and impact assessments, can cause significant tension.

The concept of 'comprehensive security' is gaining currency among policymakers in the region, and in Washington. President Clinton adopted the term 'economic security', signalling a gradual separation of economic interactions among nation-states from ideological confrontation. Others speak of post-Cold War Asia posing 'unconventional threats to America's national security,' meaning overpopulation, environmental degradation, the narcotics trade, and the quickening spread of AIDS. Indeed, the final draft of the Clinton administration's national security strategy places new emphasis on non-traditional foreign policy concerns including environmental concerns.

A corollary of both comprehensive security and increasing interdependence is co-operative security. The concept of co-operative security argues for the necessity of acknowledging the complex linkages between conventional and unconventional forms of security threats and thus focuses on creating 'habits of dialogue' and multilateral norms and instruments for building confidence, reducing threat, raising the cost and threshold of confrontation and the peaceful resolution of disputes. Co-operative security is based on three main ideas: security *with* one's neighbours as opposed to security against them; a broad interpretation of security threats to include, among others environmental degradation and resource access; and an emphasis on multilateral institutions and processes for managing regional issues and promoting habits of dialogue and co-operation.[35]

These trends are important because what is most worrying about potential conflict in North-East Asia is that it would take place in a region whose diversity and disputes have never been ameliorated by multilateral co-operation, and where security has always been defined by military might. A fundamental question for peace in the region may thus revolve around the prospects for multilateralism. This may be particularly so in the case of China, and to a lesser extent, North Korea. China's power and influence will obviously grow in the years ahead and this is itself a strong argument for involving China in emerging regional tasks and institutions.

Clearly then, the first step toward the peaceful settlement of international conflicts is the creation of a sense of international community.[36] The creation of such a community presupposes at least the mitigation and minimization of conflict, so that the interests and common needs shared by different nations outweigh the interests separating them. Common recognition that even a poor regime is better than none compels nations to collaborate to the extent of developing a minimally satisfactory solution. A functional approach can help the growth of positive and constructive common work and of common habits and interests, decreasing the significance of boundaries or conflicting claims by overlaying them with a natural growth of common activities and administrative agencies. The challenge for the region then is to develop a variety of multilateral arrangements that will demonstrate that a habit of dialogue and working together can build common—and eventually—co-operative security. In North-East Asia, continuity and persistence are critical to advancement of regionalism. Tactical learning—in which the behaviour of states towards co-operation is changed—must give way to complex learning in which values and beliefs about reaching goals through co-operation are changed.[37] While the establishment and operation of a functioning inclusive regional security dialogue appears unlikely at this particular moment, movement in this direction using innovative mechanisms is essential.[38]

Maritime issues are rising to the forefront of current regional security concerns.[39] The United Nations Convention of 1982 on the Law of the Sea has introduced new uncertainties and conflict points into the region, particularly in regard to Exclusive Economic Zones (EEZ) and continental shelf claims and boundaries. Many emerging regional security concerns such as piracy, pollution from oil spills, safety of sea lines of communication, illegal fishing, and exploitation of others' offshore resources are essentially maritime. The inclusion of issues like environmental protection, illegal activities at sea, and resource management and protection will necessitate acceptance of broader responsibilities and different priorities by military authorities, both for force structure development and their operations and training.[40] Together with the requirements for defence self-reliance and force modernization, these concerns are reflected in the significant maritime dimension of the current arms acquisition programs in the region: the maritime surveillance and intelligence collection systems, multi-role fighter aircraft with maritime attack capabilities, modern surface combatants, submarines, anti-ship missiles, naval electronic warfare systems, and mine warfare capabilities. Because some of these new systems have offensive capabilities, they can be seen as provocative, and thus destabilizing, by those countries that do not have them or lack the means to acquire them. Possession of these systems undoubtedly increases the risk of inadvertent escalation in time of conflict. It is therefore particularly important that regional mechanisms be instituted to address these maritime issues.

East Asian specialists list similar maritime problem areas for greater co-operation: piracy, smuggling, illegal immigration, transnational oil spills, incidents at sea, search and rescue, navigational safety, exchange of maritime information, illegal fishing, and management of resources in areas of overlapping claims.[41] These issues are all maritime safety problems of a civil, as opposed to a military, nature. The resulting proposals for maritime co-operation are formulated against no single adversary but rather, against common problems of crime, human depredation, pollution, and natural disaster. Progress on the harder issues may well depend on successful development of a softer, essentially civil, maritime safety regime. Successful co-operation in the marine realm can build the confidence necessary for initiatives in other spheres, and for the jump from tactical to complex learning. Such an approach has been proposed for North-East Asian seas, namely the Yellow/East China Seas and the Sea of Japan.[42]

Arguably, the most significant of the current proposed maritime confidence and security building measures is the concept of regional oceans management.[43] In practice managing North-East Asian seas will be especially complicated because they are surrounded by several countries that share the same historical and cultural background, but differ in internal political systems, external economic alignments, and levels of economic development. Indeed, relations among the states bordering these marine regions have been tenuous or even estranged for the past half century, resulting in a political environment that has inhibited multilateral ocean management programs.

Now—for the first time in two generations—North-East Asia has an opportunity for lasting peace. The end of the Cold War, the concomitant warming of relations in much of the region, the extension of maritime jurisdiction, and the coming into force of the Law of the Sea Convention in 1994[44] provide a narrow window of opportunity to forge a new order for regional seas before resurgent nationalism further complicates these issues. Needed now is a process—or first, a framework and a blueprint—for developing multilateral regional marine policy co-ordination.

Organization of the Book

That is what this study proposes to provide. It will delineate and analyse the transnational marine policy problems in the region, the alternative solutions, the realistic possibilities for co-operation in the resolution of these problems, the obstacles to be overcome, and the ways and means of doing so. In particular, it will involve the application of regime theory to the environment and fisheries sectors.

The next chapter focuses on relevant regime theory, describing general principles and definitions and the context, origin, and demand for regimes. It treats resource and environment regimes as a particular subset, and elaborates criteria for their evaluation and the factors in regime formation and change. The chapter goes on to review lessons learned elsewhere in environment and resource regime formation, and in marine policy regimes in particular.

Chapter 3 establishes the context for marine regionalism in North-East Asia. It begins with a treatment of the natural environmental setting including oceanography, geomorphology, and ecology of the Japan, Yellow, and East China Seas. The relevant parties that would be involved in regional action, the national jurisdictional claims and transnational disputes, and the integrative and dis-integrative forces influencing regionalism are identified and described. The chapter concludes with the research questions that guided the study.

After this discussion of the general regional context, national context and interests of the regional states—China and Taiwan, North Korea, South Korea, Russia, and Japan—regarding a regional maritime regime are delineated in Chapter 4. The advantages and disadvantages for each state of both a fisheries and a marine environmental protection regime are then analysed.

Chapter 5 focuses on a regime for marine environmental protection. It describes the status of marine pollution, outlines the problems and issues, reviews the existing regimes and their inadequacies, and culminates in an outline of an ideal environmental protection regime for North-East Asian seas. An appendix gives a hypothetical example of a sea use planning approach to environmental protection in the Yellow Sea.

Chapter 6 provides a similar treatment for fisheries, defining the problems and issues, and analysing the existing international regimes and their inadequacies. It goes on to present alternative regimes, including one designed specifically for the Yellow Sea, before concluding with the realistic possibilities, the necessary next steps, and the elements of a model regime for fisheries management in North-East Asia.

Chapter 7 summarizes the results of the study and recommends a course of action.

Notes

1. For the purposes of this study, North-East Asia includes China, Japan, North Korea, South Korea, Taiwan, and Russia. The marine regions include the Yellow Sea and the East China Sea taken as a unit, and the Sea of Japan.
2. Barry Buzan, *People, States and Fear: An Agenda for International Security Studies in the Post-Cold War Era*, Hemel Hempstead: Havester Wheatsheaf, 1991, pp. 105–16; 125–6.
3. Charles E. Morrison, 'Changing patterns of international relations in the North Pacific', Paper presented to the Workshop on the Russian Far East and the North Pacific, East–West Center, Honolulu, 1993.
4. Michel Oksenberg, 'The East Asian Quadrangle', as amplified in a 9 August 1993 oral presentation at the 'Asia in Transition' workshop in Seattle.
5. Robert Keohane, 'Multilateralism: an agenda for research', *International Journal*, v. 65, 4 (Autumn 1990): 733–6.
6. Robert A. Scalapino, 'The Post Cold War Asia-Pacific Security Order: Conflict or Cooperation?', Paper presented at the Conference on Economic and Security Cooperation in the Asia-Pacific: Agenda for the 1990s, Canberra, 28–30 July 1993, p. 16.
7. Morrison, 'Changing patterns of international relations in the North Pacific'.
8. Robert A. Scalapino, 'Northeast Asia: prospects for cooperation', *The Pacific Review*, 5, 2 (1992): 102.
9. Morrison, 'Changing patterns of international relations in the North Pacific'.
10. Zbigniew Brzezinski, Lecture on China at the East–West Center, Honolulu, Hawaii, 24 May 1994.
11. Mark J. Valencia, 'Preparing for the best: involving North Korea in the new Pacific Community', *Journal of Northeast Asian Studies*, 5, 13, no. 1 (1995): 65.
12. Andrew Mack, 'Multilateral security dialogue for Northeast Asia: problems and prospects', in *Forming Multilateral Security in Northeast Asia: Opportunities, Constraints, and Options*, Seoul: Kim Dae Jung Foundation, 1995, pp. 27–9.
13. Barry Buzan, 'The Post Cold War Asia-Pacific security order: conflict or cooperation?', Paper presented at the Conference on Economic and Security Cooperation in the Asia-Pacific: Agenda for the 1990s, Canberra, 28–30 July 1993, p. 16.
14. Scalapino, 'Northeast Asia: prospects for cooperation', p. 111.

15. Yoichi Funabashi, Michel Oksenberg, and Hernrich Weiss, *An Emerging China in a World of Interdependence*, Report to the Trilateral Commission, New York, Paris and Tokyo, 1994.
16. Richard Ashley, 'The poverty of neorealism', in Robert O. Keohane (ed.), *Neorealism and its Critics*, New York: Columbia University Press, 1986, pp. 255–300.
17. Bruce Cumings, 'The wicked witch of the West is dead, long live the wicked witch of the East', in Michael I. Hogan (ed.), *The End of the Cold War: Its Meaning and Implications*, New York: Cambridge University Press, 1992, pp. 88–89.
18. Stephan Haggard and Beth A. Simmons, 'Theories of international regimes', *International Organization*, 41, 3 (Summer 1987): 500–504.
19. Office of International Security Affairs, Department of Defense, United States Security Strategy for the East Asia-Pacific Region, Department of Defense, February 1995.
20. Gerald Segal, 'Northeast Asia: common security or a la carte?' *International Affairs*, 5. 67, no. 14 (1991): 765.
21. Paul Taylor, 'Regionalism: the thought and the deed', in A. J. R. Groom and Paul Taylor, (eds.), *Framework for International Relations*, New York: St. Martins Press, 1990, pp. 151–71.
22. John Gerard Ruggie, 'Multilateralism: the anatomy of an institution', *International Organization*, 46, 3 (Summer 1992): 561.
23. Boutros Boutros-Ghali, *An Agenda for Peace: Preventive Diplomacy, Peacemaking, and Peace-Keeping*, New York: United Nations, 1992, pp. 36–7.
24. Richard Solomon, 'Asian architecture: the US in the Asia-Pacific Community', *Harvard International Review*, 16, 2 (Spring 1994): 29.
25. Keohane, 'Multilateralism: an agenda for research'.
26. Richard Baker, 'Clinton to Pacific Rim: shall we talk?', *Honolulu Advertiser*, 14 November 1993, p. B-2; Richard Baker; 'APEC confirms a US shift in policy to Asia-Pacific', *Financial Review*, 23 November 1993; Intelligence, *Far Eastern Economic Review*, 26 January 1995, p. 12.
27. Close ROK–PRC–Japan cooperation stressed. *FBIS-EAS-95-041*, 2 March 1995, p. 1.
28. 'Russia to join APEC forum', *Japan Times*, 19 March 1995, p. 1.
29. Frank Ching, 'Discussing regional security', *Far Eastern Economic Review*, 12 May 1994, p. 38.
30. Lee Chang Choon, 'The Beijing–Seoul–Tokyo triangle', *Far Eastern Economic Review*, 13 October 1994, p. 36.
31. *Pac Net*, 37 (Pacific Forum/CSIS, Honolulu): 18 November 1994.
32. Richard Higgott, 'Economic cooperation: theoretical opportunities and political constraints', *The Pacific Review*, 6, 2 (1993): 105; Vinod K. Aggarwal, 'Building international institutions in Asia-Pacific', *Asian Survey*, 33, 11 (November 1993): 1029–42.
33. Morrison, 'Changing patterns of international relations in the North Pacific'.
34. Robert W. Barnett, *Beyond War: Japan's Concept of Comprehensive National Security*, Washington: Pergamon, Brassey's, 1984; John Lancaster and Barton Gellman, 'National security strategy paper arouses Pentagon, State Department debate', *Washington Post*, 3 March 1994, p. A-14.
35. Amitav Acharya, David Dewitt and Carolina Hernandez, 'Sustainable development and security in Southeast Asia: a concept paper'. Draft prepared for the first consultative meeting on Development and Security in Southeast Asia. Manila, The Philippines, 1–3 June 1995, p. 21.
36. Hans Morgenthau and Kenneth W. Thompson, *Politics Among Nations: The Struggle for Power and Peace*, New York: Knopt, 1985, p. 559; Robert Scalapino, 'Back to the future', *Far Eastern Economic Review*, 26 May 1994, p. 38.

37. Higgott, 'Economic cooperation: theoretical opportunities and political constraints' 106–107.

38. Brian Job, 'The evolving security order of the North Pacific subregion: alternative models and prospects for subregionalism.' Draft paper presented to the Ninth Asia-Pacific Roundtable, Kuala Lumpur, 5–8 June 1995.

39. Desmond Ball, 'A new era in confidence building: the second-track process in the Asia/Pacific region', *Security Dialogue*, 25, 2 (1994): 164.

40. Jack McCaffrie and Sam Bateman, 'Maritime confidence and security building measures in Asia-Pacific: challenges, prospects and policy implications'. Paper presented to the first meeting of the CSCAP Working Group on Maritime Co-operation, Kuala Lumpur, June 1995, pp. 1, 5, 11.

41. David I. Hitchcock, Jr., 'East Asia's new security agenda', *The Washington Quarterly*, 17, 1, pp. 95, 96, 103; Paik Jin-Hyun, 'Strengthening maritime security in Northeast Asia', Paper presented to the 8th Asia-Pacific Roundtable on Confidence Building and Conflict Reduction in the Pacific, ASEAN Institute for Strategic and International Studies, Kuala Lumpur, 5–8 June 1994, pp. 1–17; Stanley B. Weeks, 'Law and order at sea: Pacific co-operation in dealing with piracy, drugs, and illegal migration.' Paper presented to the first meeting of the CSCAP Working Group on Maritime Co-operation, Kuala Lumpur, June 1995, pp. 1–15; and Charles Meconis and Stanley B. Weeks, 'Co-operative Maritime Security in the Asia-Pacific Region: a Strategic Arms Control Assessment Report, July 1995, pp. 75–80.

42. For background and specific proposals for co-operation in a variety of maritime sectors, See Joseph R. Morgan and Mark J. Valencia (eds.), *Atlas for Marine Policy in East Asian Seas*. Berkeley: University of California Press, 1992, p. 152; Mark J. Valencia, (ed.), *International Conference on the Sea of Japan*, Occasional Papers of the East–West Environment and Policy Institute, No. 10, p. 239; Mark J. Valencia, (ed.), *International Conference on the Yellow Sea*, Occasional Papers of the East–West Environment and Policy Institute, No. 3, p. 165; Mark J. Valencia, 'Sea of Japan: transnational marine resource issues and possible cooperative responses', *Marine Policy*, 14, 6, pp. 507–25; Mark J. Valencia, 'Northeast Asian perspectives on the security-enhancing role of CBMs' in *Confidence and Security-Building Measures in Asia*, New York: Department of Disarmament Affairs, United Nations, pp. 12–18; Mark J. Valencia, 'The Yellow Sea: transnational marine resource management issues', *Marine Policy*, 12, 4, 382–95.

43. McCaffrie and Bateman, 'Maritime confidence and security building measures in Asia-Pacific: Challenges, prospects and policy implications', p. 10.

44. Congressional Record, *Proceedings and Debates of the 103rd Congress*, Second Session, 140, 86 (30 June 1994); Steven Greenhouse, 'US, after negotiating changes, is set to sign pact on sea mining', *The New York Times*, 10 March 1994, p. A-13.

2 Regime Theory and Maritime Regime Building[1]

General Principles, Definitions, Context and Demand for Regimes

Global interdependence generates conflict. Thus to avoid conflict, governments must adjust their policies to one another. In other words, co-operation is absolutely necessary.[2] Co-operation is a process through which governments recognize that the policies of their partners can facilitate realization of their own objectives as a result of policy collaboration or co-ordination.[3] Although co-operation can develop on the basis of complementary interests, it is more often viewed by policymakers as a means to achieve other objectives, rather than an end in itself. Co-operation requires adjustment of policies to meet the demands of others; in other words, it emerges from a pattern of discord or potential discord such as exists in the environment and the fisheries sectors in North-East Asia. Indeed, co-operation occurs when actors adjust their behaviour to the actual or anticipated preferences of others, thereby facilitating realization of each other's objectives. Co-operation is a necessary but not sufficient condition for the formation of a regime.

What is a regime, what are its functions, and why would governments want to form and participate in them? Formally, regimes are sets of implicit or explicit principles (beliefs of fact, causation, and rectitude), norms (standards of behaviour defined in terms of rights and obligations), rules (specific prescriptions or proscriptions for action), and decision-making procedures (prevailing practices for making and implementing collective choice) around which actor expectations converge.[4] This concept of international regime permits the description of patterns of co-operation and the explanation of both co-operation and discord.[5] A more applied definition treats regimes as multilateral agreements among states which aim to regulate national actions within an issue area.[6] Implicit in this definition is that regimes can define the range of permissible state action by outlining explicit injunctions. Such a definition allows a sharp distinction to be made between the concept of regime, and that of co-operation, in that regimes are subsets of co-operative behaviour and facilitate it, but co-operation can exist without a regime. Regimes are also distinguished from the broader concept of institutions, which arise from the intersection of convergent expectations and patterns of behaviour or practice.[7]

Regimes are created to solve dilemmas of collective goods. The optimal provision of international collective goods—environment, natural resources, and collective security—can only be assured if states eschew the independent decision making that would otherwise make them 'free riders', and would ultimately result in either the suboptimal provision or the non-provision of the collective goods. Two different bases of regime formation are the dilemmas of common interest and common aversion.[8] Regimes established to deal with the dilemma of common interests require *collaboration* while those created to solve the dilemma of common aversion require *co-ordination*.[9] Most issues of environmental protection and fisheries management are dilemmas of common interest and thus require collaboration.

A major function of international regimes, then, is to facilitate the making of mutually beneficial agreements among governments and thus to avoid structural anarchy and a climate of 'all against all.'[10] Specific agreements are usually neither random nor *ad hoc* but are nested within more comprehensive agreements covering a plethora of issues. This makes it easier to link issues and so facilitates trade-offs and side payments. Within these multi-layered systems, international regimes help the expectations of governments to converge. Hence regimes develop because states believe that such arrangements will enable the making of mutually beneficial agreements that would otherwise be unattainable, and that the results of repetitive *ad hoc* joint action would be worse, or more costly, than negotiation within a regime context. Furthermore, when agreements that would be beneficial to all parties are not consummated, international regimes can help to correct such a 'market failure.' Thus for regimes to be formed, there must be sufficient complementary or common interests in agreements benefiting the essential members. Clearly, governments that find their fundamental interests to be empathetically interdependent find it easier to form international regimes.

Regimes fill one or more of three critical needs: they establish a clear legal framework with liability for actions; they improve the quality and quantity of information available to states; and they reduce transaction costs. They also provide a valuable mechanism for formulating and presenting a united stand on issues *vis-à-vis* outside actors. Regimes thus create the conditions for orderly multilateral negotiations, and for different types of state action, both legitimate and non-legitimate, and they facilitate linkages among issues. By clustering certain issues together in the same forums over a period of time, they foster continuous interactions between governments, thus reducing incentives to cheat, raising the costs of detection, and enhancing the value of reputation. And the marginal cost of dealing with an additional issue is less with a regime—which is one reason why regimes often expand in scope.[11]

Indeed regimes alter the international environment so that cooperation is more likely. By establishing mutually acceptable standards of behaviour for states to follow and by providing ways to monitor compliance, regimes create the basis for decentralized enforcement based on the principle of reciprocity. 'Reciprocation'[12] is the golden rule with a reward, the belief that if one helps others, or at least fails to hurt them, they will reciprocate when the tables are turned. Implicit in this formulation is the principle that statespersons should avoid maximizing their interests in the short term for the sake of expected long run gain.

Although it depends on the issue, there are several general reasons why multilateralism can be preferable to bilateralism. Development of a multilateral mechanism is organizationally cheaper than maintaining a web of bilateral contacts.[13] Multilateral processes or mechanisms can also mute bilateral tensions and disputes. And multilateralism is not only more efficient for management of the maritime environment and mobile resources which are shared or claimed by several nations. It is politically and ecologically essential as well.

The Origin of Regimes

Regimes are created by a combination of imposition, spontaneous processes, or negotiation. There are four main theories as to why regimes develop and change: structural, functional, game-theoretic, and cognitive.[14] Each captures important features of reality but none offers a satisfactory explanation for the full range of observable phenomena. The structural theory of hegemonic stability links regime creation and change to a dominant power's existence—for example, Japan or China in North-East Asia—and the weakening of regimes to the decline of a hegemon. In a negative view, the hegemon provides coercive leadership, and enforces regime rules with positive and negative sanctions. In a positive mode, the hegemon is willing to originate and sustain a regime because it benefits from a well-ordered system, even if the weaker nations contribute little. Japan can be considered a hegemon in the fisheries and marine science sectors, and institutions that arise in these sectors are more likely to reflect its interests.

With the waning of American hegemony, leadership in the region may become joint hegemonic (Japan and China) or alternative hegemonic.[15] On the other hand there may emerge a set of like-minded regional policymakers with a common set of values who are committed to co-operation. In this situation, it is important for small powers like South Korea to make their support for co-operative endeavors genuine and credible by providing public goods, and thus refuting free-rider implications. Indeed small states can help provide the forum to mitigate major power conflict.

But international behaviour cannot be predicted or explained solely by interests and the distribution of power. It must also be understood within its functional context. Institutions determine the patterns of co-operation. Although the form and content of regimes will inevitably be strongly influenced by their most powerful members pursuing their own interests, regimes can also affect powerful state interests by altering state expectations and values. Abstract plans for morally worthy international regimes that do not consider the reality of self-interest are not practical.

The critical factor here is the state's definition of its self-interest. Although realism must be a basis for analysis of co-operation and discord, it fails to take into account that a state's definition of its interests depends not only on national interest and the distribution of world power, but on the quantity and quality of information and its access to it. Regimes thus thin the 'veil of uncertainty', and reduce transaction costs, facilitating interstate agreements and their decentralized enforcement. Asymmetrical information is an obstacle to regime building because those with less information will be reluctant to make agreements with more knowledgeable actors, or the more knowledgeable may be unwilling to diminish their information advantage by sharing. This asymmetry may be a major obstacle to agreement on a fisheries regime in North-East Asia because Japan has a preponderance of fisheries information.

Although non-hegemonic co-operation is difficult because it must accommodate states operating each in its own self-interest, the erosion of hegemony can nonetheless increase the demand for regimes. Thus, given a more or less similar array of state interests and distribution of power, a system which has a great deal of high quality information available on a reasonably even basis is likely to engender more co-operation.

Regimes are also most successful when they involve relatively few countries with similar interests in the making and maintenance of rules. This is because a small number of actors, as in North-East Asia, for example can monitor each other's compliance with rules and practices, and pursue strategies that make other governments' welfare dependent on their continued compliance. Such co-operation is almost always fragmentary and, not universal across countries and sectors, particularly at the beginning. Thus North Korea's participation is not essential to regime formation in North-East Asia. Nevertheless, the comprehensiveness of participation will have a significant impact on the types of issues that can be addressed, the strategies that are adopted, and the effectiveness of any regional organization.[16]

Governments may comply with regime rules even if it is not in their narrow self-interest to do so. In a world of many issues, such apparent self-abnegation may actually have a rational explanation. In view of the difficulties of constructing international regimes, it is also rational to seek to modify existing ones, where possible, rather than abandon unsatisfactory ones and attempt to start over. Thus regimes tend to evolve rather than die: the General Agreement on Trade and Tariffs, for example, became the World Trade Organization. However, it should be noted that commitments to regimes reduce the flexibility of governments by limiting their ability to act in their narrow self-interest. Each government must evaluate this trade-off and I will attempt to approximate their reasoning in this analysis.

The value of regimes lies not only in their service of the present national interest but also in their potential contribution to the solution of problems that cannot yet be precisely defined. Anticipation of future use of regimes by the same countries which are currently committed to them, acts as an incentive for those countries to keep these commitments even when it is difficult to do so. If a country violates regime norms and rules, it may make it difficult to conclude self-beneficial agreements in the future. Thus maintaining unrestricted flexibility can be more costly than entering and complying with regimes. This means that the *realpolitik* model of an autonomous, hierarchical state that keeps its options open and its decision-making processes closed is contradicted by the value of a reputation for reliability, and by the need to supply and receive high quality information as a basis and incentive for mutually beneficial agreements. North Korea, for example, may be slowly removing the obstacles to co-operation posed by its autarkic behaviour.

Actors will make agreements on the basis of their expectations of a nation's willingness and ability to keep its commitments. Indeed, for co-operation to succeed, each co-operative act must be perceived as a link in a chain of such acts. The essence of international regimes is injunctions which are sufficiently specific as to make violations identifiable and changes observable. Thus a good reputation is advantageous for forming, and entering, into international agreements. A government's reputation is also an important asset in persuading others to enter into agreements with it. Moreover, international regimes help governments to assess others' reputations by providing standards of behaviour against which performance can be measured, by linking these standards to specific issues, and by providing forums, often through international organizations, in which these evaluations can be made. For certain types of activities such as sharing research and development information, weak states with much to gain but little to give may have more incentive to participate in regimes than strong ones, but less incentive to spend funds on research and development. Yet without the strong states, the enterprise as a whole will fail.

Spontaneous institutional arrangements often exhibit longevity and flexibility even when the structure of relationships change. This is because the costs to individual actors, and even leading members, of not complying with, or defecting from, established regimes are generally high[17] and the alternatives are limited: what is the alternative, for example, for a particular state which objects to an emerging regime for marine fisheries management? This observation tends to undermine the realist contention that collective behaviour at the international level is a reflection only of prevailing interests and will last only as long as those interests are met. However, it should be recognized that defection does occasionally occur despite its high costs: Japan's violation of International Whaling Commission quotas and practices is a good example.[18]

Game theory argues that in the real world of complex and consecutive interactions, such as those embodied in international ocean management in North-East Asia, the costs of defecting in any particular situation must be considered in relation to the opportunity costs of foregone future interactions. Iterative interaction can limit the number of players, increase the transparency of state action, and alter the payoff structure. So game theory shows that co-operation among self-interested parties is possible even in the absence of common government, but it depends on the existence and type of international regime.

Cognitive theories of regime formation argue that the degree of agreement in ideology, values, causal relationships, and beliefs regarding the interdependence of issues, and the availability of knowledge about how to realize specific goals stimulate the formation of regimes. Cognitive theory is particularly important in explaining the content of regime rules and why they evolve.

Groups at the domestic level have regime interests, although the domestic political determinants of international co-operation are usually neglected in the regime analysis. It is clear that an inordinate preoccupation with domestic issues, as in China or Russia, provides an opportunity for medium powers, such as South Korea, to lead in regionalism. Yet interaction between domestic and international games and the role of transnational coalitions could be important factors in supporting or opposing the building of particular maritime regimes in North-East Asia. Unfortunately this interaction is not well understood[19] and this aspect of regime formation requires considerably more theoretical research. Although this study will survey major relevant domestic factors and their linkage to international relations, an in-depth analysis of their effects on, and interactions with, national decision making is beyond the scope of the work.

However, one relevant argument is that co-operation may be enhanced when central intellectual players in the region form an epistemic community which co-ordinates their activities and attempts to translate their beliefs into public policies furthering co-operation.[20] The influence on decision makers of epistemic communities grows under conditions of uncertainty such as pertain in North-East Asia today. These communities are identified by the presence of a broadly shared set of normative and principled beliefs, combined with an internalized and self-validating set of causal and methodological principles and a common policy goal operating within a set of formal, semiformal and informal institutions and networks that, in a period of dramatic historical change and uncertainty, provide the framework within which to broker a set of policy options drawn from their normative beliefs and amenable to their causal and explanatory principles.[21]

Epistemic communities have enormous value for international and regional organizations because their loyalties are more to the production and application of their knowledge than to any particular government. Through networks and 'invisible colleges', they seek to promote co-operation across national boundaries. Often, the epistemic communities are able to introduce values and visions that can capture the imagination of decision makers who, on the basis of their new understanding, may redefine strategic and economic interests so as to enhance collective human interests across national borders.

Such a community exists and is gaining strength in the marine policy arena in North-East Asia. The chances of success of an epistemic community will be enhanced when some of its members enter state institutions as officials, technocrats, or consultants, thereby affecting the political process from within. There is an ideological affinity between experts and key policymakers. Their ideas strike a balance, or find a common denominator, between competing positions within the government and in society, thus helping to break domestic political deadlocks and creating a temporary consensus that enables their implementation. Once implemented, however, those ideas may acquire a life of their own and transcend and outlast the temporary consensus that cleared the way for their promotion by the epistemic community.

The Demand for Regimes

Issue areas are sets of issues that are dealt with in common negotiations and handled by the same or closely associated bureaucracies, such as those responsible for environmental protection and fisheries management. Increasing issue density, such as in the ocean sector in North-East Asia, is likely to lead to increased demand for international regimes.[22] And the incentives to form international regimes will also be greater when different issues are closely linked. This is because *ad hoc* agreements will begin to interfere with one another and raise the costs of continually considering the effect of one set of agreements on others. Co-operative regimes in one issue area may also arise as an unintended consequence of co-operation in some other area. This growing realization of the interdependence of issues will also increase the demand for international regimes. The situation in the fisheries and environment sectors in North-east Asia is rapidly approaching this point.

Regimes are supplied when there is sufficient demand for the functions they perform.[23] Apparently there is sufficient demand for regimes in North-East Asia, including those for marine management. With the lowering of political barriers, we are currently witnessing a rapid growth of multilateral agreements in North-East Asia—particularly in economics and environmental protection. There are now ten bilateral fisheries agreements in force in the region: Japan/Russia (2); Japan/North Korea; Japan/South Korea; Japan/China; China/South Korea; North Korea/Russia; Russia/Taiwan; and North Korea/China (2).[24] In the environmental protection area, there are at least five overlapping intergovernmental or international organization-led multilateral programs: the North-West Pacific Region Action Plan of the United Nations Environment Programme (UNEP); the United Nations Development Programme/Global Environmental Facility-UNDP/GEF; the Program on Prevention and Management of Marine Pollution in East Asian Seas; the Intergovernmental Oceanographic Commission Subcommission for the Western Pacific (IOC-WESTPAC); the North-East Asian Environment Programme (Economic and Social Commission for Asia and the Pacific-ESCAP/UNDP) and one non-governmental effort supported by the Asia Foundation. As will be shown, these initiatives do not provide the comprehensive environmental co-operation needed and envisioned by this analysis. But they are a beginning.

International regimes often incorporate international organizations that provide forums and secretariats and thus act as catalysts and mediators for agreement. International organization secretariats also provide unbiased information that is made available, more or less equally, to all members.

In sum, regimes reduce asymmetries of information by upgrading the general level of available information, thereby reducing uncertainty and minimizing disagreements based on misapprehension and deception. They establish patterns of legal liability and reduce the costs of negotiations and bargaining. In short, regimes are formed because governments anticipate they will facilitate co-operation. The costs of reneging on commitments, both for a government's reputation and because of the effects of retaliation against it, are increased, while the costs of operating within these frameworks are reduced. International regimes can—and are—frequently altered, and regime rules are changed, bent, ignored, and broken. Nevertheless, because regimes are so difficult to construct in the first place, actors usually prefer to obey the rules rather than have the regime break down altogether.

Designing Institutions

Institutional design is a means of altering institutional arrangements to achieve identifiable goals or tailoring them better to serve policy programs.[25] While offering hope for solutions to difficult problems, it is also a tricky business. Opportunities to create regimes that would be beneficial to all participants are frequently lost, or when put in place, fail to yield the intended results. Worse still, they can produce unintended and unforeseen outcomes for the creators of the regime. For example, the whaling regime established by whaling nations to regulate commercial whaling has become a vehicle for those wishing to end whaling altogether. And the Antarctic regime has actually stimulated interest in exploiting the natural resources there by providing stability, and allowing the accumulation of knowledge regarding their location and development. Another problem is that institutions are frequently designed to achieve several often conflicting goals simultaneously, for example, efficiency, equity and ecological sustainability. This particular problem arises if the values and trade-offs between them cannot be measured.

Organizations evolve in response to opportunism, uncertainty, increasing information costs, and the need for management and enforcement.[26] Oran Young asserts that there are six main principles to bear in mind when designing international institutions: seize opportunities, expand the scope, multiply roles, emphasize integrative bargaining, simplify implementation, and mobilize leadership.[27] A major advantage of ongoing negotiations, however conflictual, is that they lay the groundwork for effective rule-making when the moment is ripe.

Events external to the bargaining process can precipitate a credible crisis that provides the window of opportunity needed to create a regime. Examples of crises in the resource and environment sectors which produced regimes include the precipitous decline in fur-seal populations in the Bering Sea in the early twentieth century, the alarming drop in blue-whale stocks in the 1930s, and the discovery of an ozone hole over Antarctica in the 1980s. The argument can similarly be made that the revelation of Russian and Japanese nuclear waste dumping in the Sea of Japan has produced just such a credible crisis in the marine environment sector in North-East Asia. It has already led to a modification of the London Dumping Convention to include a ban on all ocean disposal of radioactive material.[28] And the increasingly obvious decline in fish stocks in North-East Asia, the changes in the nature of the catch, and the increased incidence of poaching are producing a web of bilateral regimes, while the increasing issue density may lead to a multilateral regime.

The windows of opportunity created by credible crises do not last long, so those advocating regimes and institutions must be able to recognize such opportunities and seize them promptly. This means that it is necessary to devote time and intellectual energy to thinking systematically about the nature and relative merits of alternative regimes in specific issue areas, even when there seems little possibility of positive movement in the near future. This study attempts to do just that for the fisheries and environment sectors in North-East Asia.

When parties to a potential regime are faced with a narrow question they tend to think in terms of a zero sum game. Expanding the scope of the issue package provides greater flexibility and opportunity for side payments, thus enhancing the prospects of a win/win situation. This observation is, firstly, applicable to the nuclear waste dumping issue and supports the concept of a broader marine environmental regime as proposed by UNEP. Secondly, it is also similar to Young's maxim to multiply roles. Wherever possible it should be emphasized that parties play multiple roles, and that they cannot be certain of occupying a single, well defined role. It is also important to construct regimes which are the result of integrative rather than distributive bargaining. For instance in the nuclear waste dumping example, Russia and Japan are both dumpers and victims and the latter should be emphasized. Similarly in fisheries, all parties are both overexploiters and victims of others' exploitation.

Uncomplicated regimes that produce the desired results are much more preferable to loose or complex regimes. Loose regimes which leave much to be worked out in administrative arenas often founder on the bureaucratic politics of implementation. Regimes that require an elaborate administrative structure, such as the proposed deep seabed regime, often do not command sufficient support to come into existence. And arrangements that make verification of compliance difficult often breed distrust rather than confidence or co-operation.

Leadership is important to the formation of successful regimes. Leaders seize opportunities, initiate and structure the bargaining process, set the focus on integrative rather than distributive issues, and facilitate package arrangements. UNEP has considerable leadership experience through its Regional Seas Programme and the fact that one of these programmes has been initiated for the North-West Pacific region may bode well for a regional marine environmental protection regime.

Resource and Environment Regimes

Regimes for resources and environmental protection have been well-studied and provide many lessons relevant to this study.[29] Three conditions appear essential for effective action on environmental problems: high levels of governmental concern, a hospitable contractual environment, and sufficient national political and administrative capacity.[30] First, governmental concern must be sufficiently high to prompt states to devote scarce resources to addressing the problem. Second, states must be able to make credible commitments, jointly enact rules, and monitor each other's behaviour at moderate cost. Finally, states must have the political and administrative capacity to implement the international norms, principles, and rules, in other words, the regime itself.

Resource and environmental protection issues are usually defined and described by uncertain costs, high transaction costs, externalities, scarce common property resources, and impediments to factor mobility. Anarchical competition for common renewable resources, such as fish or the environment, often precipitates a tragedy of the commons in which everyone loses.[31] Then it becomes increasingly obvious that considerable mutual benefits can be derived from a clear unified set of rules on access and legislative authority. Nevertheless, distributional consequences still often pose obstacles—especially when market failures exist. The same factors apply to the formation of regimes which compensate for accidental or intentional marine pollution damage. Here regimes make costs more certain and reduce them for the individual state.

The result is usually a regime with definite rights and rules. Although these dimensions are the core of every international regime, they are particularly prevalent in resource and environment regimes. Examples include those limiting the freedom of fishing, and those enhancing conservation of fish stocks or the safety of shipping. Liability rules detail the locus and extent of responsibility in cases of injury to others arising from the actions of individual actors operating within the regime. The responsibility for cleaning up marine pollution from accidents is a typical example. Procedural rules describe situations requiring social or collective choices, such as the handling of disputes or the functioning of specific organizations. Some involve the allocation of factors of production, such as total allowable catches in fisheries, or the sorts of activities allowed in an area like Antarctica, or collective sanctions aimed at obtaining compliance with regime rules.

A compliance mechanism may even be developed, in the form of an institution publicly authorized to promote compliance with regime rules or collective social choices. Several factors have increased concern with compliance.[32] In the first place, the growing demands and needs of states for access to, and use of, natural resources coupled with a finite, and perhaps even shrinking, resource base lay the groundwork for increasing interstate tension and conflict. And as international environmental obligations increasingly affect national economic interests, states that do not comply with their environmental obligations are perceived to gain unfair competitive economic advantage over states that do comply.

Nevertheless, as states assume greater resource and environmental treaty commitments, there is an increasing probability that compliance will weaken. Non-compliance obviously limits the effectiveness of resource and environmental agreements, undermines the international legal process, and contributes to conflict and instability in the international system. Non-compliance can occur for a variety of reasons, including a lack of institutional, financial, or human resources, and differing interpretations of the meaning or requirements of a particular obligation. Therefore, despite the prevalence of rules in resource and environment regimes, unilateral actions, bargaining, and coercion remain central to the processes of arriving at social choices within these regimes.

There appear to be four types of national policy responses: avoidance of commitment; commitment but non-compliance; commitment and compliance; and commitment and compliance beyond that required.[33] Countries in the first two categories are 'laggards.' Laggards generally have a low level of concern regarding environmental protection and a lack of capacity to implement it, or demonstrate political resistance to collective action. Laggards also usually have weaker environmental standards and typically agree to an environmental regime in the hope of gaining assistance in capacity building. Rich laggards can be prodded to compliance by political embarrassment or pressure from their own scientists and public. Poor laggards require financial assistance to comply. Yet the implementation of environmental policies is seldom a matter of compliance alone, but more often involves a combination of binding international law, public exposure of non-compliance, normative persuasion, scientific facts, and where necessary, financial aid, and technical and scientific assistance.[34]

Resource and environment regimes vary in extent, their degree of formality and direction, and coherence. Informal understandings may be reached within the framework of the formal regime. The intensity of regime direction is a function of the degree of pressure on members to conform with a goal, such as the conservation of fish stocks or the preservation of ecosystems. There are often contradictions between use rights for resource exploiters and rights to exclude outsiders. Therefore an important dimension of resource and environment regimes is coherence, or the degree of internal consistency of the elements of the regime.

The emergence of organizations associated with international regimes raises the following questions: How much autonomy *vis-à-vis* other centers of authority in the social system should these organizations possess? What sorts of decision rules should the organization employ? How much discretion should the organizations have to make changes affecting the substantive content of the regime itself? How should the organizations be financed? Where should their revenues come from and how should they be raised?

How should the organizations be staffed? What sorts of physical facilities should the organizations have and where should these facilities be located?

Accordingly, any proposed resource and environment regime should be described in terms of the following parameters:
1. *Institutional character.* What are the principal rights, rules, and social choice procedures of the regime? How do they structure the behaviour of individual actors to produce a stream of collective outcomes?
2. *Jurisdictional boundaries.* What is the coverage of the regime in terms of geographic extent, functional scope, and membership? Is this coverage appropriate under the prevailing conditions?
3. *Conditions for operation.* What conditions are necessary for the regime to work at all? Under what conditions will the operation of the regime yield particularly desirable results?
4. *Consequences of operation.* What sorts of outcomes (either individual or collective) can the regime be expected to produce? What are the appropriate criteria for evaluating these outcomes?

Not all regimes require organizations. Indeed, if concern, capacity, and the contractual environment are all highly conducive, an organization may not be needed. The argument against organizations is that they are often costly, inefficient, wasteful, and redundant, create bureaucratic rigidities, generate unintended consequences, and sometimes impede progress toward social goals.

An alternative is an anarchical institution, of which there are several types. Zonal arrangements feature geographically or aerially delimited zones, within which individual states exercise jurisdiction or exclusive management authority. Individual states make and enforce their own rules in their zones without the consent of, or consultation with, others. With the advent of exclusive economic zones covering extensive maritime areas, zonal arrangements have gained a greater role in international society. At the same time there has been an erosion of freedom in such zones in favour of international obligations regarding, among others, protection of endangered species, pollution control, and conservation of shared fisheries stocks.

Under decentralized co-ordination, regime members agree to common rules governing well-defined activities, to be implemented by each member within its own jurisdiction. This is the model for much of the Law of the Sea Convention as well as that used for pollution control regimes in the Mediterranean Sea, the Persian Gulf, and the Caribbean under the auspices of the United Nations Environment Programme. Self-help arrangements often involve mutually agreed rules for activities that cut across zonal boundaries, such as commercial shipping or those that take place in an international commons. Individual members still assume responsibility for implementing the mutually agreed upon rules with respect to their own nationals. The regime for Antarctica even permits members to monitor each other's compliance with the regime rules.

Nations may also agree on common rules for an area or an activity but delegate administrative authority to an individual member or even an outside entity. In the case of the CITES Treaty, the designated administrator is the executive director of UNEP and he has passed administrative responsibility to the International Union for the Conservation of Nature (IUCN) and the World Wide Fund for Nature (WWF). Hence in this example, an international regime is administered by a freestanding non-governmental organization (NGO).

Despite the attributes of anarchical regimes, there are limitations on what they can achieve. They do not allocate resources, and they do not promulgate regulations needed by restricted common property arrangements like fisheries regimes. But private organizations and governments can often collect and analyse data better than international organizations. And collective decisions can be made without central authority, such as in traditional fisheries regimes. Even compliance does not necessarily require a central organization. It can be voluntary because defection has its distinct disadvantages, or it can be coerced through implied or threatened sanctions. Moreover, disputants can always resolve issues between themselves without resorting to a third party. Indeed, this is the preferred approach among North-East Asian countries. The point is that the need to establish organizations is a function of the type of institutional arrangement.

Regimes intended to deal with dilemmas of common interest must specify strict patterns of behaviour and insure that no one cheats. Because each party requires assurances that the other will also not cheat, collaboration usually requires a degree of formalization and institutionalization. Such institutionalized collaboration is often accompanied by extensive enforcement provisions for monitoring compliance.

Although anarchic arrangements are often slow to adjust to rapidly changing circumstances, central organizations may not be much better. Powerful interest groups that benefit from existing arrangements often dominate the actions of such organizations. Moreover organizations, particularly large ones, are notorious for bureaucratic obstinacy that slows or impedes change regardless of shifts in the external environment. So arrangements like the Antarctic regime or the whaling regime, which are located in the transition zone between anarchic and civil society, and which hold consultative meetings, allow for responses to changing circumstances without being bureaucratically rigid.

Instead of creating new organizations, use can sometimes be made of existing free-standing organizations, like the Food and Agriculture Organization, and the Intergovernmental Oceanographic Commission, that are not linked to specific regimes. Some free-standing organizations, like UNITAR or IIASA, specialize in service activities. Others are problem-solving agencies, while still others are focused specifically on forming new international institutions, for example, UNEP with pollution control regimes, and the International Maritime Organization with pollution of the sea from ships. Free-standing organizations are often sources of innovative ideas for negotiated regimes.

On the other hand, effective institutions can affect the political process of environmental policy-making in three important ways:

1. They can contribute to more appropriate agendas, reflecting the convergence of political and technical consensus about the nature of environmental threats.
2. They can contribute to more comprehensive and specific international policies, agreed upon through a political process conducted mainly by intergovernmental bargaining.
3. They can contribute to national policy responses which directly control sources of environmental degradation.[35]

International environmental institutions enhance the ability to make and keep agreements; promote concern among governments; and they build national political, administrative, scientific, and technical capacity. In addition they can also: enhance state concern by magnifying public pressure on recalcitrant states; influence domestic debates; provide a forum and open access to an agenda for bargaining; reduce the costs of negotiations by generating information about potential problems and zones of agreement; and, through monitoring, assure that cheating is detected. Institutions can also serve as convenient scapegoats to blame for costly environmental measures. Moreover, they can provide efficient and acceptable dispute resolution mechanisms; integrate environmental and natural resource protection with other policy goals; facilitate participation by NGOs in policy debates; create incentives to develop true cost prices; and help provide continuity and maintain momentum for future negotiations and agreements.[36]

Despite the advantages of anarchic arrangements, I argue that there is a need for organizations to administer regimes dealing with international commons or transboundary problems such as migratory fish, or pollution and maintenance of environmental quality. When protected sites need to be identified, a committee is necessary to make choices. Similarly, central organizations are needed to allocate cost-sharing for projects such as co-ordinated multilateral marine scientific research. And restricted common-property systems almost always require a central organization for such purposes as establishing effluent standards or limits on fishing effort. Organizations are indispensable for regimes which control entry or regulate the activities of individual resource users, which transfer technology and redistribute the benefits of exploitation of marine resources, or create international authorities to exploit resources directly. Furthermore, central organizations for data collection and analysis are particularly attractive because they have an aura of impartiality and credibility.

For all this, I would agree that we should be cautious about creating new organizations, and rely as much as possible instead on institutional arrangements that can cope with collective-action issues without creating organizations. In sum, organizations for environment and resource regimes should be created only when there is a demonstrable need for them.

When policymakers do choose to establish organizations, they should tailor the arrangements to the characteristics of the institution or regime under consideration. The task should start with a systematic assessment of the relative merits of alternative institutional arrangements and proceed to devising the organizations (if any) necessary to meet the requirements of the specific institutional arrangements selected. The most significant roles of international institutions—as magnifiers of concern, facilitators of agreement, and builders of capacity—do not require large administrative bureaucracies. In making such a systematic assessment, it is useful first to establish general evaluative criteria.

Criteria for the Evaluation of Regimes[37]

Political Feasibility

The most important cluster of evaluation criteria focuses on the political feasibility of regimes. These criteria relate to the implementation and maintenance of regimes. To what extent is a given regime likely to prove fundamentally acceptable to the principal actors involved? If a regime type is demonstrably unacceptable within the context of a particular region, there is little point in pursuing it in depth no matter how desirable it may seem from other points of view. Economic and political factors are by no means totally unalterable, and it would be a mistake to discard otherwise desirable regime types solely because their implementation would require certain changes. How costly would it be to achieve compliance with the principal behavioural prescriptions of a given regime? This is partly a matter of the extent to which the actions demanded by the relevant prescriptions are regarded as unusually onerous by one or more of the actors. But it is also a function of the ease of undetected violation, the probability of known violators being sanctioned, and the severity of the sanctions employed. There are wide variations in the extent to which actors comply with the behavioural prescriptions of different regimes. A regime that looks attractive on paper may be less desirable than a regime whose prescriptions are intrinsically weaker but likely to foster compliance.

Distributive Implications

The allocation of benefits among actors lies at the heart of the politics of governing arrangements. What would be the impact of a given regime on the actual distribution of benefits among the relevant actors? In other words, who gets what and how much? Which actors are favoured by a particular regime and what mechanisms would produce these results?

Goals may include the gaining or safeguarding of access to wealth (redistributing the flow of income), access to knowledge (enlightenment), increasing one's own capabilities (skill), denying access to real or potential competitors, facilitating regional stability and conflict resolution, using marine-oriented agreements as side payments for non-ocean issues, resource conservation, and management of conflict within and across various ocean issues. Costs and benefits must be thought of in short- and long-term situations and as possible trade-offs. For example, if country A desires a regional fisheries arrangement that clearly will be of more benefit to its own nationals than to those of country B, what can country B seek in exchange for participation in the regional arrangement?

The discussion of distribution naturally raises normative issues. To what extent would a given regime satisfy the claims of actors currently or potentially active in the region? Often this is essentially a matter of dividing a finite pool of resources among competing claimants. For example, what are the relative merits of Russian and Japanese claims to the salmon of the Sea of Japan? In other cases, however, allocation involves difficult value trade-offs. For example, does the provision of employment opportunities in the Russian Far East justify certain sacrifices of economic efficiency with respect to the organization of the regional fishing industry? Should oil companies operating off Sakhalin in the Russian Far East be made legally liable for all the impacts of marine oil spills even if this makes the cost of operating there prohibitive?

To what extent would a particular regime for a region like North-East Asian seas accommodate the actual or potential distributive claims of outsiders? There are in fact several distinguishable issues of this type. Should the regime include provision for new entrants wishing to utilize the region's resources, or should it have the effect of creating an exclusive club limited to the coastal states? What are the probable consequences of the regime for outsiders who are consumers of the region's products in the sense that they purchase them in the relevant international markets?

Social Consequences

When designing or evaluating a regime, planners must consider the social consequences, or the impact of any regime on the members taken as a group. That requires assessing which regime would be the most economically efficient in exploiting the region's resources. Efficiency can refer to the equalization of marginal revenue and cost, permitting the maximization of profits or rent, or it can refer to the maximization of social welfare.[38] The most widely accepted standard in this realm is Pareto Optimality.[39] Thus, a regime will be regarded as inefficient if it yields an outcome that could be improved upon for one or more of the actors in the region without damaging the interests of any of the others.

Efficiency is not the only relevant criterion in the realm of social consequences. We must also ask how extensively the natural resources of the region would be utilized if a given regime were introduced. The issues here concern the absolute level of use as well as the rate of use of various resources, in contrast to the question of whether they are exploited in an economically efficient fashion. In the case of renewable resources, the concept of optimum sustainable yield is helpful because it measures whether a given resource is being exploited below or beyond a level which is likely to lead to serious problems of depletion. For non-renewable resources, the major issues are the rate at which a given resource is depleted and the extent to which this rate conforms to various views of what is socially desirable.

A further important question is what would be the consequences of a given regime for various values that are not reflected in ordinary market calculations? These values include collective goods, the regulation of externalities or social costs, intangibles not reflected in market prices, and political goals. In resource regions like North-East Asia, such values might also include the development of neglected or backward areas, protection of pristine areas, and the conservation of living species subject to heavy exploitation.

Regime Formation and Change

Some regimes are self-generating or spontaneous, in other words, they do not involve conscious co-ordination among participants, do not require the explicit consent of their members, and are highly resistant to social engineering. They are basically a result of a trial and error learning process coupled with a form of natural selection. Other institutional arrangements are negotiated. And some are imposed by dominant powers. More common than imposition is leadership, whereby a nation with superior natural or human resources plays a critical role in designing institutional arrangements and inducing others to participate. Inducements may be negative, such as the withdrawal of aid or trade, or positive, for example, access to advanced technology or loans.

There are substantial costs of leadership in regime formation. Leaders may have to make significant concessions, or take on major burdens, to induce others to join. And failure erodes the status and stature of the leader. But those who fail to join may be denied some of the benefits. For example, if the regime is for ship design or operating procedures to prevent pollution, participants in the regime may bar ships of a non-participant from their jurisdiction. Repeat violations can often bring retaliation. But they can also lead to the collapse of the regime itself.

Regimes evolve. During the course of this century, for example, the global regime for the oceans has evolved from imposed to spontaneous to negotiated. Rapid change in the international political environment often erodes spontaneous regimes without creating conditions conducive to the formation of new arrangements. Thus negotiated and even imposed regimes are attractive and more durable during times of rapid social change. In designing regimes, one must be careful to avoid incorporating irreconcilable differences among the elements. For instance, conflict may be inevitable when unrestricted access is granted for all members to a region's resources, but sovereignty is allocated to a particular member.

The shift in the structure of power in international society as a whole, and in the region in particular will affect international maritime regimes. The power, respectively, of the United States and Russia in the region has decreased while that of China and Japan is expanding. Technological change, such as advances in fish capture techniques and equipment, and detection and understanding of the effects of pollution on human health are undermining the existing common property fisheries and environmental regimes in favour of restricted access or use regimes. And major shifts in domestic priorities caused by, among others, demand for resources due to population growth or consumer tastes can render regimes obsolete.

Nevertheless, once established, many international institutions are resistant to change even when the existing arrangements are inefficient and unfair. Common property regimes for renewable resources like fisheries and environment are particularly difficult to change. Indeed, individual agencies within governments often define their roles in terms of administering and maintaining the provisions of international regimes.

Moreover, international regimes can commonly give rise to non-governmental interest groups committed to defending the provisions of specific regimes and prepared to press governments to comply with their rules. In fact, the establishment of a regime can stimulate the growth of powerful interest groups in the member states, which then form transnational alliances in order to persuade responsible agencies to comply with the requirements of the regime. Indeed, the key variable accounting for policy change in the environment arena is the degree of domestic environmentalist pressure.[40] Although this phenomenon is highly unlikely in North Korea and might be viewed as a significant cost by that government, interest groups are emerging or gaining strength in Japan, South Korea, and Russia, and now even in China.

Marine Policy Regimes

A marine policy regime is a system of governing arrangements, together with a collection of institutions (formal or informal) for the implementation of these arrangements, in a given social structure or marine region.[41] More specifically, a marine policy regime is a set of agreements among a defined group of actors specifying: the distribution of power and authority for the marine geographical region; a system of rights and obligations for the members of the group; and a body of rules and regulations which are supposed to govern the behaviour of members.

It is thus a system of governance comprising several distinct elements: structure, objectives, functions and powers, processes, and programs.[42] Structure includes the activity for which an arrangement is designed, geographic coverage, membership, administrative framework, and institutional affiliation. The objectives of marine regional arrangements can be divided loosely into three major categories: conservation, management, and development. Conservation may include maintenance of water quality, protection of living marine resources, wetlands, marine sanctuaries, endangered species, and scenic areas. Management implies the planning and execution of programs whereby maritime activities may be carried out more effectively. It also involves managing conflicts between and among ocean users. Development focuses on the use of resources such as fish, hydrocarbons, and minerals, shoreline space, shipping lanes, and ocean space. It also involves acquiring knowledge and understanding of marine areas, as well as skills and equipment for their management. Development may also include improved systems within the member countries for marine-related action, and multistate regional systems for dispute settlement or other goals.

At the state level, there are other objectives of regional action, such as the protection of national interest, sometimes by denying access to competitors. Another objective is the provision of greater equity to developing land-locked and geographically disadvantaged states concerning access rights to the sea and its resources. Marine regional arrangements occasionally may have more than one stated objective, as well as several hidden objectives such as the strengthening of regional ties or aiding the quest by one or more states of the region for a leadership position. To be fully effective, regional organizations, including those dedicated to marine affairs, should be linked to other national, regional, and global objectives.[43]

Functions and powers are the heart of maritime regime analysis, and include the scope of the system and the degree of integration achieved. Functions operate at four levels, each of which might entail ever more politicization and controversy. A first level involves 'service,' that is, information exchange, data gathering and analysis, consultation, facilitation and co-ordination of programs, and joint planning.[44] Most maritime regimes perform this function. But regimes stuck at this level tend to avoid controversy and to concentrate on 'low priority' issues. Nevertheless, within a geographic region, a proliferation of such organizations may build the web of interrelationships necessary to move to the next higher step of successful regional arrangements.

A higher functional level involves norm creation and allocation, which includes the establishment of standards and regulations and the allocation of costs and benefits. Only a few regional marine organizations, such as the South Pacific Forum Fisheries Agency, have achieved this level. The third level, 'rule observance,' includes the monitoring and enforcement by the regional organization of the norms and agreed standards. At the highest functional level are 'operational' multilateral organizations concerned with implementing the norms for management of resource exploration and exploitation activities, technical assistance, and research analysis and development. An example might be the Enterprise, the international mining arm of the International Seabed Authority proposed in the Law of the Sea Treaty.[45]

Maritime regimes are particularly flexible and changing structures, which are influenced by the dynamics of political and economic interactions among the members.[46] Most regimes have specific provisions for their modification, whether through legislative processes or *ad hoc* negotiations. Changes may occur regardless of changes in the distribution of power and authority. In this respect, maritime regimes tend to evolve continuously, and real world regimes are often mixtures of several ideal types located between the two extremes of centralization and decentralization of power and authority. Although there will seldom be a regime type that is simultaneously optimal for all actors, one regime type may be preferred by the majority of the actors. This means that the particular regime type will be the subject of intense bargaining among the actors.

Approaches to Regional Marine Policy Co-operation

It is helpful to begin by characterizing the existing regimes in a particular region, and assessing their consequences in terms of certain well-defined criteria of evaluation. But it is also important to think systematically about alternative regimes for specific regions, and about the probable consequences of each of these alternatives. This is especially true with respect to a maritime region like the Sea of Japan and the Yellow Sea/East China Sea, in which rapidly emerging problems of resource management and environmental quality are raising fundamental questions about the adequacy of existing regimes.

There are several different approaches to regional co-operation in the maritime sphere: joint activities, regional organizations, treaty arrangements, harmonization of laws and policies, and informal contacts.[47] The most demanding of these approaches is joint activities involving sustained close interaction in specific enterprises or projects, such as joint scientific research programs, joint ventures in fisheries, or joint hydrocarbon development. Regional co-operation can proceed progressively from policy consultation, or discussion of common or mutual policy problems; to policy harmonization, or elimination of contradictory or conflicting national policies; to policy co-ordination, or national policy adjustments in pursuit of common policy goals; and finally to integration, or the implementation of common policies and structural adjustments to eliminate barriers to interaction among the region's economies.[48]

Participation in regional organizations can range from the symbolic to the substantial within the framework of larger multinational organizations such as the Intergovernmental Oceanographic Commission, the Food and Agricultural Organization of the United Nations, or the United Nations Environment Programme. When the organization undertakes extensive activities, possibly including management functions, participation can become costly, particularly for developing countries. Treaties generally are concluded only after difficult negotiations relating to specific problems. And they often result in firm obligations that are not readily adapted to changing circumstances.

Harmonization may be defined as the deliberate alignment of the laws and policies of different nations for the purpose of fulfilling their national interests in the context of co-operation. In shipping, fishing, and most other sectors, this can take the form of modest adjustments in national laws and policies that yield significant benefits to all concerned. Harmonization is likely to be less demanding than the other more obvious instruments of international co-operation. Informal contact is the least costly and the least risky method of undertaking regional co-operation, and it does not get the recognition it deserves. With informal communication it may be possible from time to time to find fruitful arenas for co-operation on a modest scale. The informal sharing of information and data bases may be a starting point in this process.

The Objectives of Maritime Regimes

Objectives or the priorities allocated to them are likely to differ between developed and developing states.[49] Most developed country governments like Japan would hope that a new fisheries regime enhanced stability in the system. Once international agreements had been concluded and conditions of operation established, they would expect that the rules would not change within some specified time frame, except through the mutual consent of all parties concerned. A second basic objective might be opportunity: opportunity for the developed country's vessels to move about freely over as large a part of the region as possible; opportunity to undertake research; and opportunity for the country's nationals to exploit the region's resources within the framework of whatever restrictions are necessary in the interests of sound management practices. Another objective of growing importance would be protection of the resources from pollution. Although the developed states in the past were largely responsible for marine pollution, these same countries are now concerned with the pollution emanating from their neighbours.

These objectives are also important for developing states, but their priorities are probably different. Their primary objectives of a new ocean regime would most likely be the opportunity to enhance their economic development, and to have a meaningful role in the management decisions of any regional institution established. Associated with these objectives would be the opportunity for receiving training, education, and mutual assistance in the marine sciences and in conservation, management, and development of their resources.

Multilateral conference diplomacy generally conflicts with the desire of governments to avoid measures that will cause their firms to lose competitive advantage. Bargaining dynamics often involves leader states, which already have advanced environmental policies and regulations, and laggard states, with undeveloped environmental measures and a desire to avoid international agreements requiring stricter measures.[50] But a potential trade-off of interests is possible between developed and developing states. A regional organization in which members of both groups participate might provide developed states with the opportunity for access, as well as stability of investments and the likelihood of improved environmental protection. Developing states would also have an interest in these issues, especially as stability might serve to encourage outside investments. But as part of the benefits of participation, developing states might first of all expect developmental assistance, training, and education from their developed partners.

The distribution of power and authority in a regime is a primary concern of both developed and developing states. Fisheries regimes, for example, can vary along a continuum which ranges between extreme centralization to extreme decentralization.[51] Extreme decentralization includes voluntary acquiescence, in which member states are free to comply with rules established by the majority of members or to ignore them. At the other extreme is consensus, which means that any member state may veto a proposal even if it is approved by a majority of the other members. In between these two extremes are various forms of 'binding majority'.

Power and authority can vary independently, although stability seems to require their approximate congruence. Nevertheless, for fisheries regimes, it should be possible to combine centralized authority with decentralized power and vice versa. Rights and obligations in a resource regime constitute a set of economic and social relations that define the position of each actor with respect to the use of the resource or environment. Variations in rights and obligations will have substantial implications for dealing with externalities, solutions for use of specific resources, and the distribution of values among the members.

Process

The term *process* as it is used here has four meanings: the interplay between forces of integration and disintegration within a regional system; the growth and/or decline of the organization; the establishment of links between one regional system and others within the same geographic location; and the impact of organizations on the nature and use of the ocean space over which their activities extend.[52]

Among the integrative forces within a region that support regional action are: the existence of other international arrangements among the member states that could contribute to the regional consciousness of the participants; ethnic, cultural, historical, or other ties among the region's countries; and clear indicators of economic benefits or other advantages to be gained through regional action, or clear indicators of common cost that might be avoided through regional co-operation.

Dis-integrative forces include: political, territorial, ideological, or other differences among countries of the region; competition for positions of leadership among two or more states of the region; and opposition to regionalism in the area by one or more regional or outside powers. Growth and decline of regional arrangements may be measured by total membership, functions and powers of the organization, new programs, and expansion of the institutional region.

For any fisheries regime to enjoy full participation, a participating entity must perceive that it will obtain advantage by the application of the regime compared to alternative action or arrangements, including non-participation.[53] Categories of benefits (advantages) and costs (disadvantages) are both sociopolitical and economic; time scales are both short and long term (Table 2.1). Each entity must decide under what type of regime the net benefit would be highest for itself.

Co-operation can be promoted by showing that the benefits would be substantial and also by showing that the costs and risks of

Table 2.1   Possible Advantages and Disadvantages of Economic and Political Co-operation in the Marine Arena by North-East Asian Nations

ADVANTAGES	DISADVANTAGES
Money saved from economy of scale	Allocation of resources to extranational projects
Contributions of extranational resources	Loss of potential benefit of unilateral approach
Increased amount and efficiency of use of external technical or financial aid	Decreased national share of technical or financial aid
Increased rate of resource development and technology transfer	
Increased efficiency of resource use	Administrative or loss-of-efficiency cost
Better information exchange and better management	Yield of competitive advantage in scientific/technical knowledge
Less strict national enforcement required	Loss of degree of control over decision making
Opportunity for bargaining for national marine interest	
Enhanced power vis-à-vis extraregional interests	Reduction of strong relationships with extraregional nations; possible collective tension with extraregional nations
Enhanced international and regional power and status of leader(s)	Yield of sovereignty to regional group or leader(s)
Strengthening of trends toward regional co-operation in other arenas	Compromise of other policy or commitments vis-à-vis intra- or extraregional entities

co-operation can be limited. But as long as regional co-operation in, for example, fisheries, is taken to mean entanglement in some large-scale regional organization of someone else's creation, nations are likely to assume that it is going to be costly and risky. It is possible, however, to cultivate co-operation without onerous organizational burdens.

The Global Maritime Scene

Over the past two decades, multilateral discussions concerning the Law of the Sea, and publicization of the importance of marine resources and uses, as well as their fragility and limits, have contributed to an increased 'marine awareness' and widespread claims to maritime space on the part of many of the world's nations. The United Nations Convention on the Law of the Sea came into effect on 16 November 1994, one year after the last of the requisite 60 signatories had ratified it.[54] The Convention changed considerably the rules of international law relating to the boundaries of the continental shelf and to movement through international straits and archipelagoes. It created the Exclusive Economic Zone (EEZ) and developed new principles relating to scientific research and the protection of the marine environment. Its plethora of rights and responsibilities in the ocean is now international law for those nations which have ratified it, including, in Asia, Indonesia, the Philippines, and Vietnam.

Although none of the major developed countries has ratified the Treaty, there are increasing signs that they are considering doing so now that a compromise agreement has been reached regarding implementation of the Convention provisions dealing with seabed mining.[55] Previously the developed countries, led by the United States, opposed ratification of the Treaty because they felt its provisions on seabed mining violated free-market principles and gave away technological advantages as well as control of the regime to the developing countries. These concerns have now been satisfied. Regardless of what the United States does, Japan is expected to ratify the Convention by June 1996, partially because it wants to nominate a candidate for the Law of the Sea Tribunal created by the Convention. Motivated by the same desire as well as wanting to legitimize its pioneer investor status, South Korea is also preparing for ratification early in 1996. And regardless of whether the major world powers, such as the United States, soon ratify the Treaty, many of its provisions have arguably already become customary international law.

Among these provisions are the right to national jurisdiction over resources and certain activities extending 200 nautical miles from coastal baselines. Scientific and technological advances have stimulated almost all coastal countries to claim such Exclusive Economic Zones, with the notable exception in North-East Asia of

China, South Korea, and Japan. Indeed, because of the sea 'rush' atmosphere which surrounds the exploitation of maritime resources, jurisdictional claims have even tended to 'creep' beyond, or develop at variance with the provisions of the Treaty.

Major scientific and technological advances also underlay the drive for the United Nations Conference on Environment and Development (UNCED), which was convened in Rio de Janeiro in 1992. The realization that the quality of the ocean environment plays an important role in such global processes as climate change,[56] decline in world fish catch,[57] and transnational distribution of certain pollutants,[58] as well as prospects for the exploitation of mineral riches on the deep seabed,[59] and petroleum in the continental slope[60] led to the inclusion of a specific focus on coastal and marine activities in UNCED's action plan for the next century. This linked the UNCLOS and UNCED processes, and the notion of sustainable development of ocean space and resources with sustainable development on land.[61] In particular, the concepts of ecosystem management,[62] large marine ecosystems,[63] and integrated ocean management[64] have emerged to address the complexity and transnationality of ocean resources and ecosystems. Some 25 regional bodies are now involved in programs aimed at protecting the world's oceans along with 14 UN agencies.[65] All these developments have, in turn, enhanced the salience of marine affairs on both national and international agendas.

Marine Regionalism: Lessons Learnt

One consequence of enhanced national maritime knowledge, expectations, and interests, and the sea enclosure movement, has been a recognition that global standards and regimes may not adequately address the special circumstances that distinguish the variety of national maritime needs and interests. The trend toward unilateralism in the seas has been partially balanced by an incipient development of marine regionalism when groups of countries have perceived that some of their mutual needs and interests, based on physical geography, complementary uses, or policy, distinguish them from other countries and are best satisfied by a regional approach.

Regionalism has been defined as sustained co-operation, formal or informal, among governments, non-governmental organizations or the private sector in three or more contiguous countries for mutual gain.[66] I use the term *regionalism* to refer to both geographical regions and subregions, as well as to the mechanisms designed to implement various activities among states. So the concept of marine regionalism has two aspects: marine regions as geographical phenomena, and regional arrangements that are designed for particular areas. Most marine regional activities to date have been oriented toward specific issues rather than discrete regional areas. The most readily definable marine areas for regional action are the semi-enclosed seas where there are often discrete marine ecosystems, and where problems of access to marine resources can be addressed more effectively than they can at a global level.[67] Only those marine-related phenomena involving three or more countries are considered *regional*.

The relationship between perceptual regions in the ocean and physically defined regions may be complex because some issues to which regional action is addressed, or for which it is contemplated, may not be confined to a semi-enclosed sea or other clearly distinguishable area.[68] Indeed, decision makers may perceive policy problems within a regional context even though they exist in a form considerably at variance with physically defined geographic units. Thus a marine region may be differentiated from other areas on the basis of coastal configuration, such as a semi-enclosed sea, but it may also be identified as a *management region* where a well-defined management problem may be handled as a discrete issue. And it may also be an 'institutional' region where one or more formal arrangements exist, for example, that area defined by the limits of competence of a regional environment protection effort. The limits of an 'institutional' region clearly should have some correspondence with the limits of the 'management' region within which exists the problem the institution is designed to address. The ideal situation is when the management and institutional regions correspond with the physical region. This is often the case in semi-enclosed seas.

A semi-enclosed sea like the Sea of Japan or the Yellow Sea/East China Seas can be considered an *international management region*, that is, a natural system or managerial unit which is not subject to the effective management of one state. In this case artificial divisions or boundaries are overlain on transnational natural systems. The outer boundaries of such regions are defined by the boundaries of natural systems and the actual or potential extent of jurisdictional claims.[69] Although some natural systems may extend well beyond jurisdictional claims, legal and political realities argue for co-operation limited to the jurisdiction of the co-operating states. Hence regional arrangements are a supplement, but not a substitute, for global administrative or managerial systems. Political differences may constrain the specific definition of the boundaries of a region, particularly if one or more parties refuse or are unable to participate in a regional regime. Fortunately, it is usually not necessary to achieve precision on the outer boundaries of a region in order to initiate co-operation. Ambiguous formulations allowing for differing interpretations can be used to overcome disputes over matters of central principle.[70] Agreement on a core area is usually sufficient, and the geographic scope of specific arrangements can then evolve on the basis of experience and *ad hoc* negotiations. The core areas of this study are depicted in Figure 2.1.

The mobility of water, fish, pollutants, and ships is a fundamental rationale for regional action within semi-enclosed regional seas.[71] Regional action can also be an asset, if not a necessity, in acquiring and interpreting data about the physical nature of the marine environment within a given area. Synchronic surveys of ocean or atmospheric conditions may be necessary to obtain a holistic understanding of natural marine phenomena, such as water mass formation and flow. The economic incentives for regional action may include joint contributions by the countries of a region to projects of high cost or to those ventures that demand a high degree of technical skill or efficiency. Regional co-operation can also be an asset in establishing rules and regulations, particularly against extra-regional actors such as foreign tankers which transit an international management region. A united stand by three or more countries is more difficult for outside shipping and other interests to assail than a unilateral promulgation of regulations by a single state. Finally, there can be a confidence-building spinoff from regional or subregional action in marine activities to co-operative activities in economic and security spheres.

Relevant Precedents

The concept of regional approaches to ocean management issues has evolved considerably since World War II.[72] The drive for regional approaches to problem solving in the global marine sphere began with the establishment of the United Nations and its special-

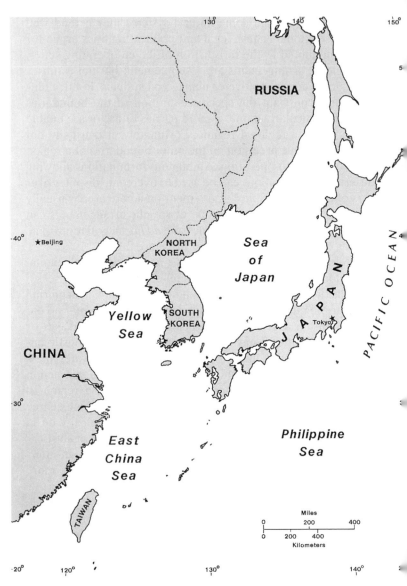

Figure 2.1 Maritime North-East Asia

ized agencies.[73] Three agencies in particular became involved in regional arrangements: the Food and Agriculture Organization (FAO), which in the 1960s inaugurated a series of regional commissions and councils, now totalling ten including the Indo-Pacific Fisheries Council (IPFC);[74] the Intergovernmental Oceanographic Commission (IOC), which has initiated ten regional scientific organizations and programmes covering such phenomena such as the Kuroshio Current off North-East Asia[75] and the United Nations Environment Programme (UNEP) with its regional seas activities, beginning in the mid-1970s with the Mediterranean Action Plan.[76]

A trend in the dynamics of marine pollution regimes has been their piecemeal pattern, combined with the gradual expansion of regimes from a focus on a single issue to encompass several issues.[77] From a regime focusing on one form of ship-generated pollution, the process of regime building widened into a more inclusive regulation of pollution from vessels and then diversified, producing other partial regimes in the fields of dumping and land-based pollution. This disjunctive approach to constructing marine environment regimes may not conform to the rather idealistic 'ecoevolutionist' or 'ecoreformist' principle of order,[78] which postulates a holistic approach to the management of the oceans, but in pragmatic terms it has been the only feasible way to proceed. Experience has also been accelerated by a spillover generated on two planes: from one functional partial issue regime to another, and from one geopolitical ocean space to another, as demonstrated by the example of the regional regimes under the auspices of the UNEP Regional Seas Programme.

Since 1972, this expansion of regimes has been accompanied by a trend toward a more comprehensive regime, first at the regional level, and then at the global level as part of the general Law of the Sea regime under the 1982 Convention on the Law of the Sea.

Full or partial marine regionalism has been attempted or proposed in a number of semi-enclosed seas, including the Baltic,[79] the Mediterranean,[80] the North Sea,[81] and the Arctic[82]—with mixed results. The Baltic Sea is similar to the Yellow Sea in physical and ecological parameters, in the disparate economic and political systems of its coastal states, and in the types and sources of pollution. But unlike North-East Asia, the Baltic countries have made great strides towards regional co-operation in marine policy, even enacting national legislation to implement regional agreements. Disputes over continental shelf boundaries have been mitigated by a stocks, and to recommend to their governments acceptable, if not optimal, levels of fishing effort and the regulatory measures necessary to achieve those levels, assuming there is no or little non-compliance. In sum, negotiations between government representatives, occasionally including cabinet ministers, have managed to produce mutually acceptable agreements, and the whole process has forced governments to co-ordinate to some degree their respective policies.

Fourth, except for the declaration of fisheries and exclusive economic zones, probably the single most important development affecting the bilateral fisheries relations between Japan and Russia and between Japan and South Korea has been the expansion of the fishery industry in Russia and South Korea since the 1970s. Although the two fisheries regimes have shown sufficient flexibility to accommodate these trends, the consequences for Japanese fishermen have been painful. Drastic reductions in fishing effort forced many Japanese fishermen to give up fishing, temporarily or permanently. Many surplus boats had to be scrapped or transferred to other areas or uses.

Nevertheless, once the governments agreed to limits on their fishermen's activities, however painful those limits may have been, the fishermen knew how much 'return on investment' they could expect from their fishing operations. Continual reductions in Japanese fishing effort in Russian and South Korean waters were expected in Japan, and the bilateral fisheries regime provided some degree of stability and predictability. Moreover, pressure from the well-organized Japanese fishing interests pushed the Japanese government to establish the now familiar array of relief measures for those affected.[24]

The Changing Situation and the Inadequacies of the Existing Regimes

The current fishery regime in North-East Asia is insufficient, fragmented, and antiquated. Although North-East Asian countries have undertaken unilateral measures to regulate foreign fishing in their coastal waters,[25] none of the regulatory regimes includes all coastal or fishing nations nor is any state a party to all the agreements. Hence there is no forum wherein all Yellow Sea and/or East China Sea fishing nations can meet to discuss the distribution of catches. Nor is there a clear dividing line for national fisheries areas. And geographically, none of the bilateral agreements takes into account all of the region. Vast areas are not covered under any formal agreement, and in some areas agreements appear to overlap.

Under the present *de facto* regime, coastal states have attempted to reserve coastal fisheries for their own fishing interests. The bilateral agreements make some modest attempts to indicate sharing of catches within jointly regulated areas, or to limit efforts in order to stay within catch quotas for designated areas. But direct discussion of allocations would entail reviewing systematically the existing *de facto* allocations, and re-examining the legal regime in the area with respect to fishing rights. Discussion of shared stock questions in the Sea of Japan would also require involvement of North Korea. This situation creates the possibility of eventual competitive bidding for quotas which could undermine stock management.

It also constrains the sharing of scientific information. The bilateral fisheries commissions established under the agreements generally do not publish their decisions or the results of their scientific deliberations for peer evaluation or general public information. Without information on the basis for decisions, the necessity of and rationale for the regulations cannot be fully understood, nor can their success be evaluated.

Despite some advantages, notably the lack of overt conflicts over fisheries until recently—due mainly to self-restraint—the present regime is fundamentally flawed in terms of fisheries management. Few species can be managed by only one country. Although the stocks are often transboundary in distribution involving two or more states, there is no corresponding multinational body to manage them. And the parties concerned often produce significantly different resource assessments. Japan is a fisheries hegemon, with a virtual monopoly of quality information—although there remain significant knowledge gaps, for example, regarding stocks in the Seas of Japan and Okhotsk.

Though perhaps inequitable, this system, comprising an interlocking web of bilateral agreements dominated by one nation, could, in theory, successfully manage the region's fisheries, particularly if hidden factors serve to make the regime more equitable. But the advent of UNCLOS, the imminent extension of jurisdiction by some of the states in the region, the development of China's and North Korea's offshore fishing capability, utilization of unconventional species, and the full or overexploitation of most stocks indicate a need for regime change. Moreover an open-access system will not lead to stable successful fisheries management.[26] The fact that many species are overfished by itself indicates that the system is not working and underscores the need for multinational monitoring and regulation of this multispecies fishery, and ultimately, of fair allocation of the resource. The extension of jurisdiction to 200 nautical miles will transform the contractual environment from a bargaining situation for an open-access common pool resource to a hierarchical enclosed resource.

Historically, conflict more than co-operation has been characteristic of fisheries relationships among North-East Asian countries. Despite its outwardly tense stability, the current balance in the regime involves serious compromises in national positions, and for this reason it is unlikely to remain stable for long. South Korea, China and perhaps, eventually, Russia and North Korea are likely to perceive gains in fisheries regulations that could be obtained at acceptable costs, particularly if prodded by their domestic fishers. But these countries may continue to allow open access in their EEZs for both domestic and foreign fishers, which would still lead to overfishing and overinvestment.

Russian Far East fisheries policy is a good example of what might happen. It is currently in disarray. The national economic crisis has resulted in shortages of fuel for fishing vessels; poor vessel maintenance; poor harbour facilities; shortages of capital and labour; disorderly privatization of state assets; assignment of quotas to local/regional entities; excessive sales of those quotas to joint ventures and foreign investors without regard to the condition of the stocks; and lack of legal protection for foreign investors. So exclusion of foreign fishermen from its EEZ would be detrimental to Russia's immediate interests: hard currency for fishing fees and use of ports, and technology or equipment transfer. But these arrangements are unlikely to hold in the longer run as concern over the condition of stocks becomes an increasingly important domestic political issue. In 1993 Russia reduced North Korea's 1994 quota of pollack and cuttlefish from 60,000 tons to 30,000 tons. Depletion of resources in the Bering and Okhotsk Seas and the needs of Russian fishermen were the reasons given by Moscow.[27]

Although South Korea still feels its agreement with Japan is, on balance, in its favour, it recognizes that the agreement also involves an unequal compromise of its fishery interests.[28] Some provisions are outmoded under UNCLOS because South Korea could have rights to fisheries out to the median line rather than only to 12 nautical miles. The present agreement makes no provision for reviewing the annual allocation relative to the maximum sustainable yield. It also appears to make South Korean delineation of its straight baselines contingent upon agreement by Japan. Another South Korean concern is the failure of flag-state enforcement of fishing regulations in the area covered by joint fisheries regulations.[29] In addition, South Korea, although theoretically within its rights to fish anywhere in the Yellow Sea and East China Sea outside the territorial seas and EEZ of North Korea, has voluntarily refrained from fishing in the areas governed by the Japan–China agreement. Seoul protests against, but nonetheless observes the fishery protection zone located in the south central part of the Yellow Sea under the Japan–China agreement, although it does not acknowledge a legal obligation to do so.[30] Nevertheless, South Korea has now expanded its fishing grounds westward and this will result in more competition and conflict between Chinese and South Korean fishermen.[31]

There is no agreement between South Korea and China. Although South Koreans are not fishing heavily in Chinese waters, Chinese vessels are expanding their activities in both North and South Korean waters, and they target the same fish as South Korean fishermen: little yellow croaker, hairtail, cuttlefish, anchovy, lefteye flounder, crab, and shrimp. In 1993 some 1,300 Chinese fishing vessels violated either South Korea's territorial waters or its Fishery Resources Protection Zone; 340 violations were reported in 1994.[32] During the previous December, negotiations were initiated to resolve the problem but there was no consensus on what to do. Nevertheless, negotiations are continuing. South Korea has proposed that the two countries declare mutual exclusion zones. But China has not yet declared a baseline, which makes it difficult to determine where such a zone should begin.

South Korea is in a difficult position because its argument with China may prejudice its argument with Japan, and vice versa. It tries to keep its separate negotiations secret. Meanwhile Japan argues that it should be included in the talks because Chinese fishing affects the interests of Japan, which itself takes about 200 thousand tons in the Yellow and East China Seas. Even so, this is only 3.3 per cent of its total take compared to 24 per cent in the 1970s.[33] Meanwhile, Japan is concerned about Chinese fishing in

Japanese coastal waters in the Sea of Japan and in the North-West Pacific.[34]

South Korea's fisheries conservation zone within the unilateral Rhee Line was initially meant to exclude Japanese fishermen. After the formal Japanese/South Korean agreement, it became insignificant. Now South Korea wants to revive the Rhee Line to protect its fish from Chinese fishermen. Indeed, to exclude Chinese fishermen, South Korea is considering expanding its jurisdiction to 12 nautical miles in the Korea Strait where it previously enforced only a 3 nautical mile territorial sea.[35] Meanwhile, China wants South Korea to respect its Trawl Prohibition Line. Beijing argues that its line is more or less consonant with its territorial sea whereas the Rhee Line extends far beyond South Korea's territorial sea.

The original purpose of the South Korea–Japan agreement was to regulate Japanese fishing because of its superior capability. However, with the growth of the South Korean fishing industry, Japan feels that its agreement with South Korea is not in its interests, primarily because the agreement does not carry enforcement powers. It is concerned about illegal fishing in its waters, depletion of its stocks, and disputes over accidents between vessels.[36] Of particular concern are large Korean trawlers operating in areas in which Japanese trawlers have been banned by domestic legislation and Korean dragnet, purse-seining, squid-angling, and conger eel (*anago*)-pot fisheries in areas where Japanese fishing is prohibited.[37] In 1987 the two countries adopted self-regulation, limiting the number of vessels, fishing period, and size of boats. But the Japanese government has continued to protest illegal activities carried out by South Korean fishermen. Indeed, then Japanese Prime Minister Morihiro Hosokawa personally raised the issue at a summit with South Korean President Kim Young-sam in November 1992. The two countries tried to renegotiate the self-regulation fishing agreement. The talks focused on South Korean illegal fishing, compensation for accidents, and mechanisms to protect marine resources. They did reach agreement, valid through 1994, on voluntary measures to reduce accidents and settle disputes. But the key issue was whether the two should renegotiate the 1965 basic fishing accord as demanded by Tokyo, or renew the agreement for three years with supplementary measures as proposed by Seoul. Japanese fisheries officials believe that neither country should establish a full 200-nautical mile fishery zone before a substitute for the 1965 agreement is in place. And Tokyo fears that termination of the current agreement might lead to South Korea resurrecting the Rhee Line.[38] Nevertheless, Japan may be considering unilaterally imposing a 40- or 50-nautical mile 'conservation zone' in the Sea of Japan.

Maritime incidents are proliferating and becoming more dangerous (Table 6.4). The detention by Chinese authorities of a Russian fishing factory ship in the East China Sea underscores another stimulant for a change in the fisheries regime—fishing by nonclaimants.[39] This incident was one in a series perpetrated by supposedly 'rogue' Chinese vessels carrying uniformed personnel and involving Russian and other countries' vessels—including even one North Korean vessel in Chinese waters. After experiencing some 17 such attacks in late 1992 and early 1993, Russia dispatched a Kara class cruiser to the area to halt 'pirate' attacks on its vessels.[40]

Korean navy and maritime police have been instructed to step up patrols in the East China Sea to cope with a series of attacks and threats on South Korean freighters by armed pirates. Also, perhaps in response, Russia detained a Chinese trawler for illegal fishing in Russian territorial waters near Sakhalin.[41] The incidents eventually had to be resolved by the Russian and Chinese foreign ministers. While it appears that some of the detained Russian vessels may have actually been smuggling cars and other goods, the tit for tat nature of the problem demands attention. Some suspect it was an attempt by China to squeeze other countries' fishing vessels out of what will eventually be its EEZ.[42]

Many of these incidents involved Japanese vessels[43] and eventually led in March 1993 to a Sino-Japanese consultation regarding maritime security in the East China Sea, in which both sides agreed to work out effective measures for co-operation in safeguarding its high seas. A second round of talks in November-December 1993 in Tokyo reached a four-point agreement which included regular consultations on combatting piracy in the East China Sea.[44] The agreement has proven very effective and the number of incidents has been sharply reduced.[45]

In August 1992, North Korean gunboats opened fire on a fleet of Chinese fishing boats in the Yellow Sea, causing casualties. The action was thought to be linked to Pyongyang's displeasure over the normalization of relations between South Korea and China.[46] The following August, South Korea's Defence Minister, Kwon Yong-hae, said that he would officially protest Chinese illegal fishing around the Northern Limit Line in the Yellow Sea.[47] The web of bilateral and multilateral agreements. And the few remaining unsettled boundaries have not prevented the establishment and implementation of conventions to protect the marine environment.

This co-operation did not come easily. It only succeeded after several aborted attempts. In the end, it was due to a general improvement in relations between the socialist and Western European nations, combined with recognition of the Baltic as a common resource and genuine concern for its living resources, particularly its fisheries. The co-operative effort has been successful in that it has been a factor in retarding the environmental degradation of the Baltic.

The Mediterranean Sea has been the focus of numerous co-operative actions by its coastal states. UNEP's strategy of a convention followed by protocols, and concurrent assessment and regulatory activities was first developed here. The umbrella instrument is the 1976 Barcelona Convention for the Protection of the Mediterranean Sea Against Pollution, supported by five protocols. These legal instruments are supplemented by various related or similar programs. The heart of the effort is the Mediterranean Action Plan (MAP), divided in turn into the Blue Plan and the Priority Action Programme (PAP). There is also an Environmental Program for the Mediterranean, the Mediterranean Centre for Research and Development in Marine Industrial Technology, and the Centre for Environment and Development for the Arab Region and Europe. Regional co-operation began with a concern about ship-generated marine pollution, evolved into a concern about the effect of pollution on living resources, and eventually developed into a more general concern about the impact on the environment of a range of human uses and thus the general pattern of economic development.

However the system is not working very well. Ratification of the convention and the protocols, especially the third protocol on land-based sources, was slow and implementation has been even slower. Control of land-based pollution requires fundamental and politically unpopular decisions in agricultural and industrial policies and practices. Naturally there is a growing concern that 'co-operation' does not erode national sovereignty. Eighty-three per cent of the contributions for the MAP come from only three countries, and those contributions are inevitably late, causing the postponement and even cancellation of activities. The Blue Plan has been plagued by organizational difficulties, lack of well-defined goals, problems with the host state, slow responses from the other governments, and financial shortages. The other programs have had similar difficulties, and suffer as well from the non-participation of key states and competing claims to ocean space. There is also a general concern that there are too many overlapping programs and activities.

The former co-ordinator of the Mediterranean Action Plan ventured some frank criticism of its implementation from his particularly pertinent perspective.[83] He felt that the Programme concentrated overly on marine pollution and was basically an amalgam of somewhat disparate activities. This is especially, and increasingly, a problem for the developing nations on the southern rim of the Sea. On the other hand, the developed Mediterranean countries are afraid of being subjected to a pan-Mediterranean 'tribunal' that might be dominated by the 'south', and thus they do not support a legal committee to supplement the scientific component of the Plan. This conundrum has lessons for North-East Asia where a legal framework which might integrate such disparate principals and socio-political systems seems unlikely.

Another fundamental problem is that UNEP headquarters in Nairobi attempts to run the Plan from there, often disregarding the suggestions and wishes of the co-ordinator's office and, more importantly, those of the Contracting Parties to the Convention which they established. There is also a belief among the Contracting Parties that an American-led world favours a global approach rather than regional initiatives which are more difficult to control.

The Mediterranean Plan process has yielded important lessons for future attempts to form international marine environmental regimes.[84] Although pollution in the Mediterranean might have been worse than it is today if the plan had not been instituted, this is partially due to factors other than the Plan itself. It was basically a political success but a substantive failure, in that co-operation resulted in a comprehensive legal and institutional plan to clean up the Mediterranean, but the environmental condition of the Sea did not appreciably improve. The Plan was politically successful because the motives of participating countries for joining were mostly not environment-related. In particular, UNEP bore most of the initial financial burden and provided financial, technical, and educational aid to strengthen the marine scientific capacity of the Lesser Developed Countries (LDCs). The Plan also provided opportunity for traditionally hostile countries to establish diplomatic ties. Moreover, signing conventions and protocols is initially inexpensive. But subsequent compliance is expensive which explains why the Program stalled in its later stages.

In the beginning there was high uncertainty, high complexity, and a perception that pollution and its prevention were common problems. There were also different incentives due to perceived asymmetry in the abatement costs and effects of pollution. LDCs feared that developed countries were trying to exploit them by making them pay for pollution that the developed countries had caused. They also worried that prevention would make them less competitive and divert resources from their economic development.

UNEP's contributions corrected the initial asymmetrical incentives. Moreover, the participation by LDC scientists reduced uncertainty regarding the distribution and effects of pollution, and therefore changed the perception of the problem. In particular, studies conducted by MedPol supported neither the hypothesis that the Mediterranean was on the verge of 'collapse' nor the assumption of extensive transmission of pollution across national boundaries. But such studies did indicate that cleaning up the Mediterranean would be very costly. Indeed, identifying land-based pollutants and rivers as the major sources of pollution meant that participating states would have to take measures which covered their entire territories. But, since pollution was largely localized and the developed countries were the main polluters, they would have to bear the highest costs. The low pollution exchange and the consequent minor interdependence between North and South also meant that no single actor had real power over others. So reduction in ecological uncertainty led to reduction of economic and political uncertainty.

France and, to a lesser extent, Italy were the major polluters of the Mediterranean and their participation was necessary to clean it up. France emerged as the scientific leader, and the developed countries had the scientific upper hand. But the false perception that the Plan was a response to a truly collective problem was used by the South to stall its implementation if the MEP did not satisfy them.

Marine regionalism has reached its pinnacle in the North Sea where comprehensive sea-use planning has been ongoing since the early 1970s. The international goal of the North Sea regime is the development of a balanced and effective battery of instruments for both national and international administration and management.

The general economic objective is the safe and efficient use of the various opportunities offered by the sea, both now and in the future. There are five specific plans: for offshore oil and gas development, for fisheries management, for water quality, for shipping routes, and for special sea uses. The North Sea regime also combines infrastructure facilities and furthers multiple or shared use of particular areas. An important feature is its small but efficient joint secretariat which facilitates co-ordination between two different conventions. The commissions established to implement the conventions are supported by technical working groups and a joint monitoring group.

A detailed proposal has been put forward for the Arctic region or 'Beringia.' Dramatic changes in the intensity of resource use, the relative interests of the principal actors, and concomitant claims and conflicts have made the existing regime outmoded. Therefore a mixed system is proposed for the governance of the Arctic. It will comprise a combination of several functional agencies for the management of fisheries with a loose overarching regional and shipping authority capable of coping with problems that cut across the domain of individual functional agencies, such as environmental protection and preservation.

From these and other examples, it is possible to deduce some general factors that favour or retard the success of marine regionalism. Clearly, the member countries must perceive that the benefits of participation in a co-operative arrangement outweigh the costs. Benefit/cost considerations include direct benefits, such as better use of marine resources, as well as indirect benefits, such as the advancement of a state's aspirations for regional leadership. In the initial stages of co-operation, the costs should be kept as low as possible.

Also relevant in benefit-cost considerations are the objectives and functions of the arrangement. Multiple use regimes have found it difficult to overcome the often contrary factors that create their need.[85] Multiple use approaches lack a vigorous political constituency because this usually forms around a single use. By their very nature, multiple use regimes partition opportunity among interest groups and thus create a more complex political milieu. They must also overcome segmented entrenched interests, which feel comfortable with single use regimes. These interests include agencies within governments, which tend to resent allocation of funds, personnel, and authority to international endeavours.

It appears, then, that sector-specific approaches have a greater chance of success than some supranational or multipurpose system that seeks to perform a multitude of functions within the framework of a single authority. This conclusion is a political reality despite the logic and apparent need in some areas for an integrated management system. A corollary is that success is encouraged by the recognition and promotion of subregional efforts rather than by an overemphasis on comprehensive regionalization. To minimize controversy and politicization in co-operative efforts, perhaps priority should be placed on innocuous sectors and combined with decentralized decision making authority.

Positive perceptions of co-operation are also necessary for successful regional efforts. For example, Western Europe, the site of the most successful marine regional efforts, has been preparing for economic unity since the late 1940s. Public and private institutions must establish regional links, and there must be a cadre of personnel committed to the regional concept. And in order to maintain momentum, expectations of progress must be reasonable and they must be satisfied. The ASEAN countries, for example, have consistently opposed a formal treaty on environmental protection as being too rigid and potentially divisive.

It is clear from the North Sea experience that formation of an action-oriented regime can take a long time. Few direct measures to reduce the level of pollution in the North Sea were agreed upon during the first decade of regional effort. Due to lack of knowledge about the level and effect of pollutants, monitoring and information gathering constituted the most important work for that initial decade. Conversely, if no progress is made within a reasonable time, the movement toward co-operation is weakened or its objectives are altered.

Many regional seas programs were only initiated after scientific revelations of regional pollution or a well-publicized environmental disaster. This is because governments tend to be forced into action only when significant political constituencies insist on it. The transition in relevant knowledge, perceptions, and values as they relate to the world's oceans is producing such a groundswell. The present generation views ocean space differently from previous generations. Although ocean ecosystems will probably never be completely understood, their complicated nature is slowly being revealed, and we are becoming increasingly aware of peoples' capacity to damage the ocean environment, with both immediate and long-term practical effects readily understandable to those without training in ocean science. For example, the loss of income to businesses dependent upon beachgoers and the loss of the opportunity to bathe during a hot summer owing to the presence of assorted waste products in coastal waters generate enormous pressures on policy makers to 'do something'. However the prerequisites of an effective public policy debate, such as clear statements of risk and the identification of costs and benefits, do not mesh well with the slowness, deliberateness, and uncertainty of the scientific process.[86]

Disagreements stemming from uncertainty can be overcome by adopting the precautionary principle, which is an agreement to take action even when cause and effect are only suspected rather than proven. Indeed, as pollution increases and becomes a publicly perceived problem, ocean management moves closer to centre stage as a significant public policy area concern meriting attention. So the combination of a thorough planning process, public pressure and a high-level political conference may be key factors in the formation of environmental protection regimes.

There are both obvious and subtle reasons why some marine regional efforts do not succeed. The obvious reasons include inadequate leadership, limited management authority, a lack of effective enforcement powers, non-binding effects on non-parties, disagreement among member states, shortages of funds, and shortage of trained personnel or equipment. More subtle problems can include limited information and/or the ability to use it, jurisdictional limitations, differing time horizons, and 'freedom' costs.[87] A lack of information regarding scientific knowledge relevant to effective ocean management can cause disagreements between governments and inhibit the generation of sound national and international policy. Boundary disputes or uncertainties retard the implementation of jurisdictional control, without which co-operation is made more difficult. Furthermore, existing jurisdiction may not encompass the entire ecosystem so that some critical elements requiring management are beyond state control. Ironically, although segmented or sectoral approaches to co-operative ocean management are much more likely to gain the political support necessary to progress, they may ultimately be unsuccessful in resolving the problem at issue, particularly if conflicting uses/users are involved.

There is obviously a time gap between the average length in

political office of ministers and the tens of years required to implement and sustain successful co-operation. Yet policy-makers need to show results in the short term. This dilemma favours multiple short-term programs which may be insufficient for the achievement of long-term results.

'Freedom' costs at both the individual and state levels may ultimately block successful co-operation as well. Individuals and states will have to surrender their absolute 'freedom of the seas,' that is, their absolute rights to use the ocean environment at any time, for any purpose, and without regard for the interests and well-being of anyone else. Such freedoms will be surrendered with considerable reluctance in the absence of anything less than a clear, overwhelming, and immediate need. But formal sovereignty is different from operational sovereignty—the legal freedom of action. The latter can be eroded in the interest of effective collective action. Indeed, environmental interdependence and related international negotiations reinforce formal sovereignty while resulting in the self-limitation of operational sovereignty.[88]

A Model Maritime Regime

These foregoing examples, read in conjunction with the work of Young and Osherenko,[89] enable the distinctive parameters of successful regional maritime regimes to be delineated. An ideal maritime regime would have many of the following features.

- First and foremost, *a positive perception of co-operation would exist throughout the region as a foundation for forming and nurturing a regime.* Public and private institutions would have established regional links, and an epistemic community, committed to the concept of regionalism, would have developed. Hence the regime would be supported by a significant political constituency.
- *The member countries would perceive that the benefits of participation outweigh the costs.* Such benefit/cost considerations would include direct benefits like better use of marine resources, as well as indirect benefits such as the advancement of state aspirations for regional leadership. In the initial stages of co-operation, the economic and political costs would be kept as low as possible.
- *The regime would be widely perceived as equitable, however measured.* The distributive consequences of the regime over

time would be difficult to predict, thus expanding the contract zone and thickening the veil of uncertainty. These consequences, in turn, would mute the positional and distributive aspects of bargaining, and encourage the singling out of a few key problems and approaches to them that each participant could accept as a package, rather than as discrete elements. Expanding the contract zone would also allow participants to perceive themselves as occupying different roles, for example, as both sources and sufferers of pollution, or as perpetrators and victims of overfishing. De-emphasizing or avoiding distributive issues would enhance regime formation by enabling the participants to focus their attention on the search for mutually beneficial solutions. If distributive issues have to be addressed, a 'cookie-cutter' approach would be perceived as more equitable, and the participants would be willing to accept this formula.

- *A shock or crisis exogenous to the negotiating process would enhance regime formation or speed it along once the process has begun, and the participants would be prepared to make the most of such 'windows of opportunity.'* Such an event would stimulate the emergence of political will to address the issues in integrative bargaining, thereby opening windows of opportunity for the regime to form. These shocks or crises might include broad shifts in values and ideas, such as a growing environmental consciousness, changes in the political system, such as the end or muting of the Cold War, or specific events, such as an environmental crisis, the 'coming out' of North Korea, or a change of key officials.
- *Disagreements due to uncertainty would be mitigated by commitments to co-operative monitoring and research to bridge knowledge gaps, and by adoption of the precautionary principle.*
- *The objectives and functions of the regime would be sector specific*, such as environmental protection or fisheries management. In order to encourage the broadening of knowledge, the agenda in the environment sector should be focused on harm rather than specific pollutants.
- *Expectations of progress would be reasonable.* Most successful environment and resource regimes have inauspicious starts. But institutions are needed early on to help create the conditions that make strong rules possible. Indeed, effective institutions seize opportunities to enhance themselves.

Progress would thus be slow but steady; and perceptible, otherwise the movement towards co-operation could be weakened or its objectives altered.

- *The regime would form in stages beginning with a limited and temporary regime—perhaps a 'maritime council'—and move eventually to a broader convention.* It would initially place priority on innocuous sectors in order to minimize controversy and politicization of the issues. Informal or supplementary interpretations of the regime would evolve over time. To prevent misunderstanding among the participants over which track they are on, the distinction between institutional bargaining and prenegotiation would be clarified.
- *Objectives and functions would be clear.* Objectives would include conservation, management and/or development; protection of state interest; or provision of greater equity. Some objectives would be hidden. Others would be linked to regional or global objectives. Functions would initially include service, information exchange, data gathering and analysis, consultation, facilitation and co-ordination of programs, joint planning, and technical assistance. Eventually functions might involve operation of an organization, management of resources, and exploration and exploitation. Ultimately they could include norm creation, compliance, allocation, and rule observance.
- *The regime would be uncomplicated, but not loose.* It would define geographic scope, membership, rules and regulations, procedures, decision-making, and, if an organization is necessary, financial matters and staffing.
- *Geographically, the regime would fit the natural system on which it focuses, although its exact boundaries would be ambiguous to mute jurisdictional concerns.* In addition, it would initially be a subregional effort rather than emphasizing comprehensive regionalization.
- *Decision-making would be decentralized.* A co-ordination regime would establish rules and procedures while leaving each member free to implement them in its own way.
- *The regime would concentrate on policy questions and be negotiated by skilled diplomats.*
- *Negotiations would result in one or more explicit agreements.*
- *The issues would be defined at the outset rather than allowing the process naturally to correct faulty initial formulations.*

- *Alternative regime designs would be available, new ways of understanding the problem would be introduced, and the participants would be willing and able to entertain new ideas.* Solutions would be creative, simple, and easily grasped by policymakers, journalists, and the public.
- *There would be no explicit or unsubtle attempts to use power in institutional bargaining.* There can be issue-specific hegemony in which a hegemon can use its dominance in scientific research and diplomatic expertise to impose its preferred outcome. Because this may backfire, power must be used with care, even by a hegemon. Nevertheless, the distribution of power and authority within the regime would be clear.
- *While there would be willing and able national leadership, some countries would also be prepared to play the role of effective follower in order to make the process work.* Strong leadership is both affected by, and affects, power relationships, and it also shapes values and ideas. But a desire for mutual control will not necessarily produce a successful regime. As long as participants sharply disagree over who will control what, regime formation will be difficult.[90] The major powers—China and Japan—would participate, and South Korea would assume a key role, perhaps by offering to host and chair meetings and, thereby, enabling the major powers to avoid appearing too dominant or assertive. Not all interested parties would need to participate in the formation of the regime. Nor must the issue be either a commonly high or low priority on all national agendas. Indeed the issue would be a high priority for one or more parties but a low priority for others.
- *A strong individual leader or leaders would provide impetus and direction for the regime.*

Notes

1. Because marine policy issues cannot be entirely divorced from the larger political context in which they are embedded, this work is informed by both realist and neorealist arguments. However, regime theory used in this study is drawn primarily from the multiple works of Keohane, Krasner, and Young cited below.

2. Robert Keohane, *After Hegemony: Cooperation and Discord in the World Political Economy*, Princeton: Princeton University Press, 1985, p. 243. (Hereafter *After Hegemony*)
3. Ibid., pp. 51–2; 63.
4. Stephen D. Krasner, 'Structural causes and regime consequences: regimes as intervening variables' in Stephen D. Krasner (ed.), *International Regimes*, Ithaca: Cornell University Press, 1983, pp. 1–21; Robert O, Keohane, 'The demand for international regimes', *International Organization*, 36, 2 (1982): 325.
5. Keohane, *After Hegemony*, pp. 52–4.
6. Oran R. Young, *International Co-operation for Natural Resources and the Environment*, Ithaca: Cornell University Press, 1989, pp. 196–8. (Hereafter *International Co-operation*)
7. Ibid., p. 103; Oran R. Young, *Resource Regimes: Natural Resources and Social Institutions*, Berkeley: University of California Press, 1982, Vol. 2, p. 16. (Hereafter *Resource Regimes*)
8. Arthur Stein, 'Coordination and collaboration: regimes in an anarchic world' in David A. Baldwin (ed.), *Neorealism and Neoliberalism: The Contemporary Debate*, New York: Columbia University Press, pp. 309–311.
9. Ibid.
10. Stein, 'Coordination and collaboration: regimes in an anarchic world'; Keohane, 'The demand for international regimes'.
11. Keohane, *After Hegemony*, pp. 103–6.
12. Robert Jervis in Krasner, *International Regimes*, p. 364.
13. Vinod Aggarwal, *Liberal Protectionism: The International Politics of Organized Textile Trade*, Berkeley: University of California Press, 1985, p. 28.
14. Stephan Haggard and Beth A. Simmons, 'Theories of international regimes', *International Organization*, 41, 3 (1987): 498; Young, *Resource Regimes*, p. 202.
15. Richard Higgott, 'Economic cooperation: theoretical opportunities and political constraints', *The Pacific Review*, 6, 2 (1993): 109.
16. Muthiah Alagappa, 'Regionalism and the quest for security: ASEAN and the Cambodian conflict', *Journal of International Affairs*, 46, 2 (Winter 1993): 442.
17. Young, *International Cooperation*, p. 203.
18. In September 1992, Norway, Iceland and other whaling nations established the North Atlantic Marine Mammal Commission as an alternative to the International Whaling Commission (IWC), and may defect from the IWC altogether. Japan is considering establishing a similar organization in the Pacific (Tsuneo Akaha, 'Balancing developmental needs and environmental concerns, domestic and international interests: Japan's Ocean Policy in the Post Cold War Era', draft manuscript, 1993, p. 84).
19. Haggard and Simmons, 'Theories of International Regimes', pp. 506, 516–17.
20. Peter M. Haas, 'Introduction: epistemic communities and international policy coordination', in Peter M. Haas (ed.), *Knowledge, Power and International Policy Coordination, International Organization*, 46, 1 (Special Issue, 1992): 31–2.
21. Higgott, 'Economic co-operation: theoretical opportunities and political constraints', p. 114; Michel Foucault may have invented the term 'epistemic community' in his *The Order of Things*, New York: Random House, 1970. However, as Ernst Haas has argued, Foucault's usage is indistinguishable from what might be called 'ideological communities.' For the meaning, definition, role, value, and examples of who may or may not constitute an epistemic community see, Burkhart Holzner and John H. Marx, *Knowledge Application*, Boston: Allyn and Baron, Inc., 1979, p. 108; Ernst Haas, *When Knowledge is Power*, Berkeley: University of California Press, 1990, pp. 40–6; Haas, 'Introduction: epistemic communities and

international policy coordination': pp. 1–36. The term epistemic community was first applied to international relations by John G. Ruggie, 'International responses to technology, concepts and trends', *International Organization*, 29, 3 (Summer 1975): 569–70. See also Peter M. Haas, 'Do regimes matter? epistemic communities and Mediterranean pollution control', *International Organization*, 43, 3 (Summer 1989): 377–403.

22. Keohane, *After Hegemony*, p. 79.
23. Haggard and Simmons, 'Theories of international regimes', p. 507.
24. Ronald P. Weidenbach and John Bardach, 'Fisheries', in Joseph R. Morgan and Mark J. Valencia (eds.), *Atlas for Marine Policy in East Asian Seas*, Berkeley: University of California Press, 1993, p. 120.
25. Young, *International Cooperation*, p. 217.
26. Haggard and Simmons, 'Theories of international regimes', p. 507.
27. Young, *International Cooperation*, pp. 230–5.
28. Mark J. Valencia, 'Engaging North Korea in regional economic cooperation', *Trends*, Institute of Southeast Asian Studies, 25–6 December 1993, p. 10.
29. Mark W. Zacher, 'Toward a theory of international regimes', in Robert L. Rothstein (ed.), *The Evolution of Theory in International Relations*, Columbia: University of South Carolina Press, 1991, pp. 119–37.
30. Robert O. Keohane, Peter M. Haas and Marc A. Levy, 'The effectiveness of international environmental institutions', in Peter M. Haas, Robert O. Keohane and Marc A. Levy (eds.), *Institutions for the Earth: Sources of Effective International Environmental Protection*, Cambridge: The MIT Press, 1993, pp. 19–20. (Hereafter 'The effectiveness of international environmental institutions')
31. Garett Hardin, 'The tragedy of the commons', *Science*, 162 (1986): 1243–8. A tragedy of the commons is a situation in which each pursues his own best interests under a regime that fosters open access to a commons, but that very freedom brings ruin to all.
32. Philippe Sands, 'Enforcing environmental security: the challenges of compliance with international obligations', *Journal of International Affairs*, 46, 2 (Winter 1993): 367–90.
33. Keohane et al., 'The effectiveness of international environmental institutions', pp. 16–17.
34. Peter Sand (ed.), *The Effectiveness of International Environmental Agreements. A Survey of Existing Legal Instruments*, Cambridge: Grotius Publications, Ltd., 1992.
35. Keohane et al., 'The effectiveness of international environmental institutions', p. 17.
36. Gary Bryner, 'Constructing international institutions to ensure effective implementation of global agreements'. Paper presented at the Annual Meeting of the International Studies Association, Chicago, 23–5 February 1995, pp. 2–3.
37. Young, *Resource Regimes*, pp. 117–37.
38. Social welfare economics is that branch of economics which brings together the pertinent theoretical relationships from positive or 'pure' economics in terms, typically, of a single end: the economic welfare of the community. Jeremy Rothenberg, *The Measurement of Social Welfare*, Englewood Cliffs, NJ: 1961.
39. The first-order conditions for Pareto Optimality in the absence of external effects, are: (1) for all consumers of any two commodities, the marginal rate of substitution between them must be equal; (2) for all producers of any two commodities, the marginal rate of production transformation between them must be equal; (3) for any two commodities, the common marginal rate of substitution must equal the common marginal rate of transformation. Rothenberg, *The Measurement of Social Welfare*, pp. 95–6.

40. Keohane et al., 'The effectiveness of international environmental institutions', p. 14.
41. Young, *Resource Regimes*.
42. Lewis Alexander, 'Marine regionalism in Southeast Asian Seas', *East–West Environment and Policy Institute Report No. 11* (1982): 8–10.
43. Alagappa, 'Regionalism and the quest for security: ASEAN and the Cambodian conflict', p. 442.
44. Eugene B. Skolnikoff, *The International Imperatives of Technology: Technological Developments and the International Political System*. Berkeley: University of California, Institute of International Studies (1972): pp. 13–16, 102–10.
45. United Nations Convention on the Law of the Sea. New York: United Nations, 1983.
46. Young, *Resource Regimes*.
47. Mark J. Valencia and George Kent, 'Co-operation: opportunities, problems and prospects' in George Kent and Mark J. Valencia (eds.), *Marine Policy in Southeast Asia*, Berkeley: University of California Press, 1985, p. 370.
48. Tsuneo Akaha, 'Northeast Asian economic cooperation: national factors and future prospects', Paper presented to the International Studies Association, Monterey, 1993.
49. Valencia and Kent, 'Co-operation: opportunities, problems and prospects', pp. 381–2.
50. Keohane et al., 'The effectiveness of international environmental institutions', p. 13.
51. Young, *International Cooperation*, p. 51.
52. Alexander, 'Marine regionalism in Southeast Asian Seas', p. 3.
53. Kent and Valencia, 'Introduction'.
54. Congressional Record, *Proceedings and Debates of the 103rd Congress*, Second Session, 140, 86 (30 June 1994).
55. Moritaka Hayashi, 'The 1994 agreement for universalizing the new Law of the Sea', Paper presented to the International Studies Association 36th Annual Convention, Chicago, 21–5 February 1995, pp. 8, 10.
56. See, e.g., W. J. McG. Tegart, G. W. Sheldon and D. C. Griffiths (eds.), *Climate Change: The Intergovernmental Panel on Climate Change Impacts Assessment*, Canberra: Australia Government Publishing Service, 1990.
57. See, e.g., 'Depleted fisheries: world crisis is now', *The Washington Post*, 20 March 1994, p. B-2; Jessica Matthews, 'Fish: The tragedy of the oceans', *The Economist*, 19 March 1994. pp. 21–4; 'The catch about fish', *The Economist*, 19 March 1994, pp. 13–14; Anne Swandson, 'World's fishermen hit bottom in pursuit of ocean's bounty', *Honolulu Advertiser*, 14 August 1994, p. A-21.
58. See e.g., *The State of Marine Environment*, Report and Studies No. 39 (UNEP 1990), *Report of the Joint Group of Experts on Scientific Aspects of Marine Pollution*, (GESAMP).
59. See e.g., Charles J. Johnson, Allen L. Clark, and Robert N. Yonover, 'The future of Pacific Ocean minerals' in Peter N. Nemetz, *The Pacific Rim: Investment, Development, and Trade*, 2nd ed.,Vancouver: University of British Columbia Press, 1990, pp. 259–76; Charles Johnson, Allen Clark and James Otto, 'Pacific Ocean minerals, the next twenty years' in Peter N. Nemetz, *The Pacific Rim: Investment, Development, and Trade*, pp. 199–222.
60. 'A driller's nightmare may be a dream energy source', *Business Week*, no. 3180, 1 October 1990, p. 69.
61. *United Nations Conference on Environment and Development*, A/CONF. 151/26, III, 14 August 1992, p. 20; K. Saigal, 'Ocean governance: a model for regional seas in the twenty-first century', Paper presented at Pacem in Maribus XXI, Takaoka, Japan, 1994, p. 2.

62. Patricia W. Birnie, 'The Law of the Sea and the United Nations Conference on Environment and Development' in Elisabeth Mann Borgese, Norton Ginsburg and Joseph R. Morgan (eds.), *Ocean Yearbook*, Chicago: The University of Chicago Press, 1993, pp. 13–39.
63. See e.g., Kenneth Sherman and Lewis M. Alexander (eds.), *Variability and Management of Large Marine Ecosystems*, Boulder, Colorado: Westview Press, 1986; Kenneth Sherman, Lewis M. Alexander and Barry D. Gold (eds.), *Large Marine Ecosystems*, Washington, D.C.: American Association for the Advancement of Science, 1990.
64. For background, see Adalberto Vallega, *Sea Management: A Theoretical Approach*, London and New York: Elsevier Applied Science, 1992.
65. World Resources Institute, *World Resources, 1992–93*. New York: Oxford University Press, 1992, pp. 68–9.
66. Muthiah Alagappa, 'Regionalism and security: a conceptual investigation', in Andrew Mack and John Ravenhill (eds.), *Pacific Cooperation*, Sydney: Allen and Unwin, 1994, p. 158.
67. Alexander, 'Marine Regionalism in Southeast Asian Seas', p. 11.
68. Ibid., p. 5.
69. Oran R. Young, *Resource Management at the International Level: The Case of the North Pacific*, New York: Nichols, 1977, p. 22.
70. Anthony Armstrong, 'Breaking the ice: initiatives to improve relations with a national adversary', summarized in United States Institute of Peace, *In Brief*, March 1992.
71. Alexander, 'Marine regionalism in Southeast Asian Seas', p. 5.
72. Ibid., pp. 1–2.
73. Probably the earliest of the modern marine regional efforts was the 1902 formation of the International Council for the Exploration of the Sea (ICES), which has the responsibility of monitoring the abundance of fish stocks in the Baltic Sea and the North-East Atlantic Ocean.
74. S.M. Garcia, 'Ocean fisheries management: the FAO programme', in Paolo Fabbri, (ed.), *Ocean Management in Global Change*, London: Elsevier Applied Science, 1992, pp. 381–419.
75. Henk Postuma, 'Marine scientific research projects undertaken by or under the auspices of international organizations', in A. H. A. Soons (ed.), *Implementation of the Law of the Sea Convention through International Institutions*, Proceedings of the 23rd Annual Conference of the Law of the Sea Institute, Law of the Sea Institute, Honolulu, 1989, pp. 509–17.
76. For a description of the origins of the Mediterranean Action Plan and an evaluation, see Aldo E. Chircop, 'Participation in marine regionalism: an appraisal in a Mediterranean context', in Elisabeth Mann Borgese, Norton Ginsburg, and Joseph R. Morgan, eds., *Ocean Yearbook 8*, Chicago: University of Chicago Press, 1989, pp. 402–16.
77. Boleslaw Boczek, 'The concept of regime and the protection and preservation of the marine environment', in Elisabeth Mann Borgese and Norton Ginsburg (eds.), *Ocean Yearbook 6*, Chicago: The University of Chicago Press, 1986, pp. 282; 285.
78. Ernst Haas, 'Words can hurt you: or who said what to whom about regimes', in Stephen D. Krasner (ed.), *International Regimes, International Organization*, 36, 2 (Special Issue 1982): 207–43.
79. For background, analysis and proposals for co-operation in the Baltic Sea, see Matthew Auer, 'Prospects for environmental cooperation in the Yellow Sea', *Emory International Law Review*, 5 (1991): 163–208; Ton Ijlstra, 'Development of resource jurisdiction in the EC's regional seas: national EEZ policies of EC member states in the Northeast Atlantic, the Mediterranean Sea, and the Baltic Sea', *Ocean*

Development and International Law, 23, 2–3 (1992); Baruch Boxer, 'Mediterranean Action Plan: an interim evaluation', *Science*, 202, 10, (November 1978): 585–90; Gunnar Kullenberg, 'Long-term changes in the Baltic ecosystem', in Kenneth Sherman and Lewis M. Alexander (eds.), *Variability and Management of Large Marine Ecosystems*, AAAS Selected Symposium 99, Boulder: Westview Press, Inc. 1986; B.I. Dybern and S.H. Fonselius, 'Pollution in the Baltic sea', in A. Voipio (ed.), *Elsevier Oceanography Series 30*, Amsterdam: Elsevier Scientific Publishing Company, 1986, pp. 351–82; ICES, *Assessment of the Marine Environment of the Baltic Sea Cooperative Research. Report XX*, Copenhagen, 1986; F.B. Pedersen, 'The sensitivity of the Baltic Sea to natural and man-made impact', in J.C.J. Nihoul (ed.), *Hydrodynamics of Semienclosed Seas*, Elsevier Oceanography Series 24, Amsterdam: Elsevier Scientific Publishing Company, 1982, pp. 385–99; Aoano Voipio, *The Baltic Sea*, Elsevier Oceanographic Series, Amsterdam: Elsevier Scientific Publishing Company, 1981.

80. For background, analysis and proposals for co-operation in the Mediterranean Sea, see Aldo Chircop, 'The Mediterranean Sea and the quest for sustainable development', *Ocean Development and International Law Journal*, 23 (1992): 17–30; Norton Ginsburg, Sidney Holt and William Murdoch, *The Mediterranean Action Plan and The Maritime Development of a Region, Pacem in Maribus 111, Proceedings of a Conference, Split, Yugaslavia, April 28–30, 1972*, The Royal University of Malta Press, 1974; Adalberto Vallega, *A Human Geographical Approach to Semienclosed Seas: The Mediterranean Case in Regional Development*, Chicago: The University of Chicago Press, 1988; S. J. Holt, 'Mediterranean: international cooperation for a sick sea', *Environment*, 16 (1974): 29–33; United Nations Environment Programme, 'Activities for the Protection and Development of the Mediterranean', in *Ocean Yearbook 1*, Chicago: The University of Chicago Press, 1978, pp. 584–97; United Nations Environment Programme, 'Recommendation for the Future Development of the Mediterranean Action Plan', in *Ocean Yearbook 2*, Chicago: The University of Chicago Press, 1980, pp. 547–54; 'Protocol for the Protection of the Mediterranean Sea against Pollution from Land-based Sources', in *Ocean Yearbook 3*, Chicago: The University of Chicago Press, 1982, pp. 489–96; 'MARPOL 73/78: The International Convention Concerning Pollution and the Mediterranean', in *Ocean Yearbook 6*, Chicago: The University of Chicago Press, 1986, pp. 572–73; United Nations Environment Programme, 'The Mediterranean Action Plan: Retrospect and Prospect', presented at the Fourth Ordinary Meeting of the Contracting Parties to the Convention for the Protection of the Mediterranean Sea Against Pollution and its Related Protocols, Genoa, 9–13 September 1985. Published as *Document UNEP/IG. 56/4*, 26 June 1985.

81. For background, analysis and proposals for co-operation in the North Sea, see Sven Andersen and Brit Flostad, 'Sea use planning in Norwegian waters: national and international dimensions',*Coastal Management*, 16 (1988): 183–200; H. Carison, 'Quality status of the North Sea. International Conference on the Protection of the North Sea', in John King Gamble (ed.), *Law of the Sea: Neglected Issues*, Proceedings of the Law of the Sea Institute Twelfth Annual Conference held at The Hague, The Netherlands 23–6 October 1978; Gerard Peet, *Techniques and methods for sea use planning and management in selected areas*, Report of a Literature Study, commissioned by Directie Noordzee, Rijkswaterstaat, 1986; Second Chamber of the State General, The Netherlands, *Harmonization of Netherands North Sea Policy 1989–1992*, 44–5, SDU uitgeverij; A.J. Smith, Elizabeth Kennet and M.B.F. Ranken, *Britain and the Sea: Further Dependence— Further Opportunities*, Edinburgh: Scottish Academic Press Limited, 1984; H.D. Smith and C.S. Lalwani, *The North Sea: Sea Use Management and Planning*, Cardiff: University of Wales Institute of Science and Technology, 1984; Elizabeth Young and Peter H. Fricke (eds.), *Sea Use Planning*. London: Fabian Society, 1975; Steiner

Andresen', 'The environmental North Sea regime: a successful regional approach' in Elisabeth Mann Borgese, Norton S. Ginsburg, and Joseph R. Morgan (eds.), *Ocean Yearbook 7*, Chicago: The University of Chicago Press, 1989, pp. 378–401; Steiner Andresen. 'The "effectiveness" of regional environmental co-operation in the Northern Seas.' Paper presented at the Annual Meeting of the International Studies Association, Chicago, 23–25 February 1995.

82. Young has devoted an entire book to a proposal for co-operation in Arctic resource management: *Resource Management at the International Level: The Case of the North Pacific*.

83. Salvino Busuttil, 'The governance of the Mediterranean', Paper presented at *Pacem in Maribus XXI*, Takaoka, Japan, September 1993.

84. Jon Birger Skjaerseth, 'The "effectivenes" of the Mediterranean Action Plan', *International Environmental Affairs*, 5, 4 (Winter 1993): 313–34.

85. Lawrence Juda and R.H. Burroughs, 'The prospects for comprehensive ocean management', *Marine Policy* (January 1990/1): 34.

86. Bryner, 'Constructing international institutions to ensure effective implementation of global agreements', p. 7.

87. Juda and Burroughs, 'The prospects for comprehensive ocean management', pp. 33–4.

88. Marc A. Levy, Robert O. Keohane, and Peter M. Haas, 'Improving the effectiveness of international environmental institutions', in Peter M. Haas, Robert O. Keohane, and Marc A. Levy (eds.), *Institutions for the Earth: Sources of Effective International Protection*, Cambridge: The MIT Press, 1993, p. 416.

89. Oran B. Young and Gail Osherenko, 'International regime formation: findings, research priorities, and applications', in Oran R. Young and Gail Osherenko (eds.), *Polar Politics: Creating International Environmental Regimes*, Ithaca and London: Cornell University Press, 1993, pp. 223–62. Young and Osherenko co-ordinated a project in which five Arctic or Arctic-related environmental and resource regimes were selected for a careful case study analysis of established hypotheses of regime formation. The regimes chosen were the regime for the conservation of North Pacific fur seals, the Svalbard regime, the regime for conservation of polar bears, the regime for the protection of the stratospheric ozone, and a comparison of the problem of Arctic haze produced by air pollutants with the regime for long-range transboundary air pollution among industrialized Northern Hemisphere countries. The project confirmed some hypotheses and disproved others, and drew important lessons for successful regime formation that are useful to this study of maritime regime building in North-East Asia.

90. Vinod K. Aggarwal, 'Building international institutions in Asia-Pacific', *Asian Survey*, 33, 11 (November 1993): 1040.

3 Marine Regionalism in North-East Asia: The Context

The Natural Environmental Setting[1]

The definition of the region to be covered by any regime is based in part on natural characteristics. The characteristics of each of the semi-enclosed seas in North-East Asia—the Yellow and East China Seas and the Sea of Japan—are different, although there are some obvious commonalities and linkages. The description that follows of the natural characteristics of North-East Asian seas illustrates the transnationality of the living resources and of pollutants, and underscores why the region's seas should be managed cooperatively as a unit.

The Sea of Japan

The Sea of Japan[2] or East Sea (Figure 3.1) is almost encircled by land. The general pattern of surface circulation indicates that pollutants, once deposited, and larval fish, once spawned, may be distributed throughout its reaches. In winter, pollutants deposited off both the Russian and South Korean coasts could be swept into North Korean waters. In summer, pollutants deposited off the North Korean coast could be transported into both Russian and South Korean waters. Pollutants in the Tatar Strait could find their way into northern Japanese waters. For purposes of regime design, then, the Sea of Japan can be treated as a separate entity. But pollutants can also reach the Sea through the Korea Straits from the East China Sea, and from the north through the Tsugaru and Soya Straits. And fish migrate both in and out of the Sea of Japan through all three straits.

The most important current in the Western Pacific north of the equator is the Kuroshio, which brings warm tropical water to the north-east, flowing close to the south and southeast coasts of Japan before turning to the east at about 36°N (Figures 3.2 and 3.3). This current and its margins are associated with high organic productivity and the resultant rich fisheries are the focus of competi-

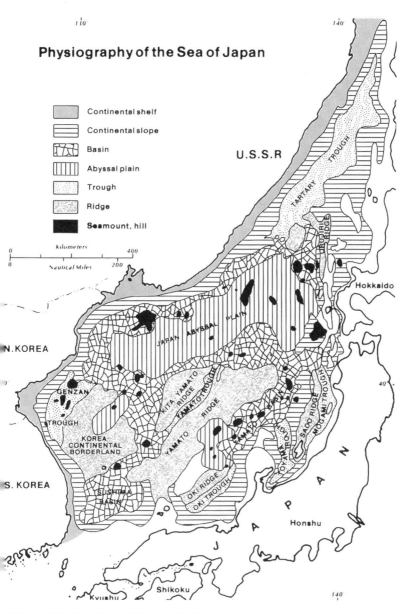

Figure 3.1 The Sea of Japan Basin
Source: Mark J. Valencia (ed.), *International Conference on the Sea of Japan*, Occasional Papers of the East–West Environment and Policy Institute, No. 10, 1989, p. 8.

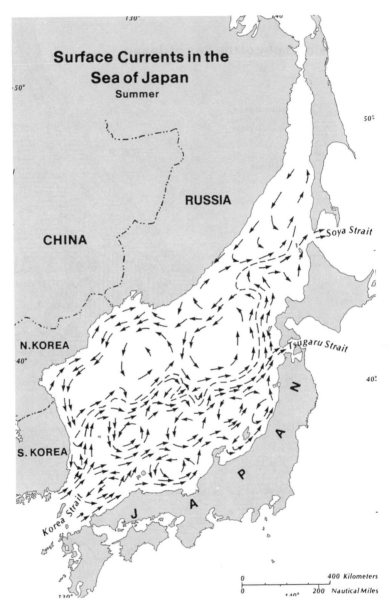

Figure 3.2 Surface Circulation in the Sea of Japan: Summer
Source: Mark J. Valencia (ed.), *International Conference on the Sea of Japan*, Occasional Papers of the East–West Environment and Policy Institute, No. 10, 1989, p. 14.

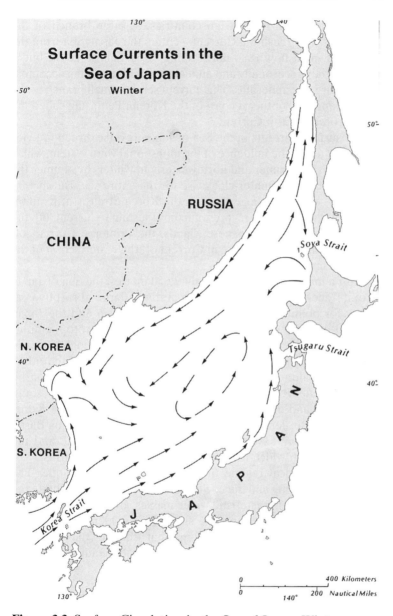

Figure 3.3 Surface Circulation in the Sea of Japan: Winter
Source: Mark J. Valencia (ed.), *International Conference on the Sea of Japan*, Occasional Papers of the East–West Environment and Policy Institute, No. 10, 1989, p. 15.

tion by the North-East Asian countries. A large branch of the Kuroshio—the Tsushima Current—enters the Japan Sea from the south. Its main flow runs north along the Japanese coast, but is highly variable seasonally and annually, sometimes being accompanied by large meanderings. This current sends a small branch to the north along the southeast coast of the Korean Peninsula called the 'East Korean Warm Current'.

The surface currents in the Sea of Japan reverse direction twice each year under the influence of the monsoonal wind system, which is southerly in summer and northwesterly in winter. In summer the current is largely counter-clockwise for the entire sea. But in winter, when water exchange through the Korea Strait is at its minimum, other currents are present: the Tsushima Current off the western coast of the Japanese islands, the Primorye Current off Primorye, and the East Korean Current off the eastern coast of the Korean Peninsula.

With a maximum depth of about 3,650 metres, the Sea of Japan is quite deep. This gives it a comparatively large absorptive capacity for pollutants. The big Sea of Japan Basin and the Japan Abyssal plain lie north of 40°30′N. Other prominent topographic features in the north include the Tartary Trough, several seamounts, including the huge Bogorov Seamount, and the ridges and troughs along the continental slope off the islands of Honshu and Hokkaido. Ridge and trough topography predominates south of 40°30′N, and includes the Yamato Rise, or Ridge, the Korea Continental Borderland (Korea Plateau), and the Tsushima Basin, all of which may have deep water petroleum potential, and the Yamato Trough which may contain metallic sulfide deposits.

The hydrological regime of the Sea of Japan is unique. The bottom topography and the enclosed nature of the sea facilitate the formation of its characteristic water masses. In general, there is a warmer sector on the Japanese side and a colder sector on the Korean-Siberian side. Though somewhat isolated, the Sea is actively influenced by the Pacific Ocean. The water column is divided into two vertical zones: the surface (0 to 200 metres) and the deep (below 200 metres). The hydrological characteristics of the upper zone vary in time and space, while the lower zone is essentially horizontally and vertically homogeneous throughout the year. Pollutants that sink below 200 metres will remain there for a long time.

The polar front separates the cold and low saline waters of the north-western Sea from the warm and high saline waters entering the Sea through the Korea Strait. The resultant vigorous mixing facilitates dilution and dispersion of pollutants. Intensive upwelling conducive to high organic productivity develops near the southern shores of Primorye in late October to early November, when strong northwesterly winds blow from the continent. Discrete instances of short-lived upwelling occur in September to early October. In late November, cooling leads to vigorous mixing and the destruction of stratification in the northwestern Sea, producing a more uniform temperature distribution at the surface.

These oceanographic features determine the distribution of biota. On the north-west shelf and continental slope of the Sea, between Olga Bay and Zolotoi Cape, organic productivity is very high. Deep-water bottom fauna, represented by gastropods (snails), molluscs (scallops and mussels), crustaceans (shrimp), and polychaetes (sea worms), form immense aggregations, exceeding $1,000 g/m^2$ (grams of organic matter per square metre of bottom area). Here the rate of accumulation of organic matter in sediments is considerably higher than on other parts of the Primorye shelf, due not only to the impact of the cold Primorye current and natural processes of sediment accumulation, but also to the descending movement of water in the convergence zone characteristic of this region. Thus pollutants at the surface could be transported rapidly to depth and incorporated into the rich biomass.

In shallow coastal areas, the water temperature in winter drops to freezing point and the coldest and saltiest water sinks rapidly, especially in stormy weather. This water spreads over the bottom from the original site to greater depths, sometimes reaching the shelf edge and creeping down along the continental slope. The areas where these waters are formed are stationary open leads, from where newly formed ice is torn and carried away by severe north-west winds. The main areas of water supply to the deep Sea of Japan Basin lie in the coastal area from Cape Povorotny to Vladimir Bay, and in central Peter the Great Bay and Posiyet Bay. Pollutants deposited here may be rapidly transported at depth throughout the Japan Sea. Primary production is especially high where the Tsushima Current and North Korean Cold Current mix.

High oxygen concentrations are present in the Sea of Japan and are a substantial contributary factor to its rich fisheries. The northeastern Sea of Japan can be divided ecologically into South Primorye (the area from Posiyet Bay to Cape Povorotny), Middle Primorye (north from Cape Povorotny to Olga Bay), and

North Primorye, the extreme northern Sea of Japan (including Tatar Strait and the southern part of Sakhalin Island). The interface between the meanders of the Tsushima Current with the cold Primorye Current is an impenetrable barrier for organisms of the Middle and North Primorye ecotypes.

There are also distinct differences between southern and northern ecotypes. Mussels that form banks in the littoral of the Sea of Okhotsk, form masses in the South Primorye ecotype at depths of 4 to 8 metres. The molluscs in rocky areas of the southern region are replaced in the north by seaweeds. The most important limiting factors governing the distribution and normal development of bottom-dwelling sedentary organisms are pollution, heavy siltation, competition from other organisms, and extreme variations in temperature and salinity.

Some organisms are transported vast distances within the region. As a result of the inflow of a strong Tsushima Current, some groups of tropical animals, including sea snakes, marine turtles, fish, pelagic squids, and jelly fish, are occasionally transported as far as the Russian Far East. But these organisms die without reproducing. Ribbon fish usually live in the deeper zone of the Western North Pacific. After heavy winter storms, they are transported by the Tsushima Current and stranded on the beaches of the west coast of Honshu. Similarly, many porcupine fish, sea snakes, marine turtles, giant squid, and even whales are stranded ashore from time to time.[3]

Because of the particular geography and oceanography of the region, some species are rare or unique. The deeper zone of the Japan Sea is characterized by northern or boreal fish which entered the Sea when communication with the Pacific Ocean and the Sea of Okhotsk was still open. At that time these invading boreal fish had no natural enemies, so they easily accomplished species differentiation. Consequently, several fish, such as *Atopocottus tribranchius* (Honma 1960), *Lycodes sadoensis* (Toyoshima and Honma 1980), *Lycodes japonicus*, and *Careproctus trachysoma* are unique to the Sea of Japan.

The Yellow Sea/East China Sea

Neither the East China Sea nor the Yellow Sea[4] alone can be considered a single semi-enclosed sea, since the boundary between them was set quite arbitrarily as a line running from just north of the Yangtze River (Chang Jiang) mouth to Cheju Island. Hence the two seas are in open communication with each other. And there is a partial opening to the Pacific Ocean to the east through the numerous straits between the Ryukyu Islands, as well as a cul-de-sac in the northern part, the Bo Hai and Korea Bay.

The Yellow Sea covers an area of about 400,000 square kilometres and measures at its maximum about 1,000 kilometres by 700 kilometres. It is very shallow with an average depth of just 44 metres, and a maximum depth of about 100 metres. The Bo Hai has a mean depth of only 21 metres and a maximum of 72 metres. These characteristics mean that this Sea, unlike the Sea of Japan, lacks a long-term sink for pollutants. The sea floor slopes gently from the Chinese continent and more rapidly from the Korean Peninsula to a north-south trending sea floor valley with its axis close to the Korean Peninsula (Figure 3.4). The dividing line be-

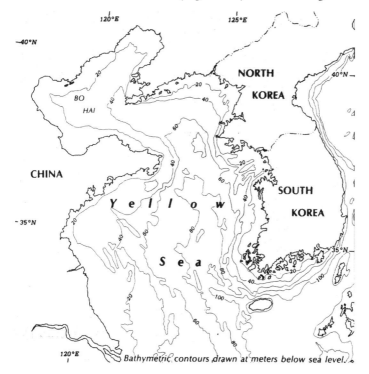

Figure 3.4 The Yellow Sea Basin
Source: Mark J. Valencia (ed.), *International Conference on the Yellow Sea*, Occasional Papers of the East–West Environment and Policy Institute, No. 3, 1987, p. 7.

tween silt derived from China and sand derived from Korea almost coincides with the seafloor valley. China argues that this natural dividing line should be the continental shelf boundary.

The East China Sea also has a broad continental shelf, with the depth of the shelf break at 150 to 166 metres. But depths increase rapidly to the east and reach a maximum of 2,717 metres in the Okinawa Trough. The Sea receives more than 1.6 billion tons of sediments annually, mostly from the Yellow (Huang He) and Yangtze Rivers, which have formed large deltas.

Winds over the Yellow Sea have distinct monsoonal characteristics and determine the variations in its water properties. Winter storms occur every three to eight days and gale strength winds usually prevail after the passage of a cold front. In summer, gale or stronger winds are associated with the passing of typhoons, which occur with a frequency of slightly less than two per year. These storms are hazards to navigation. Furthermore, any pollutants are rapidly mixed with seawater and diluted both horizontally and vertically.

The Yellow Sea is connected to the Bo Hai in the north and to the East China Sea in the south, forming a continuous circulation system (Figure 3.5). The circulation in both winter and summer is counter-clockwise, with the Yellow Sea Cold Current flowing southward along the Chinese coast and the Yellow Sea Warm Current flowing northward along the eastern side of the Yellow Sea basin. On both sides of the north-west Yellow Sea, less saline coastal water flows alongshore with the coast to its right. In summer, the circulation is very slow—less than 0.12 knots at most. There is an eastward current across the Cheju Strait. Its speed is greater near Cheju Island in winter and becomes stronger in the middle of the Strait in summer. Off northern South Korea (along the 36°N), the current flows east to north-east.

The major rivers discharging directly into the Yellow Sea include the Han, Datung, Yalu, Guang, and Sheyang. The Liao He, Hai He, and Yellow Rivers around the Bo Hai have important effects on salinity in the western Yellow Sea, whereas the Yangtze River exerts strong influence on the hydrography of the southernmost part of the Sea. All rivers have peak runoff in summer and minimum runoff in winter.

The rivers also discharge many pollutants which can be transported to adjacent countries' waters, particularly from China to the southern Korean Peninsula by the various ocean currents. There

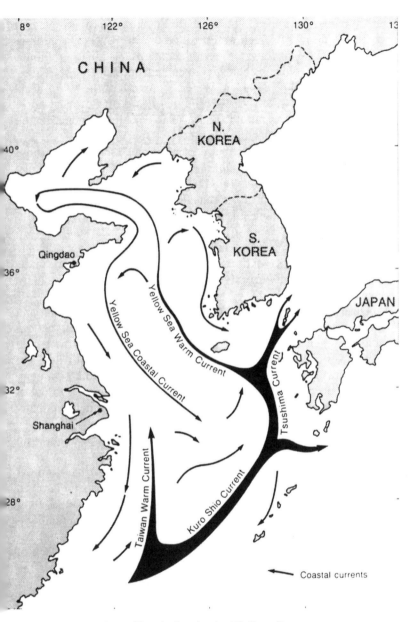

Figure 3.5 Surface Circulation in the Yellow Sea
Source: Mark J. Valencia (ed.), *International Conference on the Yellow Sea*, Occasional Papers of the East–West Environment and Policy Institute, No. 3, 1987, p. 9.

are numerous examples. The Southern Liaoning Coastal Current in the north derives its fresh water from the Yalu River. The autumn western coastal current in the south contains a significant amount of fresh water, which is believed to come from the Yangtze River. The Winter Yellow Sea Cold Water, a mixture of the Warm Current and coastal waters, thrusts southeastward offshore just north of the Yangtze River mouth and has a far-reaching influence on the hydrography of the northern East China Sea.

The tidal ranges along China's coast are much smaller than those along the coasts of North and South Korea. In China, they range from less than one metre around the tip of the Shandong Peninsula, to three metres north of the Yangtze River mouth. Tidal ranges along the Korean Peninsula are more than eight metres in northwestern South Korea producing tidal currents which usually exceed three knots and can be a hazard to navigation when approaching ports. Tidal currents on the western boundary are about two knots except for stronger currents along the middle part of the Jiangsu coast.

During the calm and warm weather months of late spring and early summer, the water column is well stratified, and little bottom sediment is resuspended into it. During the winter, the strong cold winds associated with arctic outbreaks from the north mix and homogenize the water column and resuspend bottom sediment, resulting in considerable turbidity. The net southeasterly flow of water during these arctic outbreaks means that sediments and any pollutants they may contain are pushed across the Yellow Sea. Warmer and saltier water occupies the central part of the basin where demersal species concentrate.

Concentrations of dissolved oxygen and nutrients have similar seasonal variations. In summer the concentrations of dissolved oxygen and nutrients are well stratified. The lower oxygen content in shallow water or the surface layer reflects the higher temperature there, whereas the lower oxygen concentration in the bottom layer is due to oxidation of organic matter as well as the lack of renewal. There is an oxygen maximum at and just below the thermocline. In winter the concentrations of oxygen and nutrients are vertically homogeneous.

Compared with other shelf regions in the north-west Pacific, the Yellow Sea has a relatively low primary production, about $68 g/C/m^{-2}/yr^{-1}$ (grams of carbon produced per square metre of ocean surface per year). Primary production in the northern region is usually higher than that in the southern region, and the lowest level is in the coastal waters of Jiangsu.

Biotic communities of the southeastern Yellow Sea are very complex in species composition, spatial distribution, and community structure possibly owing to the very complicated oceanographic conditions. This makes necessary co-operation in marine scientific research and living marine resource management. The diversity and abundance of the fauna are comparatively low and marked seasonal variations are characteristic of all the biotic communities.

Warm temperate species in the Yellow Sea fauna are the major component of the biomass and account for more than 70 per cent of the total abundance of resource populations; warm water species and boreal species account for about 10 per cent. Fish are the main living resource and 200 species have seen found. Of these, 45 per cent are warm water forms, 46 per cent are warm temperate forms, and 9 per cent are cold temperate forms. The number of species of crustaceans is relatively small—only 54—of which warm water and boreal forms account for 65 and 35 per cent, respectively. Because of the cold temperature, some warm water shrimps do not enter the northern Yellow Sea, while some cold water shrimps are not found in the northern East China Sea.

There are only fourteen species of cephalopods. Warm water forms and warm temperate forms account respectively for 65 and 35 per cent; there are no cold water species. Of the warm temperature species, *Sepia andreana* and *Euprymna morsei* are endemic to the Yellow Sea and do not appear in the East China Sea. About eleven mammal species are found including some which are endangered or highly valued ecologically. Except for four temperate species—minkes whale, sperm whale, humpback whale, and finless porpoise—most are cold temperate forms: harbour seal, northern fur seal, stelles sea lion, fin whale, blue whale, right whale, and gray whale. Of these, fin whales and right whales migrate into the northern Yellow Sea to 39°N in winter and spring to reproduce. Harbour seals migrate into the northern Bo Hai in winter and spring for the same purpose.

The habitats of resource populations in the Yellow Sea can be divided into two groups, nearshore and migratory. And like pollutants, the distribution of fishery resources is transnational. Nearshore species include skates, greenline, black snapper, scaled sardine, and spotted sardine. These species are found mainly in bays, estuaries, and around islands, but they move to deeper waters in winter. The migratory species, small yellow croaker, hairtail, and Pacific herring, have distinct seasonal movements and some, such as chub mackerel, Spanish mackerel, and filefish, migrate out of the Yellow Sea to the East China Sea in winter. The distribution of these two groups often overlaps, especially in overwintering and spawning periods. When water temperatures begin to drop significantly in autumn, most resource populations migrate offshore towards deeper and warmer waters and concentrate mainly in the Yellow Sea depression. There are three overwintering areas: the mid-Yellow Sea, 34 to 37°N, with depths of 60 to 80 metres; the southern Yellow Sea, 32 to 34°N, with depths about 80 metres; and the northern East China Sea. The deep water areas of the central Yellow Sea and northern East China Sea are the overwintering grounds for most species that migrate over long ranges. Clearly these areas must be protected from pollution and overfishing.

There are also several important concentrations of vulnerable resources in the Yellow and East China Seas.[5] The Ryukyu archipelago in the East China Sea harbours whale calving grounds, turtle breeding areas, coral reefs, mangroves, protected areas, and marine mammals. There is also a remarkable concentration of turtle breeding areas, mangrove stands, coral reefs, and protected areas in the Sakashima Gunto. The Japanese island of Shikoku and its southern outlier have all these resources except whale calving grounds and, in addition, harbour numerous aquaculture sites. The heavily used Korea Straits are spawning grounds for both demersal and pelagic fish, as well as whale calving areas. The island of Taiwan is enveloped by mangrove and coral reef sites, protected areas, and aquaculture sites. And the southwest coast of Korea also has a coincidence of many aquaculture sites and protected areas. These vulnerable resources would be priority areas for protection under a regional marine environmental protection regime.

The Political Context

Relevant Parties in Regional Action

The principal relevant parties comprise four groups: the national governments of littoral states, national governments of outside states with particular maritime interests in the region, international agencies, and private companies. Although these actors are arguably the most important, others include provinces, ministry and agency bureaucracies, and increasingly in Japan, Russia, and South Korea, private interest groups. The interests of these actors sometimes overlap, but they are seldom identical and frequently conflict.

The littoral states for the Sea of Japan are the Democratic People's Republic of Korea (North Korea), Japan, Russia, and the Republic of Korea (South Korea). For the Yellow/East China Sea, they comprise China, Japan, North Korea, South Korea, and Taiwan. Principal outside states with varying degrees of interest in the Sea of Japan region are China, Mongolia and, in the realm of security, the United States. For the Yellow/East China Sea, the circle widens beyond Mongolia and the United States to include principle commercial users of the sea lanes, such as Australia and ASEAN.

International agencies with marine relevance that have interests in the region include: the Asian Development Bank, with its investments in marine-related projects; the United Nations Environment Programme, which has initiated a North-West Pacific Region Action Plan; the International Maritime Organization, which promotes treaties on safety at sea and prevention of ship-sourced pollution; the Intergovernmental Oceanographic Commission, whose Subcommission for the Western Pacific (WESTPAC) includes the North-East Asian seas and their coastal countries in its coverage; and the fisheries programs of the Food and Agriculture Organization of the United Nations. In addition, there are several intergovernmental organizations whose terms of reference include North-East Asian seas. Among them are the North Pacific Marine Science Organization (PICES), which involves Canada, China, Japan, Russia, and the United States, and the Japan-East China Sea Surveys (JECSS), involving China, Japan, and South Korea. The fourth group of relevant parties comprises private companies such as shipping firms, oil companies, and banks, both within and outside the region.

Although the decisions of state-level actors are usually considered the principal factors in regional marine management and co-operation, this assumption is being eroded by recent and rapid changes in the political arena. Accepted perceptions of the operation of the political system itself are being challenged,[6] including the very ability of the traditional international structure of independent nation-states to respond to the new global challenges. The growing influence of non-governmental organizations such as national interest groups or multinational corporations is receiving increased attention. In the past, industrial actors were seen largely as national organizations which affected only the policies of their local governments. Now it is increasingly recognized that they are also independent 'transnational' actors often with interests in many states, who influence international decision-making and offer their 'flag' states new forms of leverage on the policies of other countries.

Also emerging are new conceptions of old actors. The study of 'transnational' relations has provided a new perspective on just what constitutes international politics, a perspective which has undermined the old concept of the state itself. With the diversification of actors and issues, various competing bureaucratic interests act across national boundaries, carrying on their business subnationally or 'transgovernmentally', building foreign alliances in much the same way as the new non-governmental actors. These new insights into the interactions of national and international actors are changing the concept of the nature of the international organization itself. It can be viewed as a cluster of intergovernmental and transgovernmental networks. Within the organization, there is a continuous mixing of officials dealing with such a variety of issues that the function of the agency may be as much to 'activate potential coalitions' as to engage in more formalized undertakings.

Having identified the major categories of participants in the decision-making process regarding regional action in North-East Asian seas, we should also consider the decision-making process within the various organizations. This involves, firstly, the internal mechanisms: the organizations within countries, or the public or private institutions responsible for collecting and analysing data, making recommendations, and setting policy. Associated with them are legal, political, and other constraints on a participant's freedom of choice on policy action concerning support or non-support of regional activities. Although an in-depth analysis of the internal decision-making structure of the various 'actors' in maritime regime building is beyond the scope of this work, some indication of unknowns and possible future lines of research will be made.

A second consideration is the perceptions of the leaders or decision makers themselves. These decision makers include those from within the region who actually commit, or prevent commitment of their government to 'invest' in regional activity. They also include actors from outside the region who may participate in the regional arrangement, or actively support or oppose the regional action. The interests of external and regional actors can and often do collide.

Jurisdictional Claims and Disputes[7]

There is great diversity in maritime claims made by North-East Asian countries, and some of these claims overlap (see Figure 3.6). Only Russia claims the entire suite of maritime zones permitted by the 1982 Convention on the Law of the Sea. Its claim to territorial seas 12 nautical miles wide dates from 1921; a claim to the continental shelf in terms of the *Convention on the Continental Shelf* of 1958 was made on 6 February 1968, and it decreed a 200-nautical mile EEZ on 1 March 1984. North Korea has not made a specific claim to the continental shelf but did claim territorial waters of 12 nautical miles and an EEZ of 200 nautical miles on 1 August 1977. One month before North Korea made its claims, Japan proclaimed territorial seas 12 nautical miles wide except in some critical straits, and a fishing zone of 200 nautical miles. Japan has not made a formal claim to a continental shelf but has agreed to joint development with South Korea of the shelf between them. South Korea claimed territorial seas of 12 nautical miles on 30 April 1978, except in the Korea Strait, and made a general claim to a continental shelf in a presidential proclamation on 18 January 1952.

China claimed territorial waters 12 nautical miles wide in a declaration made on 8 September 1958. There have also been announcements by the Ministry of Foreign Affairs concerning its continental shelf. On 15 March 1973, after the *Glomar IV* had been drilling in areas authorized by South Korea, Beijing claimed ownership of the seabed resources along the coast of China. On 14 February 1974, China reserved all rights over the continental shelf extending from its coast, including that under the East China Sea. That release mentioned the principle of natural prolongation apparently for the first time. China has never published precise limits to its continental shelf, but appears to claim the entire shelf in the East China Sea up to the Okinawa Trough.

Taiwan extended its claim to a territorial sea to 12 nautical miles in September 1979 and at the same time claimed an EEZ of 200 nautical miles. Taipei did not sign the Convention because it had no

84 A MARITIME REGIME FOR NORTH-EAST ASIA

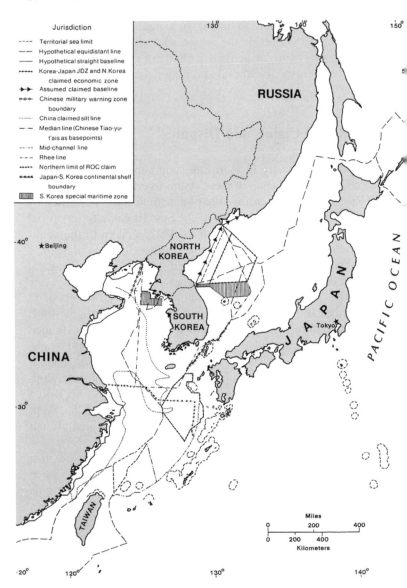

Figure 3.6 Jurisdictional Claims in North-East Asian Seas
Source: J. R. V. Prescott, 'Maritime Jurisdiction' in Joseph R. Morgan and Mark J. Valencia (eds.), *Atlas for Marine Policy in Northeast Asian Seas*, Berkeley: University of California Press, 1993, pp. 25–6.

diplomatic relations with Jamaica, and the United Nations Credentials Committee would not accept the credentials of Taiwan's delegates. Even so Taiwan will most likely comply with the Convention's terms when it comes into force.

In order to avoid confrontation and conflict, unilateral maritime claims in North-East Asian seas have generally not been explicit. Indeed, some such as South Korea, may consider that the international tension and legal, technical, and enforcement responsibilities likely to be created by its declaration of a 200 nautical miles EEZ may outweigh the resources that it would gain. Thus little progress has been made in negotiating bilateral marine boundary agreements. Only two of the eleven potential boundaries have been defined. One was the agreement between Japan and South Korea of 30 January 1974 that defined their continental shelf boundary for about 260 nautical miles through the Korea Strait and the western entrance to the Yellow Sea and, more significantly, joint development of their overlapping continental shelves in the northern East China Sea. The second was the North Korea/Russia agreement of January 1986 on their common continental shelf and EEZ boundary.[8] Although North Korea's announced EEZ claim extends only to the 'half-line', its agreed boundary with Russia apparently extends to the central portion of the Yamato Bank fishing ground and thus would not be acceptable to Japan and South Korea. Three potential boundaries are complicated by serious disputes over ownership of islands: Tok Do/Takeshima between both Koreas and Japan, the Diaoyutai/Senkaku Islands between China and Japan, and the Northern Territories/Southern Kuriles between Japan and Russia (see Figure 3.6).

Despite the lack of precision in the definition of bilateral marine boundaries, friction has so far been minimal. Although the existing marine policy regimes are weak and frequently ineffective, governments have avoided catastrophic situations by controlling the extent of their economic and technological activity. Countries seem to use marine areas that clearly belong to them and avoid zones where conflict could arise, making these zones seem more like political frontiers than political boundaries. As an example, China, Japan, Taiwan, and South Korea have generally avoided creating serious tensions by limiting prospecting for oil and natural gas to non-disputed sections of the continental margin. The exceptions of course are exploration in the Japan/South Korea joint development zone, which is opposed by China, and both Taiwanese and

Chinese exploration in the East China Sea in areas claimed by both. But the widespread depletion of fishery resources, the expansion of national fleets, 'poaching' by non-littoral countries in the East China sea, and the coming into force of the Law of the Sea Convention are putting significant pressure on China and South Korea to declare EEZs in order to legitimate their ownership and management of these resources.

Military zones have been established at various times by a number of countries, usually during hostilities or in the face of threats. But the concept of security zones extending beyond a state's territorial sea was not sanctioned by the 1982 Convention or its predecessor in 1958. The Chinese military zone at the head of the Yellow Sea and the North Korean military zones do extend beyond the territorial seas of those countries, assuming those territorial seas are 12 nautical miles wide and claimed from reasonable baselines. It is not known whether the special maritime zones of South Korea still extend beyond its territorial waters. A minor route from the Korea Strait to Nakhodka passes through a naval operations area, but this is only of concern when actual naval manoeuvres are in progress. In any case, notifications are issued before manoeuvres take place.

Integrative and Disintegrative Forces

Integrative Forces

Ideological and political adversaries in the region have hitherto pursued economic development without much concern for their neighbours, particularly those across hundreds of miles of open sea. But when all the coastal nations of the region have formally extended their maritime jurisdictions over resources and many activities to 200 nautical miles or more, almost no marine area will be left unclaimed, and many areas will be claimed by two or more countries. The nations bordering North-East Asian seas are now trying to identify and pursue their national interests in the oceans. Yet many ocean resources and activities, such as fish and fishing, pollutants and environmental protection, sea lanes and shipping, and hydrocarbon-bearing basins and hydrocarbon exploration, are transnational in character. There may be insufficient understanding and consideration of the transnational and interdependent character of the ocean environment, and the resources and activities that it harbours and supports. And the overlaying of a mosaic of unco-ordinated national jurisdictional regimes on inherently transnational resources and activities will inevitably create conflicts. But most importantly, it will create opportunities for cooperation.

What currently passes for national and particularly regional ocean policy is quite primitive, both conceptually and analytically.[9] The major impediment to progress is an inability to formulate and implement ocean policy as an integrated whole, balancing the overall interests of the nation and the region in the short and long term. The reason for the widespread fragmentation in national policy-making structures and processes for the oceans seems to have resulted from the development of ocean uses largely in isolation from each other. Different technologies have given rise to separate networks, agencies, laws, and ways of thinking and acting. And these interests have matured into fully autonomous sectors with weak links between them or no links at all. This situation was not a problem when the pace of technological change was slow, and human use of the high seas was limited. But it has now become dysfunctional because the rate of technological change is high, human use of the high seas is rapidly increasing and diversifying, and different uses and users have begun to generate adverse impacts on each other, and across national boundaries.

Despite growing interest and obvious need, North-East Asian states continue to ignore most of the opportunities currently available in the marine sphere, and to be seemingly incapable of resolving the growing multiple use conflicts in their EEZs, let alone those conflicts that are truly transnational in character. Furthermore, the ocean as a whole continues to play a role in the national and regional development process far below its potential for most of the coastal states. Improving the way national and regional ocean policy is made and practised is the principal remedy for this unsatisfactory state of affairs. In particular, to become fully effective and efficient, national and regional ocean policy must be integrated.

There is a need, then, for increased bilateral and multilateral consultations, as well as for a new degree of co-ordination, to meet the challenge of changes in marine use patterns and concepts. By a curious dialectical process, the extension of national jurisdictions increases the need for international co-operation. When nations extended their jurisdiction, they assumed that they had also simultaneously extended their ability and authority to manage unilaterally the seas they acquired. But nations found instead that they had to manage what had been left unmanaged in the past, and that purely national management was in many cases, frustrated by the overlapping of political and ecological boundaries, and by the high costs of exploitative and managerial technologies and infrastructures. This realization should encourage regional co-operation. Besides creating an opportunity for the re-examination of national ocean management, extension of jurisdiction also presents an opportunity for re-examining a nation's relationships with its neighbours, with a view to moving towards a more ideal structure of international relations.

With the signing of the historic Convention on the Law of the Sea by 119 nations in December 1982,[10] and its coming into force in November 1994,[11] the venue for addressing issues of ocean law and policy has moved from the global to the regional and bilateral level. Hence the Convention heralds a new era of transnational rule making regarding national rights and responsibilities in the world's oceans. It already serves as a framework within which nations carry out their ocean management rights and responsibilities. The Convention and customary international law acknowledge sovereignty over resources out to 200 nautical miles, which means that most valuable marine resources have now been encompassed by national jurisdiction.

Article 122 of the Convention defines an enclosed or semi-enclosed sea in two ways: geographically, on the basis of its narrow physical connections with nearby bodies of water, and legally, on the basis that it consists entirely of the territorial seas and EEZs of two or more countries. The North-East Asian seas are semi-enclosed on both grounds. The East China/Yellow Sea consists of the territorial seas and EEZs of China, North Korea, South Korea, and Taiwan, while the waters of the Sea of Japan are completely under some form of jurisdiction from Japan, North Korea, South Korea, and Russia. The importance of enclosed or semi-enclosed seas in the management of marine regions is emphasized in Article 123 of the Convention, which holds that:

> States bordering an enclosed or semi-enclosed sea should cooperate with each other in the exercise of their rights and in the performance of their duties under this Convention. To this end they shall endeavour, directly or through an appropriate regional organization:
> (a) to co-ordinate the management, conservation, exploration and exploitation of the living resources of the sea;

(b) to co-ordinate the implementation of their rights and duties with respect to the protection and preservation of the marine environment;
(c) to co-ordinate their scientific research policies and undertake, where appropriate, joint programmes of scientific research in the area;
(d) to invite, as appropriate, other interested States or international organizations to co-operate with them in furtherance of the provisions of this article.

The Convention provides the international legal framework within which regional problems should be solved. Unfortunately it is mute as to what sorts of regional bodies might be brought into being for this purpose.

Another integrative factor may be the United States' new position that multilateralism is the key to Asia's future. Washington now supports multilateral security dialogues, including such a forum in North-East Asia, based on Track Two Initiatives.[12] And the United States maintains that it will remain engaged as an active player in the economic, security, and political affairs of the region. Indeed, America's interests compel any administration in Washington to help shape the emerging Pacific Community.[13] As it is, the United States still has significant influence in the region and its new policies may have a demonstration effect for powers there. Its encouragement and support of regional co-operative efforts may enhance the likelihood of their taking root in the region and growing in degree and kind.

The Clinton administration's support for ratification of the amended Law of the Sea Convention is reinforcing this support for multilateralism.[14] Japan, South Korea, and China are also likely to ratify it. This will legitimize regional co-operation in marine matters and put pressure on Russia at least to abide by the Convention's provisions. The resulting political environment could have positive implications for marine regionalism in North-East Asia.

One integrative by-product of the sometimes divisive negotiations lasting more than ten years on the Law of the Sea was the development of a cadre of law-of-the-sea experts from the region itself, who had frequent contacts and hence opportunities to discover their mutual interests—in short, an epistemic community. Although real regional co-operation is still in an early stage of development, marine policy problems will come to play an increasingly important role in the international relations of East Asian states. Already there are numerous indications that the countries concerned are being drawn slowly but surely into what could become a continuing dialogue, through which constructive and mutually beneficial marine policies may evolve within the North-East Asian region.[15]

Dis-integrative Forces

There are also serious obstacles to marine regionalism in North-East Asia. And the immediate benefits of international functional arrangements may fade in the light of the immense political conflicts that still divide the region. North Korea remains a steadfast socialist country juxtaposed to the capitalism of Japan, South Korea, and Taiwan. Mongolia and Russia, and perhaps China, are in varying degrees of 'transition'. In political terms, there are four countries with six governments, each with little history or experience in multilateral co-operation.

In this context, a primary obstacle to truly regional co-operation is the difficulty of involving China and Taiwan in a multilateral marine policy regime covering areas claimed by both, particularly given the increasing tension in their relationship.[16] Still, China and Taiwan are in the process of forming a united front regarding South China Sea issues, as well as jointly exploring for petroleum in the East China Sea.[17] Another difficulty is the isolation and pugnacious, non-participatory stance of North Korea. Since it borders and claims continental shelf, 'security zones', and EEZs in the Japan and Yellow Seas, North Korea's eventual participation in functional marine policy regimes is important.

Then there is the problem of a destabilized and near-anarchical Russia. It will not be clear for some time who speaks for Russian Far East maritime policy, and how stable and steadfast that policy is or will be. And although the participation of China and Japan is critical to a successful regime, both may be reluctant to take part unless they can dominate. In general, most big powers prefer to avoid multilateral regimes in which the smaller nations can form blocks against them. It will be necessary to present a convincing argument that the major powers can gain more benefit from a multilateral regime than from bilateral agreements which they can dominate.

Another complicating factor is the island and concomitant maritime boundary disputes that plague the region, the southern

Kuriles/Northern Territories, Tok Do/Takeshima, and Senkaku/Diaoyutai disputes, as well as the overlapping continental shelf claims of China/Japan/Taiwan in the East China Sea. Maritime issues are generally only a ripple in the great ebb and flow of economic and political relations in North-East Asia. But many national frontiers are now maritime in nature, which can elevate these issues into symbols of national pride and integrity. Some maritime issues may be so crucially situated in time or substance *vis-à-vis* the balance of much greater issues that they could act like a rogue wave or surge which significantly disturbs political relations in the region. Disputes over islands or boundaries in areas of great petroleum potential could belong in this category. Considering the tenuous or even hostile relations between most of the states in the region and the likelihood of petroleum in disputed continental shelf areas, maritime issues could become the 'tail' that wags the 'dog' of international relations in North-East Asia.

Last but not least is a conceptual dichotomy inherent in the Law of the Sea Convention. On the one hand the Treaty enjoins its ratifiers to co-operate in managing semi-enclosed seas. On the other, Article 56[17] gives the coastal state sovereign rights over the natural resources, whether living or non-living, and over other activities for the economic exploitation of the zone, as well as jurisdiction over the establishment and use of artificial islands and other structures and installations, marine scientific research, the preservation of the marine environment, and other rights and duties provided for by the Convention. Because all of the waters of North-East Asian seas will be partitioned into the territorial seas and EEZs of the littoral states, the provisions of Article 56 could represent a stumbling block to regional co-operation unless and until the littoral states agree to yield some of their newly won rights to a regional body.

The absence of a multilateral maritime regime in North-East Asia reflects political calculations by its nation-states about the rewards/risks and losses/benefits of maintaining the *status quo* versus developing regimes acceptable and beneficial to all sides involved. North-East Asian countries are simply not yet sufficiently aware of how serious the need is for a multilateral maritime regime that focuses on the management of fisheries resources and maritime environmental protection. For China, environmental protection has generally been viewed as a domestic concern or, at most, as a worthwhile but trivial aspect of international co-operation.[19] In terms of maritime affairs as a whole, China's priority seems to be the South China Sea, although even this is debatable. A similar case can be made for Japan, which seems to pay more attention to the fate of its 'Northern Territories'. South Korea, like Japan, tends to be more concerned about air pollutants that originate in the deserts of western China. And North Korea has yet to demonstrate convincingly that its interests extend beyond its land borders. The North Korean example illustrates the tendency of North-East Asian countries, when thinking maritime, to think about boundary disputes, not protection of the deteriorating marine environment or management of fisheries. It is these perceptions that must change. And they are changing.

Taken together, these disintegrative forces argue strongly for an *ad hoc*, issue-specific evolutionary process towards a multilateral approach to maritime regime building. In this context, Robert Scalapino's concentric arch approach may have some relevance.[20] For maritime regimes in North-East Asia, such an approach would first involve North and South Korea, then China, Japan, and Russia, and finally, the United States and other external actors.

Research Questions

Marine policy problems present particular challenges to the advocates of international co-operation in North-East Asia. Given the indifferent relations among states in the region, their governmental inexperience, and the inherent uncertainty and sensitivity of marine policy issues, it is necessary to develop arguments that politicians can and will use to convince their governments to participate in regimes. These arguments must be specific and reach beyond generalities and obvious platitudes, such as the sharing of a semi-enclosed sea endangered by pollution, or the less-threatening and less risky nature of a multilateral forum as compared to bilateral discussions. Arguments must be tailored to each nation for each issue because the politics and the negotiating approach will differ for each nation depending on whether the issues are to be dealt with individually or comprehensively.

The overall objective of this study is to determine if and how the resolution of ocean policy issues can improve relations between nations. More specifically, in the context of extended national jurisdiction and the overlaying of national policies on fundamentally transnational resources and activities, the research will determine the causes and consequences of, and solutions to, significant actual and potential maritime policy conflicts in East Asia.

Questions which will guide the study include the following: What mutual or collective gains can be obtained by transcending national perspectives and thinking of regions as unified managerial units? What are the emerging and potential modes and means of, opportunities for, and constraints to regional co-operation? Which specific issues and aspects of marine policy really can be addressed more effectively by an international approach, and at what level and degree of formality? What is the appropriate 'region' for each of the opportunities for co-operation? What are the various national interests with regard to these opportunities, and what are the likely advantages, disadvantages, and trade-offs for each nation? What would the costs of a co-operative solution be and how should they be allocated? Which recurring issues might benefit from a regional institutional arrangement? What might be expected of such institutions and what would be their appropriate structure and function? Should the arrangements be formalized within the established framework of an intergovernmental organization, or should they be allowed to evolve gradually through a series of governmental and non-governmental initiatives within the much looser framework of a 'network'? Should a single, complex arrangement be structured around a variety of ocean resources and uses, or should the outcome consist of separate arrangements, such as fish, hydrocarbon, shipping, and environment for each ocean use or type of resource? Should the arrangement or arrangements be limited to a single function, for example, data exchange, co-operative research, consultation, enforcement, or conflict avoidance, or designed to serve a number of functions? Should the arrangement be designed as a 'regime' based on agreed upon 'rules' for the use of resources and the management of the shared ocean environment? What relevant lessons have been learned elsewhere, and what aspects should and could be adapted to the region?

Notes

1. Joseph R. Morgan, 'The Natural Environmental Setting', in Joseph R. Morgan and Mark J. Valencia (eds.), *Atlas for Marine Policy in East Asian Seas*, Berkeley: University of California, 1992, pp. 5–6.

2. Mark J. Valencia (ed.), *International Conference on the Sea of Japan: Transnational Ocean Resource Management Issues and Options for Cooperation*, East–West Environment and Policy Institute Occasional Paper No. 10, 1989, pp. 11—27.

3. Y. Honma and T. Kitani, 'Records of the marine mammals in the waters adjacent to Niigata and Sado Island in the Sea of Japan, based partially on the old documents', *Bulletin of the Biogeographic Society of Japan*, 36 (1981): 93–101; Y. Honma, T. Kitani and R. Mizusawa, 'Records of Cephalopoda in the waters adjacent to Niigata and Sado Island in the Japan Sea, based partially on the pelagic squids stranded ashore', *Bulletin of the Biogeographic Society of Japan*, 38, 12 (1983): 23–9.

4. Mark J. Valencia, *International Conference on the Yellow Sea: Transnational Ocean Resource Management Issues and Options for Cooperation*. East–West Environment and Policy Institute Occasional Paper No. 3, 1987, pp. 6–25.

5. Joseph R. Morgan and Mark J. Valencia, 'Integrations', in Joseph R. Morgan and Mark J. Valencia (eds.), *Atlas for Marine Policy in East Asian Seas*, Berkeley: University of California Press, 1992, pp. 142–4; 150–1.

6. Robert O. Keohane and Joseph S. Nye, *Power and Interdependence: World Politics in Transition*, Boston: Little, Brown, 1977, p. 240.

7. J. R. V. Prescott, 'Maritime jurisdictions', in Morgan and Valencia, *Atlas for Marine Policy in East Asian Seas*, pp. 25–35.

8. Daniel J. Dzurek, 'Deciphering the North Korean–Soviet (Russian) maritime boundary agreements', *Ocean Development and International Law*, 23 (1992): 35–6; 50–54.

9. Edward L. Miles, 'Concept, approaches, and applications in sea use planning and management', *Ocean Development and International Law*, 20 (1989): 215.

10. *United Nations Convention on the Law of the Sea*, New York: United Nations, 1983.

11. Steven Greenhouse, 'US, after negotiating changes, is set to sign pact on sea mining', *The New York Times*, 10 March 1994, p. A-13.

12. Winston Lord, 'A New Pacific Community: Goals for American Policy', opening statement at confirmation hearings for Assistant Secretary of State, Bureau of East Asian Affairs, United States Congress, 31 March 1993; 'Group Therapy', *Far Eastern Economic Review*, 15 April 1993, pp. 10–11; Department of Defense, Office of International Security Affairs, *United States Security Strategy for the East Asia-Pacific Region*, Washington, February 1995.

13. Richard H. Solomon, 'Asian architecture: the US in the Asia-Pacific community', *Harvard International Review*, Spring 1994, 60.

14. Greenhouse, 'US, after negotiating changes, is set to sign pact on sea mining', *New York Times*, 10 March 1994, p. A-13.

15. Examples of relevant dialogue include the following meetings; *International Conference on East Asian Seas: Cooperative Solutions to Transnational Issues*, Seoul, 21–3 September 1992; *The Soviet Far East and the North Pacific Region: Emerging Issues in International Relations*, Honolulu, 20–3 May 1991; *East China Sea: Transnational Marine Policy Issues and Possibilities of Cooperation*, Dalian, China, 27–9 June 1991; *International Conference on the Japan and Okhotsk Seas*, Vladivostok, Russia, September 1989; *International Conference on the Sea of Japan*, Niigata, Japan, 11–14 October 1988; *International Conference on the Yellow Sea*, Honolulu, 23–7 June 1987; 'Japan to seek regional meeting to look at water pollution, other problems', *International Environment Reporter*, 4 December 1991; *Northeast Asian Conference on Environmental Cooperation*, Environment Agency of Japan and Niigata Prefecture, 13–16 October 1992. Japan has established a center

to elaborate the concept of regional cooperation and to prepare specific proposals for cooperation around the Sea of Japan. See RA Report No. 15, July 1993, Center for Russia in Asia, University of Hawaii, Honolulu, p. 44.

16. *Asia Wall Street Journal*, 11–12 August 1995, p. 1; Julian Baum, 'Pressure cooker', *Far Eastern Economic Review*, 24 August 1995, pp. 16–17.

17. 'China and Taiwan plan cooperative exploration of the South China Sea', *World Journal*, 17 January 1994, p. 1; Mark J. Valencia, 'The South China Sea issues: context, conjecture and a cooperative solution', Paper presented at the VIII Asia-Pacific Roundtable, Institute for Strategic and International Studies, Kuala Lumpur, June 1994.

18. *United Nations Convention on the Law of the Sea*, New York: United Nations, Article 56.

19. This is implied by the recommendations of Han Guogang et al. 'China's environmental protection objectives by the year 2000', *International Journal of Social Economics*, 18, 8/9/10 (1991): 180–92.

20. Robert Scalapino, 'Challenges to the sovereignty of the modern state and their political and security implications', Paper presented at the VII Asia-Pacific Roundtable Confidence Building and Conflict Resolution in the Pacific, Kuala Lumpur, 6–9 June 1993; Robert Scalapino, 'Historical perceptions and current realities regarding Northeast Asian regional cooperation', North Pacific Cooperative Security Dialogue, Working Paper No. 20, October 1992.

4 National Interests: Advantages and Disadvantages of Participation in Regional Maritime Regimes

The absence of maritime regimes in North-East Asia reflects a Cold War calculus of the rewards/risks and costs/benefits of maintaining the status quo, versus developing regimes acceptable and beneficial to all parties concerned. This calculus is changing. This chapter analyses the national context and the advantages and disadvantages for each polity of participating in a regional regime for marine environmental protection and for fisheries management.

China and Taiwan[1]

Chinese National Reunification and Maritime Regime Building in North-East Asia

The unfinished civil war between China and Taiwan presents the most acute problem for maritime regime building in North-East Asia. Because Taiwan has been stripped of its sovereign nation-state status since the 1970s it is ineligible to enter into formal bi- or multilateral treaties. Yet it is continuing the struggle for acceptance as a full-fledged sovereign nation-state. But given the enormous challenges facing the present and future mainland Chinese leadership, it is highly unlikely that China will change its attitude toward Taiwan for some time to come. No Chinese leader would want to be accused of 'losing' the country's largest piece of offshore territory.[2] So a relevant question for regional maritime regime building is whether Chinese reunification politics is an incentive or a hindrance.

The Beijing–Taipei diplomatic rivalry[3] is intense and relentless, as demonstrated by China's reaction to Taiwan President Lee Teng-hui's 'unofficial' visit to the United States.[4] Nevertheless changes in global and regional political, economic, and military dynamics, and ensuing external policy adjustments by China and Taiwan have led to dramatic shifts in cross-Taiwan Straits (hereafter cross-straits) politics.⁵ Then Taiwan President Chiang Chingkuo's decision in 1986 to lift the ban on private travel to the Chinese mainland thawed the frozen cross-straits relationship. Subsequent developments have brought China and Taiwan so close together that it is difficult to imagine that either would be willing to risk a return to the pre-1987 era of no contact, no compromise, and no negotiations. Except over the issue of international diplomatic recognition of Taiwan, the two sides have a wide range of common interests. And those common interests may be incentives for them to participate in a regional maritime regime as a further step in the confidence-building process.

The interests of China and Taiwan converge in several areas. The most important are nationalism, national pride, and the resurgence of the Chinese nation, followed by trade and investment.⁶ Cross-Straits trade, albeit indirect, has grown from around US$47 million in 1978 to about US$7.4 billion in 1993.⁷ Taiwan's economic planning, transportation, and foreign trade leaders have repeatedly emphasized the Chinese mainland factor in formulating their respective policies, and called for greater relaxation of the remaining restrictions on cross-straits contacts. Indeed, in January 1995, Taiwan announced plans to designate the southern port of Kaohsiung and one airport as an offshore shipping and an aviation centre respectively, so that ships and planes can move directly between Taiwan and China without technically violating Taiwan's ban on travel links. Direct shipping links will be next.⁸ Academics in Taiwan are now more convinced than ever that China's policy of forging strong economic ties with Taiwan is meant as a means of leverage against Taiwanese elements who are seeking formal independence, and as such, the policy will strengthen with time.⁹

Another area of convergent interest is the need to find agreement over the management of fisheries resources in the Taiwan Strait. The domestic fisheries policies of China and Taiwan have differed in focus and emphasis. For China, the chief management emphasis has been on controls over fishing activities in coastal waters in order to protect spawning adults and larval and juvenile fish. Local entities such as co-operatives or communes, appear to play a much stronger role in the day-to-day management of coastal fisheries than the central or even the provincial governments. Although conservation of stocks is considered a 'good,' there appears to be less concern about individual stocks than there is about total yield. The policy seems to be based on the notion that a decline in the population of one species need not cause alarm or changes in management regulations, as long as the decline is offset or compensated for by an increase in the catch of another similar stock.

Of Taiwan's total fishing output of approximately 1 million metric tons, at least one-third is caught by deep-sea fishing boats in waters outside its own jurisdiction, another third in inshore areas about 30 nautical miles from shore, and less than 5 per cent along the coasts. Culture activities produce the rest. The rapid expansion and modernization of the Taiwanese fishing fleet after World War II led to the widening of the scope of fishing operations well beyond the South-East Asian region to the Pacific, Indian, and Atlantic oceans. Many of Taiwan's fishing grounds are now within the recent declared EEZs of other countries as well as their territorial and archipelagic waters. For example, Taiwan boats take up to 40,000 tons of fish from the Yellow Sea. Taiwan's major interest in fisheries is the exploration of new grounds, meaning either finding unclaimed physical areas in which to fish or collaborating with other countries to gain access to fishing grounds now under their jurisdiction. Fisheries are an important aspect of the local economy of both Fujian, the Chinese province bordering the Strait, and Taiwan.

Thus the fisheries interests of China and Taiwan are now converging. But the formal state of unfinished civil war between the two makes fishing in each other's claimed waters an explosive issue. In the past, any mainland fishing boat that wandered across the hypothetical median line was confronted, and even fired upon, by Taiwanese naval vessels. In March 1994, Taiwan formally decided to confiscate mainland vessels if they were detained twice.[10] This action prompted protests from Beijing. Taiwan, in turn, complained about its fishermen being mistreated by their mainland compatriots.

Because of the traditional ties of kinship between the peoples of Fujian Province and Taiwan, such fisheries disputes exact a long lasting political toll. Fujian is the starting point for the competition between Beijing and Taipei for the hearts and minds of the Chinese people[11] living on both sides of the Taiwan Straits. Now that shelling across the Straits by the two armies has ceased, hostile treatment of fishermen can only prolong the hostility between the two Chinese political centres. It damages the fragile goodwill of the people closest to the Taiwan Straits—a goodwill both Beijing and Taipei have ardently courted. Although the two sides have agreed in principle to allow official vessels from either side to intervene in disputes,[12] conflict over the management of fisheries resources in the Taiwan Straits area is likely to continue and, in the short run, even escalate. Yet co-operation is necessary and perhaps inevitable. The two sides are currently negotiating an agreement but the sticking point is a dispute over the term 'territorial waters' as applied to Taiwan.[13]

There are several reasons for this prediction of increasing co-operation. In Taiwan's case, the first is that its fishing fleets have been banned from fishing in their traditional areas in the North Pacific, making the waters of the Taiwan Strait more valuable. Second, Taiwan's fishing fleet now faces a more assertive Russian enforcement of fisheries regulations as well as less friendly waters close to Japan and Korea, whose fleets were also expelled from others' waters. A third factor is that Taiwan is not currently recognized as a sovereign entity by any of the North-East Asian sovereign states, so that disputes over fishing activities are often settled to its disadvantage. There have been a number of incidents in which Taiwanese fishing boats were required to fly the national flag of the People's Republic of China when they ventured into South Korean[14] and even Singaporean waters.[15]

For China, there is an increasing convergence of interests with Taiwan on ocean policy issues in general, and on management of fisheries in the Taiwan Strait in particular. First, the mainland has to deal with the consequences of Taiwan's behaviour as a sovereign political entity, while legally and publicly denying it that status. This creates confusion and problems in the maritime sector that demand resolution. For example, the tacit agreement between China and Taiwan to use the hypothetical median line in the Taiwan Straits as the boundary separating areas of jurisdiction accords fully with the norms adopted by sovereign states. And since China claims Taiwan as one of its provinces, and Taiwanese boats are increasingly employing mainland labour, Beijing may have to shoulder the diplomatic and other costs of settling complaints about infractions committed by Taiwanese fishing vessels which enter other countries' waters.[16]

A second challenge for Beijing derives from the growing political assertiveness and economic needs of the coastal localities under its control. The economic reforms pursued in the last decade and a half have led to tension between Beijing and the capitals of the coastal provinces.[17] This coastal-interior dichotomy has been recognized as a domestic strategic concern by the central government—so much so that the *People's Daily* sees the growth gap between the eastern and western regions of China as 'a problem,' which, if not solved, 'is going to affect the unity among different nationalities and the solidification of our national defence.'[18] The people of Fujian Province are the embodiment of this dichotomy. They have extensive ties of kinship and culture with the residents of Taiwan and have vigorously pursued Taiwanese investment.

In the drive for economic development, China's coastal provinces are increasing their utilization of ocean resources. Traditionally, China's coastal and inland provinces have relied heavily on fresh-water fish rather than marine fish. This preference is already changing, and Chinese marine scientists and local economic planners are increasingly arguing for a 'return to the oceans' for natural resources.[19] Thus as consumers' incomes grow, so will their demand for seafood and with it, more investment in marine fisheries.

A third area of converging interests is the pressing task of having jointly to combat piracy, drug trafficking, and the smuggling of illegal immigrants in the Taiwan Strait and the East China Sea. The transportation of illegal immigrants from the mainland, in which criminal elements in Taiwan are known to play an active role, is a thorny issue that hurts the relationship between both China and Taiwan, and the principal destination countries, Japan and the United States.

Finally, both China and Taiwan claim sovereignty over large areas of the continental shelf in the East China Sea, and the Diaoyu (Senkaku) islands. But Japan is the single most powerful actor in the Yellow Sea and the East China Sea in terms of technology, economic power, and scientific knowledge. So it would be logical for China and Taiwan to co-operate tacitly to strengthen their position. In disputes over maritime areas, claims of sovereignty become hollow unless they are backed by actual exploration and utilization of the disputed area. Thus it is logical that Taiwan's wholly owned US company, the Overseas Petroleum Investment Company, and the mainland's China National Offshore Oil Corporation have begun discussions to explore jointly for oil in the Taiwan Strait and the East China Sea.[20] Oil exploration is an area of cross-straits co-operation that offers an excellent opportunity to yield economic and political benefits to both countries.

Thus, if the issue of Beijing/Taipei diplomatic competition can be shelved, convergence of interests could lead both China and Taiwan to participate in regimes such as APEC.[21] Co-ordinated management of marine resources and protection of the maritime environment clearly would bring them economic and political benefits. The building of trust between the two countries, and with the rest of North-East Asia, would be an additional mutual benefit of regime participation.

The single most serious obstacle to participation in a regional maritime regime is Taiwan's continued striving for diplomatic recognition, and the context of that quest: whether it is recognized as representing only itself or the people on both sides of the Taiwan Strait. By way of domestic legislation, Beijing and Taipei have formalized their pledges to pursue national reunification. The difference between them lies in how that goal is to be reached. Taiwan is not ready to accept dual recognition before reunification becomes a reality, while China views dual recognition as a step toward permanent separation. With the establishment of full diplomatic ties between China and South Korea in August 1992, the tug of war between Beijing and Taipei for diplomatic recognition in North-East Asia ended in Beijing's favour.

Beijing is likely to have the upper hand for some time to come. Its triumph is not a result of political ideology but of the new geopolitics in the region, and economic interdependence. Nations co-operate because there are rewards to be gained and the cost of confrontation is not sustainable.[22] Given the potential for further growth of the Chinese market and its significance for Japan and South Korea, it is unthinkable that either would risk undoing their relationships with Beijing by switching their diplomatic recognition back to Taipei.

Beijing's 'zero-sum' approach to the issue of Taiwanese diplomatic representation on a series of regimes effectively renders impossible an intergovernmental regime that includes both countries. But in the case of fisheries management, and particularly in fighting marine pollution in North-East Asian seas, it is crucial to have both China and Taiwan share membership and obligations. Therefore, in forming a maritime regime, membership for China and Taiwan should not be entangled with the politics of Chinese national reunification. This means that a regime must not be defined as one between nation-states. Incentives should be given instead to both China and Taiwan, so that they can each sell the concept of regime membership to their populations by arguing that the regime is only a consultative group addressing non-diplomatic resource and environmental issues that affect all members sharing North-East Asian waters.

The establishment of such a regime would also necessitate setting aside sovereignty disputes over the Diaoyu (Senkaku) islands, and the Sino-South Korean-Japanese continental shelf disputes. There are many precedents showing that the shelving of sovereignty disputes can lead to close co-operation and consequent direct benefits. This suggests that each member should consider exercising restraint by forgoing its preference for a pre-emptive agreement on boundaries in the East China Sea and the Yellow Sea.

Regime Membership for China and Taiwan: Advantages and Disadvantages

China

For China, regime participation could have several advantages. In the area of fisheries management, one advantage might be a fairer international legal environment in which the parameters for fishing and fisheries protection would be set. China currently has bilateral fisheries agreements with Japan and is negotiating a similar arrangement with South Korea, but no formal and binding fisheries agreement exists between China and Taiwan.[23] In the absence of such an agreement, the only constraint on Taiwan would be the voluntary one of compliance with the terms of China's agreements with third parties. It is under no compunction to do so, which effectively puts Taiwan in a position to undermine the fruits of bilateral maritime co-operation among China, South Korea, and Japan. Furthermore, a multilateral legal framework would help offset China's relative disadvantage *vis-à-vis* Japan in scientific data on, and understanding of, the Yellow Sea and the East China Sea. China would also be in a strong position to influence the nature of the duties and obligations that it would have regarding fisheries management and protection.

Conversely, if China joined a maritime regime it would have to use its limited monetary resources to preserve fish stocks in marine areas further offshore that it does not yet heavily utilize. It would also have to divert attention from its pressing terrestrial problems to ensure that its own fleets complied with regime rules. Still, these disadvantages pale alongside the benefits to be gained from regime participation, particularly if that participation contributed to better overall relationships in other areas.

In the area of marine environmental protection, regime participation would allow China to benefit from the technological knowhow made available to it by Japan, South Korea, and Taiwan. China has already adopted comprehensive marine environmental protection laws and carried out extensive pollution surveys and research.[24] But enforcement is a problem. Some vessels, platforms, and cities continue to discharge pollutants into the sea.

Yet China has set itself ambitious environmental protection objectives for the year 2000. The objective for marine environmental protection is that coastal waters should meet national seawater quality standards, which means that first and second grade sea areas should maintain good water quality, and third grade waters should meet the national standards by 1995. China plans to reach this goal through: strict enforcement of the Marine Environmental Protection Law and other regulations; scientific functional zoning of sea areas based on their ecological loading and environmental carrying capacities, and the establishment of appropriate water quality standards; establishment of natural reserves and a coastal forest belt to protect rare tree species, endangered animals, special natural ecosystems, topography and landforms; controlling pollutant discharge into the sea, particularly land-based pollutants, and promoting rational use of the sea's purifying capacity by selecting the best locations for discharge outlets and using sophisticated discharge methods; making marine dumping areas safe, scientific, and economic by strictly controlling waste dumping, including shipbreaking; strengthening marine pollution monitoring and surveillance; and establishing an emergency system of marine oil pollution damage control.

China recognizes that international co-operation and the transfer of scientific knowledge and technology will be necessary to implement this plan. Access to increased knowledge enabling the source of pollutants to be located, will help China better to argue its principle that the source country should pay the clean up cost. Although China is increasingly aware of its environmental responsibilities, it wants to be sure it shoulders a *fair* share of the burden.

Of broader significance, regime participation would provide an opportunity for China to express egalitarianism and collectivism in the international arena. Regime participation can also be an effective confidence building measure among China, Japan, South Korea, and Taiwan. And it can demonstrate that China is a responsible member of the international community. Japan is increasingly tying its significant Official Development Assistance loans to China to a reduction in the airborne environmental air pollutants from China that impact on Japan. Through regime participation, China could demonstrate to Japan that it is genuine and serious about improving the environment, and about its responsibilities to the region in this regard, thereby placating the neighbours that are so important to its economic development.

A disadvantage for China is that it would have to clean up marine pollution in its own as well as international waters. Given China's limited financial resources, this could have major economic implications. The coastal provinces in China have traditionally relied on labour-intensive production of light industrial goods as their major source of revenue. Should the sources of marine pollutants be traced to factories on Chinese soil, the Chinese government would be obliged to pressure its local governments, and particularly state-owned pollution-generating factories, to comply with costly measures which it had negotiated.

Taiwan

For Taiwan, the first advantage of regime participation is the opportunity to reach beyond the current framework for negotiating with China through the one officially designated channel: the Straits Relations Fund and its mainland counterpart, the Association for Cross-straits Affairs. Since the April 1993 'summit' between the heads of the two agencies, the agenda has become so filled with more pressing issues that fisheries disputes and other ocean-related issues have not been adequately addressed.[25] Because this is the only channel of official communication, and its government is faced with an aggressive media both at home and abroad, Taiwan, more than the mainland, is under immense pressure to take a single-issue, result-oriented approach, thus deferring maritime issues.

Through regime participation, Taiwan could not only be a player of equal status with the mainland. It would also be able to deal with the complex issue of maritime affairs out of the media spotlight and with less chance of unrelated points of contention with Beijing emerging. Taiwan would like to be admitted to the United Nations Convention on the Law of the Sea, and regime participation might bolster its chances of at least a *de facto* recognition of its accession to the Treaty and the benefits that would accrue therefrom.

The primary disadvantage for Taiwan is that as a full member of a maritime regime, it would have to share its technological know-how and monetary resources with China. Taiwan might perceive that its assistance could be utilized by the mainland to threaten Taiwan's security. On the other hand, by sharing such relatively innocuous data, Taiwan would be taking a giant step forward in fostering confidence building with China.

Conclusion

The idea of China and Taiwan sharing equal status and shouldering equal responsibility in a North-East Asian maritime regime is not new. The two sides already share equal seats on such important multilateral bodies as the Asian Development Bank and in the Asia-Pacific Economic Co-operation process. The underlying key is that a maritime regime for North-East Asia cannot be intergovernmental or have diplomatic power. This principle applies not just to the China-Taiwan recognition issue. It would also help to put on the back burner the sovereignty disputes between China/Taiwan *vis-à-vis* Japan, and between China/Taiwan *vis-à-vis* South Korea. Given the dynamics of post-Cold War North-East Asian diplomacy, there is no alternative. The interlocking importance of overall bilateral relations among the North-East Asian countries is so great that none of the four players can afford to risk a bruising diplomatic battle over such sovereignty disputes before they enter into a maritime regime.

While it is true, then, that any maritime regime in North-East Asia would be incomplete without the full participation of both China and Taiwan, the politics of Chinese national reunification does not have to be a hindrance to maritime regime building in the region. It is not clear that regime participation by both China and Taiwan will enhance the process of Chinese national reunification. Nevertheless, by including both China and Taiwan as members of a maritime regime, North-East Asian countries can better co-ordinate their efforts to manage their marine environment and, in the process, help create a more lasting security, in the comprehensive sense of the term.

Japan[26]

Context

Japan may be in a good international political position to lead the formation of a fisheries or environmental protection regime. It has long had bilateral fisheries agreements with all nations of the region. It also has diplomatic ties with every nation except North Korea, and is considering normalizing relations with Russia. In addition, only Japan is free of the 'split personalities' of the other countries of the region: China and Taiwan, South and North Korea, and Russia and the other former Soviet republics. And it is steadily improving relations with most of its neighbours. In 1992, Japan held its first informal summit meeting with South Korea (Prime Minister Miyazawa and President Roh Tae Woo), and made its first cabinet-level contacts with Taiwan.[27] Russia, South Korea, and China are already important fishery counterparts, and a maritime regime would strengthen these bonds.

The threat of another country's leadership could prod Japan into action. Discussions within the Economic and Social Commission for Asia and the Pacific (ESCAP) have indicated that South Korea, not Japan, should take the initiative in forming multilateral maritime regimes due to its central geographic location, which borders on all three regional seas.[28] And Kim Young Sam's proposal at his summit meeting with Prime Minister Hosokawa to forge a China-South Korea-Japan environmental initiative caught Japan's Ministry of Foreign Affairs and its Environment Agency by surprise. And although UNEP's North-West Pacific Action Plan suggests regional centres for monitoring, compiling and exchanging data, and maintaining a task force of experts, none of these centres is slated to be based in Japan.[29]

A Japan-led fisheries or environmental regime could take advantage of Japan's lead in scientific knowledge and technology. In November 1992, UNEP opened two International Environmental Technology Centers in Osaka and Shiga, Japan, in order to give developing countries access to environmental technologies and strategies employed by industrial nations.[30] And since Japan already leads the region in fisheries research, a pre-emptive move towards co-operative leadership could be the creation of a regional fisheries data centre, which would make Japan the region's 'information broker' in both environment and fisheries.

In addition to enhancing its status and providing a means of exerting influence, a regional regime for environment or fisheries could have political side benefits for Japan. Even after the Law of the Sea comes into effect, such a regime could postpone implementation of EEZs by Japan's neighbours, and thus the delimitation problems which will attend them. With a regime in place, Japan's bargaining position might be improved once the drawing of such lines became inevitable.[31]

Fisheries and Marine Environmental Protection Regimes

While they are related, the specific incentives for Japan's leadership in regional fisheries and marine environmental regime formation have notable differences. The two regimes are therefore considered separately.

Fisheries

Japan's distant-water fishing industry is developing a siege mentality.[32] Japanese fishing has been phased out of the US and Russian zones, pollack fishing in the Bering Sea has been stopped between 1995 and 1996, the high seas salmon fishery has been halted, and the squid driftnet industry has been closed. Japan's 'research' on minke whales has been severely criticized. Moreover, labour costs are rising rapidly, and now comprise 40 per cent of production costs.

Although Japan has itself declared a 200 nm EEZ, it does not apply it to South Korean and Chinese nationals, or in the eastern Sea of Japan. But Japan has specified that foreign states may fish in its fishing zone at the discretion of its Ministry of Agriculture, Forestry, and Fisheries. Japan does not set catch quotas except for Russian fishermen, for whom the quotas now equal Japanese quotas in Russian waters for fish other than salmon. Japan prefers instead to rely on co-operatives to control resource utilization in coastal areas, and depends on governmental licenses at the prefectural and national levels to control effort in the offshore and distant-water fisheries.[33] Enforcement in the zone of extended fishing jurisdiction is an expansion of pre-existing programs, except with respect to foreign fishing vessels. In the legislation which established the fishing zone they received special attention, including maximum fines, forfeiture of vessels, gear, and catch, and notification of the foreign government.

Japan is increasingly concerned about the sustainability of its fish stocks and its supply of fish. The 1992 annual report of the Japanese Fisheries Agency underscored the importance of sustainable fisheries 'through the sound conservation, management and rational use of marine resources on the basis of scientific knowledge and through the practice of fishing harmonized with the environment,' especially through better monitoring and the use of more discriminating fishing gear and methods.[34] A government white paper issued in April 1993 recommended that Japan's fishery industry sustainably manage its resources 'to cope with decreasing catches in regulated international waters', by transcending the divisions within the industry.[35] In addition, the government and the coastal fishing industry are presently developing together a national fishery resource conservation system that will be managed locally.[36] To stimulate the supply of fish, the Overseas Fisheries Cooperation Foundation provides financial aid and technical assistance to overseas fisheries industries, including those of such 'developing countries' as South Korea and Taiwan.[37]

Fish comprises 44 per cent of what Japanese eat, giving Japan the world's highest per capita fish consumption. Not surprisingly, its fishing operations are the most extensive in the world. Japan probably has a virtual monopoly of knowledge regarding *regional* fisheries. And it has successfully overcome—if only temporarily—the constraint of its neighbours' extension of jurisdiction by negotiating, or allowing its fishing industry to negotiate, separate access arrangements with every nation in the region. Hence it has the broadest geographic access to fisheries of any nation in the region.

In spite of these efforts, Japan is at a critical juncture regarding its fisheries in North-East Asia. Japan and other countries have so overfished North-East Asian seas that not only have stocks significantly dwindled, but the entire ecosystem has been altered.[38] Competition between North-East Asian nations for scarce resources is rapidly increasing. For example, Japan's catch in the Yellow and East China Seas is now only about 3 per cent of its total. As other nations become better able to harvest their own resources, bilateral fishery ventures will appear less beneficial to them, leaving Japanese fishing companies in a position of declining power, reducing in turn Japan's leverage in joint ventures.[39] In fact, the complex web of bilateral fishing agreements in the region, so painstakingly crafted by Japan, is now inadequate. It may even be in danger of collapsing, and with it Japan's existing access arrangements.[40]

But as a distant-water offshore fishing nation, Japan is also concerned about the level of protection coastal states give to spawning populations of fish and to juvenile fish in nursery areas, because of the possible effects of overfishing in coastal areas on the status of offshore stocks. Further, many fish stocks upon which the Japanese fleet depends are being overfished, or are in danger of becoming so, because of expanding fishing pressure from the host country. So Japan is becoming interested less in access than in conserving the stocks, particularly those which it shares. South Korean and Chinese fishing in Japan's waters beyond its 12 nautical miles territorial sea has significantly degraded crab, squid, and flounder stocks there, and caused Japan's coastal fishermen to call for implementation of the fisheries zone or a full EEZ in the Sea of Japan.[41] A regional fisheries regime would provide a forum for ironing out such disputes, and reduce or eliminate the need annually to renegotiate bilateral fishing quotas as well.

There is, therefore, considerable incentive for Japan to set up a multilateral regime now, when it could use its relatively powerful fisheries position to influence the regime's form. Such an organization could divert attention from Japan's regionally dominant position and encourage an aura of legitimacy for the institutions it creates or allows to be created. It could also lead to sustainable development of the region's fisheries, which would be of long-term benefit to Japan. But this opportunity must be pursued before the situation erodes much further.

In such a regime, Japan could combine its fish catching/processing technology and surplus vessels with manpower and resources from other countries, while providing its own markets for the products.[42] This could help to keep Japan's fishing economics competitive while maintaining supply. Processing could be spread along Japan's west coast, enhancing its development and thus reducing the disparity in development levels within the country.

The alternative is bleak. If China and South Korea declare EEZs, Japan will lose a major portion of its fishery grounds, including all of the Yellow Sea and 50 per cent of the Sea of Japan. Japan is the most extensive user of the fisheries resources of the Sea of Japan and has the longest of this sea's coastlines. It has the most to lose from an unstable situation.[43] Many Japanese fishermen will be displaced if Japan's neighbours establish EEZs, or if a regional fishery regime enforces more severe restrictions on fisheries.

If a regime were in place, not only could these fishermen be compensated by Japan's time-tested array of relief measures, but the new opportunities created, including expanding aquaculture operations, would further ease the shock.[44] Through scientific research access to other countries' waters as well as their data, Japan might gain an enhanced ability to assess the region's fishery stocks overall, as opposed to the present situation in which the various countries produce different and competing stock assessments.[45] Japan's potential role as a conduit of scientific information is easily envisaged. Moreover, fishery activities by extra-regional nations could be eliminated.

Considerable momentum is building in Japan for an enhanced international leadership role in fisheries. In general, Japan has gone further than any other nation in the region to ensure that any regional fishery management systems that may develop can be implemented without amending present laws.[46] Japan is already well ahead of other North-East Asian nations with its 200 nautical mile Fishery Zone Law, which stipulates that Japan must respect recommendations by international organizations concerning the conservation and management of fishery resources.

Japanese fishing interests face a classic dilemma. Should their fleet continue to fish the region's stocks as fast and as intensely as it can before the stocks collapse or other fishing nations, such as South Korea or Taiwan move in? Or should Japan participate in a regime designed to manage and limit the catch in order to be able to fish for longer but at a reduced level, hoping that competitors will join and comply with the rules? Positions of policymakers on this issue are likely to depend on whether they take advice from their leading scientists, and whether they think in the short or long term. Local politicians may listen to their fishing constituencies, which are in dire straits[47] and want continued unregulated access to other's waters. On the other hand, the Foreign Ministry may take a broader, far-sighted approach in the interest of both conserving stocks and the stability of the regime, and thus in the interests of Japan's longer-term relations with its neighbours. The challenge to advocates of regional co-operation is to develop an argument that Japan's politicians can use to justify their support for, or at least acquiescence in, regional management.

Regarding such regional arrangements, some Japanese fisheries experts feel that the best solution to the Sino-South Korean and Japan–South Korea fisheries disputes is to renegotiate a fisheries agreement among Japan, China, and South Korea. Or perhaps these countries could conduct joint scientific studies of fisheries resources, collectively determine the total fish quota for each of the countries, establish joint resource conservation zones and joint fishing zones, and carry out joint fisheries propagation projects. Foreign fishers might be given a share of the Japanese trawl fisheries in the Sea of Japan in return for Japanese access to their neighbours' coastal waters.[48] There is some movement in this direction. Some 50 delegates representing the fishing industry in the three countries met in Japan in August 1993 to discuss prevention of accidents in the shared waters of the East China and Yellow Seas. Still others advocate dissolving the existing bilateral commissions but using their framework to build an 'East Asian Fisheries Co-operation Regime' which would include all North-East Asian governments.

In what was probably the most notable regional fishery initiative up to 1994, Japan, Russia, China, and South Korea agreed, at Japan's urging, to set up an international committee to protect the fishery resources of the Sea of Japan. The group also agreed to urge North Korea to participate.[49] This forum could serve as the catalyst for a multilateral regime. Japan's growing interest in playing a more prominent role in international politics could make it more than just an intermediary, the role it currently fills. Japan clearly understands the need to co-operate with its neighbours and has considerable experience with such institutional relationships. The need to co-ordinate fishery policies could facilitate the improvement of relations overall.

Protection of the Marine Environment

In addition to the rapid decline in their fish populations, North-East Asian seas are suffering from the onslaught of pollution, which is expected to increase as the developing countries follow a 'develop first, clean up later' philosophy. Already, squid from the Sea of Japan have some of the highest levels of PCB, TBT (an organic compound used in protective ship paints), and radioactive elements in the world.[50] The recent revelations of nuclear waste dumping in the Sea of Japan by both Russia and Japan have created a crisis atmosphere, and could be the exogenous event needed to stimulate the formation of a coherent environmental regime, especially one monitored and enforced with Japanese technology.

With its advanced pollution control measures, effective energy conservation technology, and successful combination of economic growth and energy conservation, Japan is a natural choice as an environmental model for neighbouring countries.[51] Furthermore, mounting domestic concern over the marine environment and a more vocal public are pressuring the Japanese government to take steps to reduce pollution in North-East Asian seas.[52] Indeed, Japan and its public are increasingly concerned about the effects on its own fisheries and aquaculture facilities of pollutants from the region's rapidly industrializing countries, and about the rapidly declining industrial safeguards in Russia.[53]

Japan has a solid legal and policy foundation on which to base its international initiatives. It has subscribed to, and complied with, more pollution control treaties than any other North-East Asian nation except the former Soviet Union.[54] It has some of the strongest marine environmental legislation in Asia, including the 1970 Marine Pollution Control Law and the 1978 Special Law for the Conservation of the Environment of the Seto Inland Sea. Japan has put pressure on industry to boost environmental standards through the new 'Basic Environmental Law', which was implemented in 1994. This Law has a special section on international co-operation for environmental protection, and stipulates that Japan take initiatives in this direction.[55]

Japan's Council on Ocean Development was formed in 1991 and serves to combine Japan's myriad ocean interests under a single umbrella organization and to articulate their views in order to inform and focus national ocean policy debates.[56] Consisting of representatives from government agencies, ocean industries, and academic and scientific experts, the Council constitutes a major step towards formulation of a national ocean policy. Its 1991 report called for more active Japanese initiatives through international organizations in order to 'reduce and prevent' land-source and ship-source marine pollution.[57]

In the last few years, Japan has made considerable efforts in environmental diplomacy. In 1989 it hosted the world's first international environmental forum for legislators, attended by lawmakers from eighteen countries. In 1990 the Environment Agency established a Global Environment Department to strengthen Japan's international environmental efforts.[58] The Environment Agency also convened a North-East Asian Conference on Environmental Protection in October 1992 and again in 1993.[59] Japan's stated goal is to convene the conference annually, and to upgrade it to ministerial-level meetings. Japan has also helped to foster global networking for the conservation of wetlands through the Ramsar Convention, and committed 10 million yen to wetland conservation at the Ramsar meeting in Japan in June 1993.[60]

In October 1993 the Japanese Organization for Economic Co-operation and Development produced a report urging the country to 'boost its financial contributions to the international fight against environmental problems to a level more commensurate with its economic strength.[61] And in a move of considerable significance, Japan broke ranks with the West at the 1992 UNCED conference, and announced an increase of 50 per cent in its international aid for protection of the environment.[62] The Japanese delegates went on to state that sustainable development would become the watchword for Japan's entire aid programme, the budget of which is currently larger than any other nation's, and is set to double in the next three years. With a long record of success from its policy of 'economics driving politics', Japan now appears to be considering the advantages of industrial opportunities created by the politics of environmental change.

Perhaps more important, the Environment Agency obtained approval in February 1993 to establish a 'Global Investment Fund' to assist environmental non-governmental organizations (NGOs) in Japan and abroad.[63] In June 1993 the Agency announced that Japan would donate more than US$10 billion over the next five years for environmentally conscious regional development programs, noting that the Asia-Pacific region should set an example for the world by co-operating on sustainable development.[64] And in September 1993, government ministries and agencies were seeking 592 billion yen for environmental protection, representing a total budgetary increase of 8 per cent over the current fiscal year.[65] This was a marked improvement over 1986, when an Environment Agency panel produced a document entitled, 'Basic Directions for Environmental Considerations in Development Assistance' to developing countries. It received little attention, with ministries such as MAFF never bothering to read it.[66] Even business organizations have started to put emphasis on environmental issues in their overseas ventures. In 1990 the Federation of Economic Organizations (*Keidanren*) submitted to its members a recommendation suggesting a considerable greening of its overseas project guidelines and demanding a substantial improvement of the Japanese government's environmental decision-making system.[67]

In August 1993, Prime Minister Morihiro Hosokawa announced that global environmental issues demand 'immediate attention.' He went on to state that 'I intend to take new [international] initiatives drawing fully upon Japan's own experience and abilities' in this area.[68] Wakako Hironaka, the Environment Agency director under Prime Minister Hosokawa, suggested in October 1993 that Japan's Self-Defence Forces should put international 'greenkeeping operations' on their agenda, a move which would aid Japan's goal of being a benign leader. She then stated her desire to host a meeting of environmental ministers from industrialized countries on forms of environmental aid to be disbursed to developing nations.[69] Morihiro Hosokawa's successor, Tsutomu Hata, echoed these remarks, saying, 'Rather than waiting for other countries' initiatives, Japan should actively make proposals based on the knowledge it has accumulated through peace and prosperity.'[70] The current prime minister, Tomiichi Murayama has continued to support this theme. In October 1994, he gave a speech to an international conference on the global environment, in which he sought to revive momentum for Japan's contributions towards conserving the global environment.[71] All of these efforts are establishing the foundation for Japan's leadership in environmental regime formation in the region.

Japan can also build on its considerable experience in regional environmental co-operation with South Korea, Russia, and China. As early as 1974, Japan and South Korea had an 'Exchange of Notes on Preventing Pollution of the Sea', which they annexed to their boundary and Joint Development Zone agreements.[72] Although this agreement pertains only to pollution from petroleum, it does state that exploration and exploitation shall be carried out in such a way that 'fisheries will not be unduly affected'. The agreement also stipulates measures to prevent and remove pollution and to deal with the damage, making the concessionaires jointly liable for compensation.

There is growing optimism in Japan that environmental protection is an idea whose time in the region has come. Japan perceives that China has changed its attitude and now discloses that its environment is polluted and that its environmental management system does not work very well. In May 1993 Japan and South Korea agreed to exchange experience and 'know-how' for joint ventures 'at' third parties, and find ways to extend assistance to new problem areas such as the environment.[73] In June 1993 they signed a draft convention on joint environmental co-operation that included co-operation on policies for the prevention of air, marine, freshwater, and soil pollution, the protection of biodiversity, and the prevention of global warming.[74] A committee of experts will choose joint projects, exchange scientists and research, and promote seminars. Japan has signed similar accords with the former Soviet Union and with China.[75] In August 1993, Japan signed a wastewater treatment pact with South Korea providing for Japanese technology transfer to South Korea.[76]

In 1992 Russia called for twelve economic co-operation schemes to be undertaken in conjunction with Japan, including the disposal of highly toxic industrial waste.[77] In the aftermath of Russian dumping of radioactive waste in the Sea of Japan, Tokyo has agreed to provide Russia with technological assistance for the storage and treatment of radioactive waste.[78] In April 1994 Japan, South Korea, and Russia conducted a joint study of the impact of radioactive waste dumping in the Sea of Japan.[79] At the same time, Japan announced it would no longer tie aid to Russia to a solution to the Kuriles dispute, and promised about US$3.5 billion in aid, some of which will go towards environmental protection projects.[80] In October 1993 the two countries signed a joint economic declaration which included co-operation in fisheries and the environment.[81]

Regarding China, Japan opened an 'Environment Center in East Asia' in Beijing, with $80 million in startup funds in 1992.[82] Japan and China plan to sign an agreement 'checking' the acid rain emanating from China's industry, possibly with the help of South Korea. Jiro Kondo, chairman of the Japan Science Council, praised the agreement as 'basically good,' but warned that 'co-operation will only be fruitful if we avoid viewing Japan as the country suffering from pollution, while terming China the polluter.'[83]

Perhaps more significant is the agreement between Japan and the United States, under the concept of 'global partnership', to initiate an intergovernmental forum for protection of the environment in the developing countries of Asia. The forum will include discussion of methods for supporting environmental NGOs, and the training of environmental experts in these countries.[84]

International Constraints on Japan's Leadership of Regional Maritime Regimes

Arrayed against these numerous incentives are several important constraints, or disincentives, which may prevent Japan from leading or even participating in multilateral fisheries or environmental regimes. The legacy of 250 years of isolationism under the *shogun*s prevents Japan from feeling comfortable in its international relations. When forced to open its ports in 1853, Japan quickly saw the economic advantages of imperialism and committed itself to acquiring its own empire. Vestiges of this mindset remain today. Moreover, Japan still suffers from a 'legitimacy deficit' overseas which impairs its leadership capabilities.[85] Sources of the problem include the legacy of its militarism and colonialism, its mercantilistic reputation, and a perception that Japan is unable to articulate universal values and principles.[86] Japan's recent efforts to build trust, including Prime Minister Hosokawa's statements in China and Korea regretting Japan's aggression in World War II, and Emperor Akihito's visit to China in 1993, were seriously undermined by careless statements by the justice minister in April 1994, and the Environment Agency chief in the following August.[87]

Japan must also overcome a perception that it believes the only worthwhile overseas investments are those which will benefit Japan either directly or indirectly, regardless of the socio-environmental consequences for the countries receiving the investments.[88] Indeed Yoshiji Nogami, then head of the Japan Institute of International Affairs, stated:[89] 'People here don't want a global partnership. They see the world through an economic window and will continue to do so even in the post-Cold War world.' This perception has been somewhat mitigated by Japan's 1993 ODA guidelines setting conditions on recipients in the areas of military spending and human rights. Nevertheless, while the United States stresses making Asian Development Bank loans dependent on basic human needs and environmental protection, Japan prefers loans which will help to build economic infrastructure and boost manufacturing.[90] Basically, Japan prefers bilateral to multilateral arrangements because they are easier to control, and because it fears that a multilateral organization will simply spend its money in projects over which Japan has little control.

There are also particular factors working against Japanese leadership of a maritime regime. The dispute over the southern Kurile Islands/Northern Territories is one of them. It presents a formidable obstacle to regime formation.[91] In any case Japan has general doubts about Russia's ability to comply with a long-term, multilateral regime. Moreover, the present situation in fisheries *vis-à-vis* Russia favours Japan, since the Russian Far East is currently selling access to its resources to Japanese fishermen for foreign exchange.[92] Multilateral fishing quotas would reduce Japan's take within regional EEZs and further constrain its 'high seas' fishing. And Japan would have to supply most of the money to a fisheries regime. The thought of pumping more investment into developing countries in order to put them on a more equal footing for multilateral consultations is a difficult pill to swallow. The *quid pro quo* for Japan would probably have to be a larger slice of the fishery pie.

The diffuse style of Japanese foreign policy formulation, which requires strong party leadership to overcome the bickering of the *habatsu* (factions), continues to be an obstacle to Japanese efforts to play a more forward role in regional politics. The combination of the *habatsu* system and the deeply felt need for at least a semblance of consensus produces massive gridlock, preventing proactive regional leadership by the Japanese government.

The recent domestic turmoil has also not been conducive to Japanese regional leadership. And for the foreseeable future, a multiparty system, with shifting coalitions and alliances seems probable.[93] Japan has responded to the uncertainty of the post-Cold War environment by watching-and-waiting.

Although Japan has been expanding its multilateral role in Asia—for instance, in the Asian Development Bank, ASEAN, and APEC—its activity has fallen short of policy pronouncements. Basically Japan's Asian diplomacy is still reactive and is thus conditioned by other countries' expectations and requests. As Japan expands its international role, it will probably pause at each stage to assess other countries' reactions before making further advances. What policy changes do occur will likely unfold through the gradualist, reactive, risk-minimizing processes of the past. During the current transitional period of fragile coalition governments and politicians merely hoping to survive the next election, Japan's leaders of whatever stripe will be less inclined than ever to break with the past to launch assertive action in domestic or foreign affairs. Accordingly, one can expect few diplomatic initiatives and much cautious circumspection. As Japan continues to test the international waters for an appropriate post-Cold War role, its probable approach will be that of active supporter, within established structures and relationships. Any abrupt shift or reversal of direction seems improbable, even as Japan becomes increasingly prominent in global, and especially Asian regional affairs.

There is also a conceptual dichotomy among Japanese marine policymakers as to how and where Japan should focus its activities. One view holds that Japan is part of a unique region, the Sea of Japan Rim; that special treatment should be accorded this region; and that Japan should be taking the lead in doing so. This view is prevalent among politicians in western Japan who argue that the complementarities and the different levels of economic development of the nations bordering the Sea of Japan mandate a bright future. The opposing view argues that the emphasis on the Sea of Japan Rim is basically a negative reaction to the Pacific/Western focus of Japan, and that Japan should instead be open and outward looking, not isolated and inward focused. In this alternative view, Japan should embrace a broader regime for ocean policy, which includes ASEAN and Australia/New Zealand, and not create a fortress.

With regards to a regional regime, the conservative view in Japan is that there is at present a fundamental understanding on Law of the Sea and maritime zones, and that a change in regime would be complicated, confusing and conflictual. Indeed, the hegemon usually prefers a network of bilateral relations dominated by itself, in which the parties interact on a superior/subordinate basis with the hegemon but have little opportunity to interact with each other and thus to band together. This approximates the present situation in the fisheries sector in North-East Asia, with Japan playing the dominant role. In the conservative view, it is better to maintain the *status quo* and to use subtle negotiation to resolve conflicts. The problems that do arise are invariably between two or three countries and not the entire region, and are thus better handled by bi- and tri-lateral agreements and diplomatic dialogue.

The fisheries community in Japan is dominated by interests more or less supporting the status quo. Domestic divisions exist between distant-water and coastal fisheries' interests, between regions, and between companies which employ different fishing methods.[94] The largest fishery organization is the Japan Fisheries Association (JFA), which comprises 414 enterprises representing all aspects of the fishing industry. The JFA exerts substantial influence over Japan's fishery policy, participates in private-level international fisheries negotiations, and has sent representatives to all UNCLOS III negotiations since 1973. Because the JFA is dominated by distant-water fishery interests, the interests of coastal and small-scale fisheries are represented nationally by the less influential National Federation of Fisheries Cooperatives. This division has led to wide differences within the fishing industry on international policy issues, and has bent Japan's fishery policy towards distant-water fishery interests.[95]

Ministries funnel money back to the various interest groups in the form of sanctioned research projects such as the assessment of fishery resources, giving Japan by far the region's largest database on fisheries. If Japan entered a multilateral fishery regime, it would have to acknowledge that Japan's own fishermen are often the most egregious transgressors of other countries' waters.[96] That admission would lead to an alteration of Japan's fisheries policies and perhaps damage the exceptionally close relations between fishery interests and government leaders.

Finally, Japan lacks an articulate national consciousness. While recent public opinion polls demonstrate growing public awareness in Japan of the need to take a more active role in world affairs, many Japanese seem content to see their country remain passive on international issues. This sentiment works against any prime minister who concentrates on foreign policy issues. Indeed, it is sometimes said that Japan actually depends on outside pressure to overcome entrenched domestic lobbies.[97] For their part, NGOs have grown stronger since UNCED but they still lack sufficient scientific and technical expertise, and are not well-organized. More problematic is that NGOs are not yet fully accepted in Japanese society and their leadership seems more or less to accept the status quo.

This passive attitude among Japanese is reinforced by the mutual security treaty with the United States, which may be the only major active foreign policy choice Japan has made in the post-war period. The treaty effectively creates more inertia and helps to prevent Japan from formulating a new role for itself in a post-Cold War world, especially regarding the Asian neighbours it has alternately invaded and shunned over the last century. Since modern Japan has never experienced any other security relationship, it may have forgotten that alternatives exist. The result is that Japan seems not to identify with the rest of Asia[98] but continues instead to look at the United States as a role model. So the visible foot-dragging of the United States at UNCED and elsewhere has also impeded Japan's taking an active role in the environmental future of East Asia.

Conclusions

There are several factors in favour of a prominent role for Japan in a future North-East Asian maritime regime for fisheries and environment: its economic dominance, its technology, knowledge, and experience in environmental regulation, its web of bilateral fisheries agreements with neighbouring countries, and its lack of a split political personality like China and Korea. And leadership of such a regime might benefit Japan by delaying the implementation of EEZs in the region and the escalation of the boundary disputes they would inevitably bring. Other benefits would include conservation of fisheries resources and the environment, elimination of the transaction costs of annually renegotiating fishing quotas, and enhancement of Japan's status in the world. In many ways, Japan has the least to lose and the most to gain from leading a multilateral maritime regime in the region, and a sea change in domestic politics over the last year may bode well for Japan's eventual regional leadership.

For the foreseeable future though, there are strong factors mitigating against Japan assuming a leadership role. In the international context, the Kurile Islands dispute, memories of Japan's expansionist wars, and its perceived propensity for placing economic gain above all else constrain acceptance of its regional leadership. Bureaucratic inertia, resulting from the need for consensus among Japan's many domestic political entities, and buttressed by its Confucian social system, forces Japan into a reactive rather than proactive role when maintaining the status quo is no longer tenable. This obstacle is reinforced by Japanese uncertainty and caution towards foreigners, and their reluctance to become entangled in world affairs.

Although there is general agreement that Japan should continue to shelter under the security and political umbrella of the United States, domestic lobbying is increasing for it to conclude more direct and comprehensive maritime agreements with its neighbours. With the growth of the middle class and its awareness of its potential for influencing both domestic and foreign politics, including environmental issues, will come conceptual support for a global perspective. Leadership on policy issues may then be dragged out of the bureaucracy and back into the Diet, ultimately forcing Japan's leaders to take the initiative on international issues.

The rapidly increasing density of unilateral, bilateral, and multilateral pronouncements and agreements involving Japan distinctly suggests that the nation is increasingly comfortable with a larger role in the region. But the Japanese government is still in the midst of its greatest political upheaval since World War II. It remains to be seen how pervasive or permanent this restructuring will be, and how it will influence Japan's regional role.

So Japan faces the double task of having to overcome its domestic haggling and its bullying past before it can become a leader in any regional fisheries/environmental regime. An important asset would be a strong streamlined government which did not simply follow the Western lead. If the government is to achieve this goal, it must replace its single-minded economic strategy with a multifaceted, values-oriented policy, and it must redefine the prime minister's role to give him a mandate strong enough to overcome Japan's massive *habatsu* system.[99]

In terms of international perceptions, Japan must exercise leadership which is wise and non-threatening, and which does not arouse the ire of neighbouring countries. Japan must demonstrate that it has the neighbouring countries' best interests at heart, and is not taking the initiative solely to further its own ends. In other words, Japan must use its power very carefully, otherwise resistance will be considerable, both within and outside the Japanese government. Rather than attempt to impose a system from above, Japan's best course might be to promote the conditions necessary for the spontaneous formation of a North-East Asian maritime regime. It could accomplish this through hortative and financial contributions, through further strengthening of bilateral and multilateral ties, by fostering loosely structured consultative bodies in the region, and by building on the growing sense of an East Asian identity as well as the regional tendency towards consensual politics.

In recent years, Japan's international maritime agreements have become more multilateral and equitable.[100] Although its visible commitment to effective fisheries and environmental management is a fairly recent phenomenon, Japan is nonetheless slowly leading the way towards a North-East Asian maritime regime. But to lead successfully in regime formation, Japan must assume the 'foreign policy of a major power with an unassuming posture.'[101] Otherwise another regional country may very well steal the regional leadership spotlight, at least in the ocean policy realm.

North Korea

The Political and Economic Context

Since the end of the Korean War, North Korea has willingly been a closed society dedicated to socialism and self-reliance. In this context, its angling for foreign investment and diplomatic recognition may be the silver lining in the dark cloud of pessimism surrounding the North Korea nuclear controversy.[102] The nuclear issue has increased tension between North Korea and the United States as well as the rest of the region. However, it should be remembered that before this issue arose, the general trend was towards an incipient economic opening of North Korea, and an improved political relationship with the United States, Japan, and, most dramatically, South Korea. Indeed, there is an influential school of thought in South Korea that believes Pyongyang's fundamental attitude is gradually becoming more positive, and that to enhance long-term regional security, North Korea's tentative economic opening should be supported.[103] This view is consistent with the commonly held Asian perspective that to change a society, one must engage it and influence it through a wide spectrum of multilateral initiatives.

With the resolution of the nuclear issue, the general dissipation of Cold War tension and the incipient trends toward multilateralism in North-East Asia present opportunities for involving North Korea in regional regimes. To build North Korean confidence and experience in the norms of behaviour in international society, efforts to engage Pyongyang might begin in relatively innocuous fields such as environmental protection and economic development. The United States, Japan, and South Korea should be ready to reach beyond symbolism to specifics in these areas.

North Korea needs assistance to overcome its economic crisis. It has been isolated economically and politically by the sweeping reform in former communist countries.[104] Its GNP decreased by 5.2 per cent in 1991 and by 7.6 per cent in 1992. Although the worst may be over, there are still widespread food and energy shortages. Pyongyang has publicly acknowledged that its survival depends on gaining foreign exchange and technology, and thus it is striving to rebuild the trade relations that were severed by the breakup of the Soviet Union.

Before the nuclear issue intensified, North Korea was open to improved economic and political relations with South Korea, Japan, and the United States. Desirous of Japanese yen, but without too many strings attached, it had begun to negotiate colonial reparations of US$5 billion from Japan. And South and North Korea had agreed on measures for reconciliation and non-aggression, and exchanges and co-operation. The chairman of Daewoo, Kim Woo Choong, had held discussions with North Korean officials which focused on setting up light industrial plants in Nampo in the North, building a gas pipeline from Yakutia to South Korea, using North Korean labour on Daewoo's overseas projects, and constructing road and rail links between the two countries.

In December 1992, a major party and cabinet change reaffirmed North Korea's policy of seeking a limited accommodation with Western countries and a cautious economic opening. The members of the new cabinet are reportedly more moderate and internationalist.[105] A revised constitution has added clauses encouraging joint ventures, guaranteeing the rights of foreigners, and establishing a basis for expanded ties with capitalist countries. North Korea has also promulgated laws on foreign investment, joint ventures, and foreign enterprises which allow 100 per cent foreign ownership.

North Korea has also moved forward with plans for a free-trade zone in the Najin-Sonbong area, a free-trade port in Chongjin,[106] and with infrastructure development for its portion of the UNDP-co-ordinated Tumen River Area Development Program. So it can be said to have embarked on a very tentative program of economic reform. Nevertheless, any economic opening is likely to be gradual, and tempered by ideological and social discipline.

Clearly, productive North Korean involvement in any regional economic or environmental initiative would require a sea change in Pyongyang's attitude and openness, as well as a massive training and development effort to modernize North Korea's economic capacity. If the necessary intergovernmental agreements prove unacceptable or difficult for North Korea, then co-operation might be channeled through non-governmental bodies.

North Korea is already involved in two regional economic activities:

- the North-East Asia Economic Forum:[107] a nongovernmental organization devoted to facilitating research, dialogue, and dissemination of information on economic co-operation in North-East Asia, and
- the Tumen River Area Development Project (TRADP):[108] an international free trade zone at the trijunction of Russia, North Korea, and China, which proposes to combine complementary factor inputs such as Russian and Mongolian resources, Chinese and North Korean labour, and Japanese and South Korean capital, technology, managerial expertise, and markets.

Many more opportunities for co-operation exist, particularly if the nuclear issue is peacefully resolved.

Fisheries

Since 1978, North Korea's total nominal marine catch in the North-West Pacific has increased by an estimated 25 per cent. It was 1.6 million metric tonnes in 1987. In 1989 a target of 5.0 million metric tonnes, including harvests of seaweeds and freshwater fish was set by the government for all North Korean fisheries.[109] But it is doubtful it was achieved. The current 7-year plan sets a catch target of 3.0 million mt plus 8.0 million mt of aquaculture production.[110] Major fishing ports are found along both coasts, although the larger fishing fleets probably set sail from east coast ports. Fishing occurs in national waters and in Russian waters, where it is based on the reciprocal fisheries agreement between the two countries.

North Korea has a large industrial fishing fleet with motherships, and factory/freezer fishing by both trawlers and purse seiners similar to those of the former Soviet Union. The prime target species are Alaska pollack, sardine, anchovy, Pacific cod, various flounders, and possibly Pacific herring.[111]

The North Korean catch, like Russia's, is dominated by Alaska pollack, which accounts for an estimated 70 per cent of the country's total marine catch. The Alaska pollack is mostly caught in the Sea of Japan during the winter months (November to February), when large schools migrate to shallow coastal waters to spawn. Sardines and anchovies are caught in quantity in the Sea of Japan during the warmer summer and fall months (May to November). Catches in the shallow and seasonally warmer Yellow Sea include both demersal and pelagic fishes, shrimps, crabs, clams, and oysters. Catches in the Yellow Sea are smaller than those in the Sea of Japan, accounting for only an estimated 20 to 30 per cent of the total catch.

The current condition of North Korea's fleet may be indicated by a rare glimpse at one part of it: the crab fleet. This fleet is a 'rusting, leaking collection of hulks'[112] that sells its snow crab catch in Japan for ten times less than the Japanese equivalent. In 1993 Sakaiminato town bought US$12 million in crabs and other seafood from North Korean boats. But in 1994, sales were more than 50 per cent lower than the 1992 level because North Korea's fishermen were literally running out of fuel for their vessels.

North Korea reportedly has large-scale marine aquaculture.[113] Production along the east coast probably includes oysters, mussels, clams, abalone, and various seaweeds, as well as brown algae such as *wakame* and sea tangle. West coast production probably includes oysters, clams, mussels, laver, various brown algae, and recently, fleshy prawn.

North Korea's fisheries goals are to expand and modernize its fleet, modernize its fish processing industry, increase exports of fish and fishery products, and expand technical facilities for the marketing of fish and fishery products.[114] North Korea has also attempted to protect the fish in its declared jurisdictional zones from foreign fleets. In 1978 it enacted a law on 'Regulations on the Economic Activity of Foreigners and Foreign Shipping and Aircraft within the Economic Zone', which prohibited foreign fishermen from taking certain species of fish. North Korea has enforced its EEZ and military zones against South Korea, Japan, and China, even to the extent of firing upon Chinese fishing boats.[115] Because of its declared EEZ, the limited portion of the resource that it regulates, and its limited distant-water fishing, North Korea is the only nation likely to remain unaffected by regional allocation discussions. Still, shared and migratory stock management requires consideration of stocks in North Korean waters and North Korean fishing activities there.

The main advantage to North Korea of participation in a regional fisheries regime would be the opportunity to engage in discussions with regional entities in which it stood to lose little. North Korea could also avoid diplomatic strife and enhance protection of its own fisheries, assuming China and Japan abided by regime regulations and ceased poaching in its waters. It could receive technology, training, and increased knowledge of the stocks, as well as of other economies, in the region. In a later stage, greater market access in the region could increase its foreign exchange.

The main disadvantages of such a regime for North Korea would be the exposure of its perhaps embarrassingly outdated technology, capability level, and practices, and the possible sharing of information on coastal installations, enforcement capabilities, consumption levels, and tides and currents, that it considers important to its security. However, since the nuclear impasse has seemingly been resolved peacefully and satisfactorily for all concerned, such defence-oriented issues are less likely in the future to outweigh economic and political needs.

Protection of the Marine Environment

Industrial pollution, particularly on the east coast, remains the single most serious marine environmental problem in North Korea.[116] For many years it invested in heavy industries which discharged most of their untreated and inadequately treated effluent directly into rivers and coastal waters, including the Yalu River. This chemical effluent contained mercury, cyanide, arsenic, pesticides, and other organo-chlorine compounds. The major culprits include steel mills, electronic power generation, fertilizer plants, petrochemical plants, synthetic fibre factories, and cement plants, most of which have been operational for many years, and some for as long as thirty to fifty years. Old industries inherited from the Japanese colonial era are particularly problematic. The coke plants at the eighty-two-year-old Hwanghae Iron Works at Songrim, for example, produce highly toxic wastes containing phenols, cyanides, and naphthalene, which are discharged into the Taedong River. These wastes may already exceed current water quality standards.

The Sinuiju Chemical and Fibres Complex loses little in comparison. It releases 100,000 tonnes of waste daily into the Amnok/Yalu River in the course of producing viscose rayon, paper, and cardboard from reed by treatment with caustic soda. This effluent probably contains lignite, sodium, and zinc, all of which are of concern to the Chinese as well as the North Korean authorities. The four-decade-old system of primary sedimentation tanks is not working, and expensive processing chemicals are not recovered before the effluents are released, resulting in inefficiency as well as a degraded river system. This loss of valuable raw materials in waste streams is a story repeated in many North Korean industrial complexes.[117] On top of all this, North Korea has engaged in massive land reclamation and river modification.

These problems are caused by the lack of treatment facilities as much as by obsolete or overused equipment. For example, the vinalon and fertilizer complexes at Hamhung on the east coast have basic waste water treatment facilities, but cannot recover trace metals and other dangerous chemicals that are contained in the waste water. These wastes are released into a drain and marine outfall, and thence into the coastal marine environment. The waste stream also includes organic compounds, sulphides, various dissolved solids, urea, ammonia, cyanides, and arsenic. Industrial complexes like Hamhung also lack second lines of defence against equipment failure, such as guard ponds. North Korea's largest mine, the Maoshan iron mine, is adjacent to the Tumen River main channel. It has no tailings pond and discharges voluminous amounts of waste material directly into the river. The Awudi chemical plant also contributes severe water pollution to the lower Tumen River, reportedly giving fish a 'kerosene' smell.

Although there is no data quantifying the amount of industrial effluent entering the marine environment, and no reliable information on the concentrations of toxic substances, available reports suggest that the long-term effects could seriously impair the quality of the coastal environment and possibly cause human health problems, particularly in areas close to the discharge points. Given the large number of factories along the coast, the cumulative effects of toxic substances could be significant.

Major efforts will be required to help North Korea improve the treatment and production facilities in its chemical complexes. The lack of national financial and technical capability will make the process even more difficult. And although North Korea has signed the London Dumping Convention, it has not yet provided port discharge facilities to receive oily wastes, sewage, or garbage from visiting vessels, nor does it monitor and enforce its rules in this regard against foreign vessels. Similarly, North Korea has basic oil spill control boats and equipment but these are old and inadequate, and contingency planning and practice is not implemented. During the 1980s there were incidents of oil contamination from tanker spills.

North Korea has begun to show more interest in marine pollution, particularly after the 1992 Earth Summit, officially known as the United Nations Conference on Environment and Development (UNCED), brought environmental awareness to the highest level of government.[118] After UNCED, Pyongyang quickly signed the major agreements on climate change, forestry, biodiversity, and the action plan, as well as a variety of other global and regional environmental treaties (Table 4.1). It also reorganized its administrative structure to strengthen the planning and administrative organs, so that they could be more responsive and effective in addressing the environmental issues of the country. The establishment of the State Environment Commission was the result of those efforts. It consists of ten departments:

1. Environmental Monitoring and Development,

Table 4.1 North Korean Environmental Treaty Commitments or International Participation, mid-1992

Treaty/Commitment/Participation

Global

Climate Change Convention
Biodiversity Convention
Forestry Principles
Agenda 21
Rio Declaration
Vienna Convention and Montreal Protocol
World Heritage/MAB
Ramsar Convention on International Wetlands
Convention on International Trade of Endangered Species
Bonn Convention on Migratory Species Conservation
International Tropical Timber Agreement
International Undertaking on Plant Genetic Resources
Convention on Prohibition of Military, etc.
Environmental Modification
Annex 16, Environmental Protection, International Civil Aviation
Treaty of Exploration/Use of Outer Space, etc.
International Whaling Commission
Indo-Pacific Fisheries Commission
Prohibition of Biological Weapons
Chemical Weapons Convention
Code of Conduct on Pesticide Use; UNEP Chemicals Information Exchange; Basle Convention on Hazardous Waste Transboundary Movements
Protection of Victims in International Armed Conflicts
International Convention for Prevention of Pollution from Ships
International Convention on Oil Pollution from Maritime Transport of Oil
Conventions on Early Notification, Assistance, and Liability from Nuclear Accidents
Treaty Banning Atmospheric Nuclear Testing
Nuclear Non-Proliferation Treaty

Regional

North-East Asia Environmental Consultations
North-West Pacific Action Plan (UNEP)

Table 4.1 *Continued*

IOC/UNESCO WESTPAC
Antarctic Treaty
Agreement on Network of Aquaculture Centers in Asia-Pacific
Tumen River Development Project Environmental Guidelines
UNDP/GEF/ADB Greenhouse Gas Inventories Project
UNDP/GEF East Asian Marine Pollution Project
UNDP/GEF Investment Strategy
UNDP Subregional Energy-Environment, Clean Coal Technology, New and Renewable Sources of Energy, and Agriculture Projects

Sources: Peter Sand (ed.), *The Effectiveness of International Environmental Agreements*, Cambridge: Grotius Publications, 1992; P. Hayes and L. Zarsky, *Cooperation and Environmental Issues in Northeast Asia*, Nautilus Institute, Report to IGCC, San Diego, September 1993.

2. Environment Supervision,
3. Ecological Conservation,
4. Meteorological,
5. Hydrological,
6. Oceanographic,
7. Science and Technology,
8. Planning,
9. External Relations, and
10. Communication.

The Commission reports to the Committee of Environment, which is made up of various cabinet ministers and chaired by the deputy prime minister.

The recently enacted 'Law of Environmental Protection of the Democratic People's Republic of Korea' includes provisions prohibiting pollution from ships and coastal and marine development projects. It also requires vessels to be equipped with pollution protection devices, and the inspection thereof, as well as onshore waste disposal.[119] Specific legislation regulating the discharge of oils, solid and liquid wastes from vessels, port management, industrial waste treatment, and agricultural waste has been drafted. Although North Korea has promulgated water classification standards, emission standards and maximum permissible levels, and procedures for applying to set up an industrial enterprise, and requires permits for discharges, land development, and reclamations, little is known about how these work in practice. The Government has also recently developed an Environmental Action Plan to implement UNCED Agenda 21 which includes coastal and marine pollution prevention, control, and management.[120] And it is taking steps to implement the framework convention on climate change and the UN Convention on Biodiversity, reserving 20 per cent of its territory for the protection of biodiversity.

North Korea has also begun to show more interest in transnational environmental matters, particularly after it was revealed that Japan regularly disposes of nuclear waste in the Sea of Japan, and, Pyongyang claims, the former Soviet Union has dumped nuclear submarine reactors there as well.[121] The news that the former Soviet navy has dumped eighteen decommissioned nuclear reactors and 13,150 containers of radioactive waste since 1978, most of it in the Sea of Japan, created an uproar in the world environmental community and drew a rare comment from North Korea.[122] It severely criticized Russia for posing a threat with its nuclear arms and radioactive waste dumping, while 'having the cheek' to press North Korea to accept nuclear inspections.[123] And while it may have little direct connection to environmental protection, North Korea has used both geographic and ecological arguments in its diplomatic efforts to halt Japan's transport of irradiated and other nuclear materials to and from Europe. At one point, North Korea even offered to host an international seminar on regimes for pollution control.[124]

Ongoing regional co-operative environmental initiatives which involve North Korea include:

- *the United Nations Environment Programme's North-West Pacific Region Action Plan (UNEP/NOWPAP)*[125] for the wise use, development, and management of the coastal and marine environment. This project has been hampered by considerable wrangling over the plan's priorities and the allocation of costs and responsibilities, and thus needs to be revitalized.
- *the United Nations Development Programme/Global Environmental Facility (UNDP/GEF) Program on Prevention and Management of Marine Pollution in East Asian Seas*,[126] which includes China and North Korea in its efforts to support the participating governments in the prevention, control, and management of marine pollution at both the national and

regional levels. North Korea is particularly interested in participating in the proposed network of information management and marine pollution monitoring centers, and wants assistance to upgrade the equipment and facilities of its West Sea Oceanographic Research Institute.
- *the Intergovernmental Oceanographic Commission's Subcommission for the Western Pacific (IOC/WESTPAC),*[127] which defines regional problems and implements programs for regional marine scientific research, and facilitates regional exchange of scientific data, training and education; and
- *the North-East Asian Environment Programme,*[128] which promotes frank intergovernmental policy dialogue on environmental problems of common concern to the region as a whole, and information sharing, joint surveys, and collaborative research and planning.

North Korea has also been invited to join the South Korea/China project on the prevention of pollution of the Yellow Sea.[129] It certainly has shipping and fisheries interests in the region. North Korea is also a member of the International Maritime Organization and the United Nations Food and Agriculture Organization.

A serendipitous opportunity for co-operation involving North Korea might be an offer of assistance in monitoring and/or retrieving the dumped Russian nuclear submarine reactors in waters under Pyongyang's jurisdiction. This could even be a joint US–Japan initiative under the environmental wing of their global partnership.[130]

In summary, the advantages to North Korea of participating in a regional marine environmental regime include:
1. a low-risk opportunity to 'feel out' potential regional partners in more difficult areas and, in so doing, give an appearance of responsibility and reasonableness;
2. the opportunity to use the nuclear waste pollution issue as a political weapon to ward off criticism by Japan and Russia of its own nuclear practices;
3. increased capacity in terms of financial support, technology, training, and knowledge, particularly of the sources and distribution of pollution; and
4. ultimately, a cleaner marine environment.

The disadvantages include:
1. further opening of its society, and exposure of its citizenry to foreigners and foreign standards, practices, and life-styles;

2. possibly embarrassing revelation of its own serious pollution problems; and
3. sharing in the costs of control and clean-up.

South Korea

The Policy Context

There is a current convergence of natural and political circumstances in South Korea which is forcing marine affairs onto the domestic and foreign policy agendas. Geographically, South Korea is virtually an island state since its only land border is with its arch enemy, North Korea.[131] South Korea is also encircled by three major powers, Japan, China, and Russia, which have always regarded the Korean peninsula as having vital strategic value. So Korea has long been the focus of strategic politics by its neighbours, and has sought to break out of the big-power encirclement by establishing friendly relations with distant countries, and by diplomacy designed to restrain the hostile behaviour of its neighbours.

The principal means for South Korea's 'reaching out' has lain with the sea. Flushed with the success of its Northern Policy (or Nordpolitik—Roh Tae Woo's initiative to establish relations with China and the Soviet Union), South Korea sees ocean affairs as offering a new opportunity for it to take the initiative in regional affairs. Because Japan and China are suspicious of one another, it is more acceptable for South Korea to take the initiative, for example in beginning APEC[132] and CSCAP. Indeed South Korea's political independence and territorial integrity have been closely bound up with its capacity to keep its sea lanes of communication open.[133] And the North Korean threat meant that security added a critical dimension to South Korea's marine policy.

Second, the lack of arable land and natural resources has spurred South Korea's emphasis on marine policy. In 1991 the population of the southern 45 per cent of the Korean Peninsula, an area of 99,299 square kilometres, was 43.3 million. The maritime area under Korean national jurisdiction on the basis of delimitation by the median-line-equidistance formula is, however, about 447,000 square kilometres. Since only about 26 per cent of the total land area consists of lowlands and plains, this maritime space is about seventeen times larger than the arable land area. With its high population density (436 people per square kilometre), lack of natural land resources, and pressure for economic development, effective management of living resources and efficient utilization of non-living resources from South Korea's own maritime jurisdictional areas are of critical importance.

Third, Korea's marine policy occupies a prominent place on its agenda for socioeconomic development. Marine policy in South Korea has traditionally concentrated on two major issues: coastal fisheries and defence. However, since the mid-1960s, rapid industrialization, increased need for space, growing population density, increased foreign trade, technological advances, increased recognition of environmental values, and the changing international maritime order are challenging this traditional hierarchy of national marine interests. One result has been the development of distant-water fisheries, shipping, shipbuilding, ports, and the offshore and coastal construction industry, as well as the regionalization of marine policy.

In the early 1960s, South Korea's economic development policy changed fundamentally from inward-looking import substitution to outward-looking exports. Because of its lack of natural resources and limited domestic market size, export expansion was the strategy it chose to achieve speedy economic growth. In the three decades after 1962, South Korea's GNP increased from US$2.3 billion to $281.7 billion, with per capita GNP soaring from US$87 to US$6,518. In 1990 the Korean ocean sector contributed about 7 per cent of the total GNP, or US$14.7 billion. Environmental issues, both global and local, also began to loom large on the policy agenda, and the concept of sustainability was introduced into national economic policy. The importance of multiple-use plans in ocean management is increasingly recognized, and the philosophy of ocean management is changing from a zero-sum to a positive-sum game.

Within this broad context, three specific factors will shape South Korea's future marine policies: governmental reforms which promote deregulation, liberalization and globalization; foreign policy; and a forward-looking perspective regarding environmental concerns in coastal areas. The new economic, environmental, and technological trends such as the Uruguay Round, Green Round and Technology Round will have a critical effect on the sustainable growth of the Korean ocean sector. Marine policies will also take account of future threats or opportunities, and will seek to control or reduce their undesirable impacts. In 1973 and again in 1978, for example, Arab oil embargos had a catalytic impact on efforts to achieve energy self-sufficiency. This led to: accelerated oil and gas exploration in South Korea's own waters as well as on foreign continental shelves; intensified efforts to acquire a pioneer investor status for deep seabed minerals in the Clarion-Clipperton Zone of the Pacific Ocean; an expansion of South Korea's research scope to the Antarctic with the establishment of a permanent research station in 1988 on King George Island; and revitalization of the technology necessary to harness tidal power energy on the western coast of the Korean Peninsula. In each case, the goal was to ensure access to resources and their availability to future generations.

South Korea spends less than most governments on environmental protection—only 0.51 per cent of its GDP in 1995—although the amount has been increasing in recent years. Nevertheless, marine environmental policy has emerged as an important social issue in South Korea because of massive coastal reclamation and landfilling, increased coastal uses, and sporadic oil spills. The total reclaimed area is now about 1,100 square kilometres, representing 15 per cent of the total potential area in water depths less than 20 metres—about 7,305 square kilometres.[134] The reclamation of the total potential area would add new arable land corresponding to about 24 per cent of the total existing arable area.

Recognizing the new international legal order, the South Korean government from the 1990s focused on sectoral legal regimes instead of integrated legal regimes concerning the marine environment. Accordingly, the Environment Preservation Act of 1977 was divided into seven laws including the Basic Environment Policy Act (Law No. 4257, approved on 1 August 1990 and amended on 31 December 1991), the Water Environmental Preservation Act (Law No. 4260, approved on 1 August 1990 and amended on 8 December 1992), and the Natural Environment Preservation Act (Law No. 4492, approved on 31 December 1991). In addition, the MPPA (Marine Pollution Prevention Act) was amended in 1986, after South Korea's accession to MARPOL 73/78, and again in 1991 to meet the standards of Annex V of MARPOL 73/78. The MPPA focuses on pollution prevention and control from vessels, offshore structures, and ocean dumping. It also deals with pollution from seabed drilling, but does not cover pollution from radioactive materials and from naval vessels.

The formulation and implementation of a national marine policy is under way. The Prime Minister's Office has set up a Working Group for an Integrated Marine Policy (WGIMP), representing every sector of the Korean government concerned with maritime affairs. The Working Group has identified eight agenda items:
1. reinforcing ocean technology development, oceanographic research, and national ocean services;
2. enhancing the international competitiveness of the shipping and shipbuilding industries through a self-controlled open-door policy;
3. maintaining fish stocks at a maximum sustainable yield in adjacent seas, and securing distant-water fishing grounds by intensifying bilateral and multilateral co-operation;
4. establishing a new direction in the exploitation of oil and gas from the continental shelf and obtaining pioneer investor status in the Pacific Ocean under the UN Law of the Sea Treaty;
5. managing the coastal area for multiple use;
6. protecting the marine environment through contingency plans and the mitigation of ocean pollution;
7. strengthening marine diplomacy, in the wake of the new international regime of the Law of the Sea; and
8. reorganizing the governmental operating arm for marine policy.[135]

To implement this agenda, the Korean government has created eleven ministries with subordinate agencies (Table 4.2). These governmental organizations set their own policies and priorities and neither conform to, nor are responsible for creating or enforcing, a national master plan for marine policy. Korean governance for marine policy is thus somewhat decentralized, although the Economic Planning Board (EPB) has overall responsibility for producing economic development plans and presenting the yearly resource budgets. The influential ministries, those in charge of Agriculture and Fisheries, Transportation, Foreign Affairs, Home Affairs, Commerce and Resources, Construction, and Science and Technology, are allowed a great deal of discretion in the application of rules and regulations.

The current President, Kim Yung Sam of South Korea, is the son of a fishing magnate and is familiar with the industry and its problems.[136] During his election campaign, President Kim pledged to set

Table 4.2 Governance of Korean Marine Policy, 1993

Function	Ministry
Shipping	MPA
Commercial Port	MPA
Fishery	NFA
Maritime Surveillance	MPD
Naval Defense	MOND
Diplomacy	MOFA
Pollution	MPA, MPD, MOE**, NFA
R&D, Education	MOST, MOE*
Shipbuilding	MOTIE
Reclamation	MOC, MOAF
Hydrography	HAO
Minerals, Oil, and Gas	MOTIE

Notes: MOFA (Ministry of Foreign Affairs), MOND (Ministry of National Defence), MOE* (Ministry of Education), MOAF (Ministry of Agriculture and Fisheries), MOTIE (Ministry of Trade, Industry and Energy), MOC (Ministry of Construction), MOE** (Ministry of Environment), MOST (Ministry of Science and Technology), MPA (Maritime and Port Administration), NFA (National Fisheries Administration), MPD (Maritime Police Department), HAO (Hydrographic Affairs Office), MOTIE (Ministry of Trade, Industry and Energy).

up a 'Ministry of Ocean Industry', integrating the sectoral marine administrations into one institution. The basic framework of a Ministry of Ocean Industry would incorporate the Maritime and Port Administration, the Fisheries Administration, and some other marine-related agencies. The WGIMP recommended that a new Ministry of Ocean Industry should be given the following basic functions:
- ocean industry promotion to strengthen international competitiveness in the world ocean industry;
- contingency planning to cope with marine pollution and accidental spills, and safety and salvage for disabled ships;
- coastal and distant-water fisheries and aquaculture;
- management of ports, seamen, and ship operations;

- shipping;
- marine monitoring and scientific research;
- marine technology development;
- coast guard;
- marine policy planing;
- marine policy co-ordination;
- coastal zone management.

The WGIMP further recommended that the marine-related ministries should take steps to strengthen the following functions:
- maritime diplomacy in the Ministry of Foreign Affairs (MOFA);
- marine energy (tidal power) and minerals (oil and gas exploitation, deep seabed mining) development in the Ministry of Trade, Industry and Energy (MOTIE);
- permission and authorization of marine activities (fisheries, recreation) in the local offices of the Ministry of Home Affairs (MOHA);
- environmental assessment relating to coastal reclamation in the Ministry of Environment (MOE).

Proposals to reduce the overall number of ministries have delayed the implementation of these recommendations.

Fisheries

Fisheries are particularly important to South Korea. Fish are the major source of animal protein in the Korean diet while fisheries provide Korea's coastal communities with an important source of income and employment, and foreign exchange for the country as a whole.[137] South Korea's fishing fleet had fewer than 50 vessels in the early 1960s, but its size increased to over 800 vessels by the late 1970s, when the value of fish exports exceeded US$400 million, or over 3 per cent of Korea's total exports.[138] In 1981 South Korea became the world's third largest distant-water fishing nation.

In the 1990s, total fisheries landings in South Korea have been about 3 million tons annually with over 80 per cent of the total catch coming from its neighbours' waters, presumably in the Yellow and East China Seas.[139] The marine catch in the Yellow Sea and East China Sea is comprised of a diverse assortment of species including the largehead hairtail, croakers, the Japanese sardine, the Japanese chub mackerel, crustaceans, and shellfish.[140] Catches from the Sea of Japan, by contrast, are limited to a few species, which include the Alaska pollack, the Pacific saury, the Japanese anchovy, and the Japanese flying squid. South Korean fishing activities around northern Japan are regulated under a bilateral fisheries agreement with Japan.

In the early 1980s, South Korea's fishery industry suffered from the widespread extension of jurisdiction to 200 nautical miles, declining fisheries resources, and increased labour costs.[141] The fisheries industry is currently under more stress because of coastal land reclamation, coastal pollution, the trade liberalization movement, and the United Nations ban on driftnet fishing. Fishermen are fewer and older. Many stocks in the Yellow Sea are already overfished, and China is expanding its fishing fleet and fishing area in both the Yellow and East China Seas. Forced out of US and Soviet waters by the EEZ declarations, South Korean distant-water vessels have gradually moved their focus to the 'Doughnut Hole' in the Bering Sea, the 'Peanut Hole' in the Sea of Okhotsk, and Japanese waters. But now Korean vessels are being severely regulated in the 'Doughnut Hole' by a Russia/United States agreement and have been barred from the 'Peanut Hole' altogether. And Japan is striving to eliminate South Korean fishing in its waters off Hokkaido.[142]

Consequently, South Korean imports of fishery products have increased rapidly since the early 1980s. Total fisheries imports in 1990 were nine times greater than in 1980, and this trend is likely to continue.[143] But, these statistics are *somewhat* misleading because South Korea exports high-priced species, and its overall fisheries production still exceeds domestic consumption by some 600,000 metric tons.

Faced with a persistent decline of its distant-water fishing industry, South Korea quickly adjusted its fisheries policy and sought bilateral fishery arrangements with neighbours, searched for new fishing grounds and species, explored new methods of operation, and began to emphasize aquaculture. South Korea's present fisheries policy is to: enhance fish stocks through the construction and operation of fish hatcheries, artificial reefs, and mariculture; reduce fishing efforts on overexploited species in North-East Asian seas; halt the degradation of fisheries habitats and nurseries; strengthen fishery technology, marketing, and processing to increase valued-added products; and improve ongoing diplomatic efforts to increase access to foreign fishing grounds and to develop international management regimes.[144] The objectives of South

Korean fisheries management include: the rational management of scarce resources; the protection of the jobs and income of fishermen; and the restructuring of national fisheries to match capacity with available resources.

The 1982 UN Convention on the Law of the Sea set certain standards of performance in the national management of fisheries. It recommended a system for licensing domestic and foreign fisheries (if allowed), and for the setting of catch quotas, season and gear restrictions, age and size restrictions, and terms for joint ventures and enforcement measures. The task of implementing most of these standards is relatively simple for South Korea. With a territorial sea claim of only twelve nautical miles, and an agreement not to apply any extension of its jurisdiction against Japan, it can leave its existing regulations virtually unchanged. But some standards, like that for the setting of catch quotas, are hard to apply given the limited area involved.

Although implementing the Convention is relatively simple, South Korea remains concerned about the relationship between its coastal fisheries and Japanese catches offshore. Overfishing offshore could reduce the spawning population, thereby reducing populations of juvenile fish. The issue is complex because South Korea also fishes offshore, and is thus a potential perpetrator as well as a victim of overfishing. Furthermore, the fisheries relationship between South Korea and Japan in the Yellow Sea is distinctly influenced by their relationship in the Japan Sea and Korea Strait where South Korea fishes. Nevertheless, South Korea has established management regulations in its bilateral arrangements with Japan. And licensing of domestic fisheries is well established. The South Korean regulatory system bears some similarities to Japan's, but the two systems are rapidly diverging because of substantial modifications both countries have made with respect to distant-water operations, coastal fisheries modernization, and market structure.[145]

South Korea also has considerable fisheries problems with China. Korean fishermen are lobbying Seoul to declare a 200 nautical mile EEZ, or at least an exclusive fisheries zone, because of Chinese fishing in Korean waters. South Korean authorities recorded 840 Chinese violations in 1993, down from 994 in 1992 and 1,112 in 1991. Chinese illegal fishing around the Northern Limit Line in the Yellow Sea is particularly dangerous because South Korean naval vessels patrolling the area cannot always readily distinguish between an intruding North Korean vessel and a Chinese fishing boat. In August 1995, it formally asked China to strengthen measures to prevent Chinese fishing boats from intruding in the sensitive area.[146] South Korea wants to declare mutual exclusion zones but China has not yet declared a baseline, making the starting point difficult to determine.

South Korea is expanding its fishing grounds. In 1992 it revised regulations governing the operations of fishing boats in the East China and Yellow Seas to allow the expansion of the fishing grounds for South Korean vessels. The fishery restraint line was moved an average of 48 kilometres westward, almost coinciding with the China-Japan fishery agreement line and creating an additional 72,000 square kilometres of fishing grounds.[147]

South Korea supports a regional management regime as a medium to long-term goal. In the short run such a regime seems unlikely because China does not seem interested. The problem stems, in part, from a lack of leadership. Who should or could lead? And then for South Korea, there is the issue of unsettled boundaries which must be overcome. But one way to get around the boundary issues would be to do as the North Sea countries have done: apply 200 nautical mile EEZs only against third parties and not against each other.

Another problem for the implementation of a regional management regime is that the countries in the region do not apply a quota system, but limit the number and size of fishing boats instead. The traditional fishing rules originated in Japanese bilateral arrangements and it would be difficult to change them because Japan is advantaged by the existing system. A multilateral regime, on the other hand, would have to allocate total allowable catches (TACs).

The potential benefits for South Korea of a multilateral fishery regime in North-East Asia include the following:
1. South Korea can avoid politically and economically costly fisheries disputes with Taiwan, China, North Korea, Japan and Russia, and thus save time and negotiation costs.
2. South Korea would benefit from a regional regime which effectively eliminates fishing by extra-regional states.
3. South Korea would benefit from the better conservation and management of fishery resources resulting from information exchange, and co-operative scientific research.
4. Co-ordination with neighbouring countries on the management of transnational demersal and pelagic stocks in the

south and southeastern waters of Korea could substantially increase South Korea's fish production over the long-term.
5. Since South Korea fishes more extensively in North-East Asian seas than any nation except for Japan, the greater access to neighbour's EEZs that might be gained through such a regime would benefit it.
6. If the regime led to joint ventures, South Korea could combine its superior fishing technology with potential cheap labour from China and North Korea to fish stocks in, say, Russian waters. This would enhance fish production while keeping costs low, enabling South Korea to compete effectively with other transnational fishing companies for the region's huge market.
7. If South Korea led the formation of the regime, it would enhance its international stature and diplomatic reputation.
8. South Korea could use the regime as another point of diplomatic contact with North Korea, thereby helping to build the confidence necessary to deal with more difficult issues.

Marine Environmental Protection

Because of its location, South Korea is the only coastal state affected by pollution of the Yellow Sea, the East China Sea, and the Sea of Japan. Seoul began to express its concern in the mid-1970s over marine pollution and its effects on fishery resources in the Yellow Sea. With limited arable and inhabitable space and the push for economic growth, priority has been placed on development over conservation. Indeed, there is a concern for precise environmental assessments of development programmes and construction decisions to ensure that South Korea's industrialization and modernization are not critically impeded by the groundless outcries of environmentalists. Nevertheless, South Korea has become more serious about implementing its marine pollution control law, and determining the extent of marine pollution as well as the difficulties in enforcing pollution regulations. It now experiences an average of 200 marine pollution accidents a year with some 50 being major incidents. For example, in July 1995 the Sea Prince carrying 83,000 tons of crude ran aground off Yosu and spilled about 1,000 tons into the sea.[148] South Korea now recognizes that pollution could adversely affect its fishery resources and tourism/recreation,[149] with serious, perhaps irreversible, economic consequences.

Given South Korea's location and the effects of pollution, it also recognizes that it would be in its interest to participate in a regional regime for the protection of the marine environment. Indeed, South Korean policymakers appear to have realized that there is no alternative if the environment and fishery resources in North-East Asia are to be preserved. Environmental degradation, depletion of fishery resources, maritime anarchy, and political conflict are not in the interest of any regional nation.

South Korea's high-profile participation at the APEC forum summit in Seattle in November 1994, and remarks by South Korea's then foreign minister, Han Sung-jo that North-East Asia's security concerns can best be addressed by a multilateral forum, which would include China, the United States, and Japan, indicate Seoul's growing willingness to participate in and even lead North-East Asia's future multilateral arrangements.[150] South Korea continued to promote the establishment of a North-East Asia Security Dialogue at the ASEAN Regional Forum.[151] South Korean initiated marine environment protection and fisheries management regimes could be the first step.

South Korea has shown considerable interest in the construction of a regional regime for managing the environment. In June 1992, during the UN Conference on Environment and Development in Rio de Janiero, the Prime Minister, Won Shik Chung, declared Korea's support for establishing a North-East Asia Environment Co-operation Organization to protect the marine environment, and prevent air pollution and acid rain.[152] Since September 1992, South Korea has taken an active interest in establishing an informal co-operative environmental protection network in North-East Asia. And at the UNEP-sponsored regional environmental conference in February 1993, South Korea showed a keen desire to co-operate with other countries in the region to build pollution control and waste disposal facilities.[153]

In response to extensive Russian and Japanese radioactive waste dumping in the Sea of Japan, South Korea has proposed joint Russian/Japanese/South Korean surveys at specific dump sites, and surveys of water quality throughout the Sea of Japan.[154] To Seoul's delight, North Korea initially showed interest in participating in the joint surveys.[155] South Korea has even gone a step further, calling for joint funding of a survey vessel to enforce a ban on radioactive dumping in the sea. And it has also asked the International Atomic Energy Agency to take measures against such dumping.[156] Seoul's objective is to persuade Russia to stop the marine dumping of nuclear waste.

It seems clear that South Korea, more than any other nation in the region, would like to exert regional leadership in the protection of the marine environment of North-East Asia. In furtherance of this goal, China and South Korea signed an Agreement on Environmental Co-operation on 28 October 1993, which included control of air and water pollution, municipal and industrial wastewater treatment, control of agricultural runoff and pesticides, and control of coastal and marine pollution.[157]

South Korea puts more emphasis on the Yellow Sea because it is more polluted, it is semi-enclosed, and Korea's western coast is more developed than its eastern shore. The Yellow Sea is already one of the most polluted in the world. And South Korea is particularly interested in helping China curb its pollution because pollutant input is increasing rapidly in the air and in the rivers flowing into the Yellow Sea. China is planning to build nuclear power plants sited on the sea and these could present an environmental security risk for South Korea. Accordingly, South Korea has proposed a joint program with China on the Yellow Sea. The first step is to determine the degree of pollution.

South Korea could also use the regime to broaden its diplomatic contacts with North Korea and perhaps advance the prospect of unification, a long-cherished national aspiration of the Korean people.[158] Both North and South Korea participated in an international conference on environmental protection in Japan in October 1992. And the prospects of co-operation between North Korea and South Korea appear to have improved despite the recent nuclear tensions between them. Meanwhile, co-operation in a relatively innocuous sphere like protection of the marine environment under a broader multilateral regime could help South Korea forge and maintain closer links with the North.

The North-South thaw is due, in part, to South Korea's improved relations with China. In the past, China was a close ally of North Korea. With the end of the Cold War, the North's strategic importance to China has declined and Chinese policy toward South Korea has undergone a drastic positive change. Like Russia, China is now openly seeking technological and entrepreneurial assistance from South Korea. The first prime ministerial talks were held in

1990. At the end of 1991, a breakthrough was achieved when the two sides formally signed a reconciliation and non-aggression pact.[159]

South Korea is not concerned with possible duplication of effort. In its view, duplication is a fact of human existence and it is good to have several simultaneous initiatives which can compete. The separate efforts can eventually be combined or co-ordinated. So South Korea is launching its projects on a bi- or tri-lateral basis as well as participating in the NOWPAP process. Unlike other participants, it has already ratified most of the conventions called for by NOWPAP.

South Korea feels it should lead any initiative to promote regional co-operation in environmental protection because of its geographic location and its status as a medium power. This is why President Kim has proposed co-operation in environmental protection among China, Japan, and South Korea. Seoul can use these initiatives to build its status and diplomatic clout. The approach must be step-by-step: first the Yellow Sea and then the Sea of Japan. Land-based pollution sources are the most pressing problem and South Korea also welcomes NOWPAP as a means of addressing them. The present stage is one of defining the problem: how much pollution from where? From this starting point, countries will decide how to co-operate. Tripartite co-operation will commence with a working-level meeting in 1996.

Russia[160]

Russia has much to gain politically and economically from co-operation regarding the management of the Sea of Japan. The Sea of Japan rim is being viewed with increasing interest by Japan and South Korea, both of which need access to new markets, new sources of raw materials, and secure overseas investments. Indeed, with the emergence of close ties between Primorsky Krai (the southernmost maritime province in the Russian Far East) and North-East Asian nations has come talk of a Sakhalin-Kuriles-Hokkaido natural economic territory.[161] The Japan–Russia joint economic declaration of October 1993, for example, lists eleven areas for bilateral co-operation, including fisheries and environmental co-operation.[162]

The RFE is in many ways the most technologically advanced

region in Russia due to its military technology and industries. Much of this sophistication could be applied to fishery and environmental management. Investment in fisheries and environmental technology could also provide alternatives to the current dependence on the export, legal and illegal, of Russian arms by an RFE military-industrial complex which has lost its huge Cold War subsidies.[163] Eventual integration of the RFE into management regimes for North-East Asian seas is thus likely. Indeed, the fisheries and environmental policies now emanating from Moscow display a greater awareness of the need to preserve and properly manage maritime resources. But Russian participation in management regimes will be a long and difficult process, with much backsliding and resurgent isolationism.

Fisheries Management: A Raft of Problems

During the last years of its existence, the Soviet Union passed legislation consistent with provisions in the Convention on the Law of the Sea mandating the conservation of fishery resource. A decree passed in 1984, 'On the Economic Zone of the USSR' (hereafter the 1984 Decree), made the Ministry of Fisheries (MOF) responsible for fisheries conservation. The 1984 Decree was a significant step forward in that it mandated annual licensing, and quotas for territorial seas fisheries within the allowable catches set for each species and geographic area. The permits were to be issued by regional fisheries administrations, which supposedly would have more knowledge of the conditions of the local stocks. The 1984 Decree also specified fishing seasons and gear restrictions, and included special provisions for the conservation of anadromous stocks, marine mammals, endangered species and sensitive areas. The enforcement of the regulations *vis-à-vis* foreign violators was by means of fines, confiscation of catches, gear, and vessels, and notification of the flag state.

The effect of the 1984 Decree was diminished by a bureaucratic reorganization that resulted in an unclear allocation of authority. In 1988 the State Committee on the Environment (Goskompriroda) was established, becoming the Ministry of the Environment (Minpriroda) in 1991. Though much of the responsibility for supervision and enforcement of fishing regulations was transferred to this new ministry, the MOF stubbornly refused to yield its power. Moreover, responsibilities for resource exploitation and for environmental protection were never clearly separated in the former Soviet Union.[164]

Again in 1991, the staff of the MOF was halved and its mandate greatly reduced to induce state-owned fishery fleets and joint ventures (JVs) to take charge of their own day-to-day operations. But the state-owned operations are quickly losing ground as independent operations, based mainly in Vladivostok, gain expertise in the market system. Currently there are plans to establish a Regional Fisheries Committee in Khabarovsk to oversee these firms and reduce their influence over fishery policy and management.[165] And in an attempt to prevent the enterprises consolidating their control over fisheries, the former Soviet Cabinet of Ministers introduced a system of payments for fishing rights in 1991.

In August 1991 the smaller MOF was transferred to the Ministry of Agriculture, where it languished until early 1992, when fierce resistance allowed it to become independent once again. At this time the MOF was transformed into the fairly weak Russian Federation Committee on Fishery Management (CFM). During the governmental reorganization of January 1994, the CFM was left with its committee status.[166] The CFM is unlikely to be very effective since its mandate is too broad for its funding and it maintains a rigid, closed bureaucratic structure. In sum, there is still no comprehensive fishery law in Russia nor a formal channel giving fishermen input to decision making. Enterprises must instead thread their way through a network of outdated and overlapping legislation and presidential decrees.

In late 1991 President Boris Yeltsin ceded control of the continental shelf to regional authorities. But the central government retained jurisdiction over exports, and until mid-1993, marine resources remained government property. Yet few of the recommendations made by research institutes such as the Pacific Ocean Fishery and Oceanography Research Institute (TINRO) were accepted.[167] Since 1993, a vaguely worded privatization edict, the 'Law on Property', has placed responsibility for natural resource management in the hands of the soviets, also called the people's deputies.[168] However, there are soviets at the regional, *oblast*, district, and city levels, resulting in inter-soviet battles that have hastened the collapse of the fishery management bureaucracy and further clouded the issue of who has the authority to set and allocate harvest quotas. Quota allocation has even taken place at the municipal level, and in more than one instance a soviet has objected to fishing rights granted at another level, forcing fleets which had already put to sea to return to port.[169] Still, these are signs that the resource management framework may be starting to solidify. The powers of the municipal soviets have been decreasing, while nearly all of the regions have set up their own Fishery Departments with mandates to manage the fishery resources, including exports, and to regulate fishing in territorial waters.[170]

Fisheries

Management problems have been exacerbated by the current depression. Many economic targets for 1989 and succeeding years were not even remotely met because of a mounting inflation rate, withdrawal of federal subsidies, the failure of the consumer market, and an expanding budget deficit.[171] At present, the equipment necessary to monitor fishery violations in and near Russia's eastern EEZ is limited to two vessels, built in 1968 and 1971.[172] This has prompted Moscow to appoint a former KGB officer, Evgheni Slavsky, to lead the Russian team responsible for high seas fishery patrols. Inflation has also made fines meaningless. In January 1994, illegal fishing in the high-seas 'Peanut Hole' in the Sea of Okhotsk cost an offender 10,000 roubles, or about seven US dollars, while seal poaching costs only 1,000 roubles per hide, or about 70 cents.[173]

Even if monitoring equipment were available, the desperate state of the Russian fishing fleet, black marketing, and the diversion of large amounts of undeclared catch to Japan and elsewhere for hard currency limit the ability of regional authorities to control harvest levels.[174] Furthermore, the relevant government departments have largely avoided their responsibilities under the pretext of hardships imposed by the transition to a market economy.[175] So Russia's harvest of fisheries from its EEZ is declining, from 48 per cent in 1985 to 44 per cent in 1991 to an expected 41 per cent by the year 2000, and its overall fishing harvest has declined by 35 per cent since 1988.[176] And severe cash shortfall has forced the fishing industry to seek external funding, to the point of outright sales of fishing rights for hard cash, in order to keep its ailing fleet afloat and fueled. And foreign partners in fishery JVs, though a promising source of economic revival, have discovered that under-the-table money can allow them to heavily poach Russian fish.[177]

The desperate drive for these 'floating dollars' has undermined fishery conservation measures. It was recently revealed that Soviet and Russian fishing fleets have, over the last 30 years, routinely killed tens of thousands of whales, including several highly endangered species, in violation of International Whaling Commission quotas, and sold the meat to Japan.[178] And depletion of such resources as caviar, sturgeon, and salmon is likely to accelerate,[179] even though these resources have already declined to the point of preventing payback of government loans. About 10 billion roubles in loans to new fishery enterprises for supplies and equipment in 1993 has gone unpaid because of the apparent collapse of the Russian salmon stocks.[180] Even the more established Russian fishery enterprises have been forced to abandon the quest for modern technology in their desperation to obtain the hard cash necessary for continued operation.

As a result of these difficulties, the former Soviet Union and now Russia have been less able and willing than Japan and South Korea to practice fisheries and overall marine environmental management in North-East Asian seas. Even so, Russian fishermen insist that they have limited their take in order to preserve spawning populations, and accuse other countries, especially Japan and South Korea, of unlimited fishing, including within the Russian EEZ. This, in turn, has led to strained fisheries relations as Russian and Chinese, and Russian and Japanese enforcement officials harass each other's fishing fleets.[181] And now Russia has banned foreign fishing in the Sea of Okhotsk's 'Peanut Hole'.[182]

Chaotic, unco-ordinated exporting of fish products after 1990 soon led to a sharp decline in fish prices. In response, the RFE created a fisheries 'association' which enforced strict sanctions for any violation of domestic fishery regulations.[183] But compliance was weak. The self-styled 'governor' of Sakhalin *oblast* under Gorbachev, Valentin Fyodorov, who had garnered an unprecedented amount of autonomy, allowed his local industry to keep a larger percentage of fish catch than stipulated by Moscow and to sell the fish without Moscow's consent.[184] While such actions can have a destabilizing effect on the central government, they nevertheless show that the RFE is very keen on increasing its international trade connections.

A further problem in the RFE fishing industry is a shortage of labour—15–20 per cent in some coastal fish processing enterprises, and 50 per cent in some supporting enterprises. The ports of Nakhodka and Vladivostok have particularly acute labour needs.

Over 10,000 people are recruited annually for seasonal labour, many of them from North-East Asian countries. Although this increases Russian interaction with North-East Asia, the influx of hardworking and market-savvy immigrants also fuels deep-rooted xenophobia.

Some progress is being made in addressing these problems. In 1993 limits were placed on exports of Russian fishery products which had reached 50 per cent of total production. At the same time, the central government wrested some control from the local, especially the municipal, governments, and imposed some control over the foreign capital earned through JVs. Moscow also renationalized nine of its eleven main seaports in 1993, including Vladivostok and Nakhodka, which had essentially been transformed by then from free economic zones to free-fire zones as Mafia-style gangs pirated many cargoes and terrorized entrepreneurs.[185] A gradual stabilization of Russian decentralization regarding fisheries is possible, as is the eventual reallocation of funding to the modernization and revitalization of the Far East fishery industry.[186]

Despite the debilitating domestic dilemmas, there has been some progress in international co-operation. The main marine management link between Russia and North-East Asian countries is their mutual interest in joint exploitation of fish stocks with habitats which straddle maritime borders, and it is here that increasing co-operation may have the greatest political and economic effects. The main commercial species in the southern Sea of Okhotsk and in the Sea of Japan include walleye pollack, cod, and saury in the southern Kurile islands, herring and sand lance in Aniva Gulf, ivasi in Terpeniya Gulf and the Sea of Japan, and the Greenland halibut on the continental slope.

Recognizing that many of these stocks are shared with its southern neighbours, the former USSR recommended co-operation with Japan and North and South Korea on the further study of the biology, distribution, and general state of several fish species. By the year 2000, it was projected that the expected reduction of walleye pollack, cod, and ivasi stocks would seriously affect the entire ecosystem. TINRO has now worked with research institutes in Japan and North Korea for ten years studying Japanese squid and giant octopus migrations, exchanging information, and participating in joint meetings in Tokyo and Nakhodka. These studies will form the basis of recommendations for catch limits on squid and octopus which, it is hoped, will be adopted and enforced by all three governments. A further goal is the inclusion of South Korea in these decisions.[187]

Russia is considering mechanizing its fish production process and further increasing its catch of secondary species. For this it will require both aid and markets. There is already a booming regional market in such secondary products as fish stuffing (surimi), brown algae, and chitin.[188]

In spite of high taxes and their tendency towards corruption, fishery joint ventures hold great promise for economic growth in the RFE over the long term. In 1991–2, fishery products through JVs accounted for over half of the RFE's total exports and about 80 per cent of its export volume. In contrast, exports of other commodities have been decreasing.[189] The Soviet Joint Fishery Law, enacted in 1987, allows enterprises to initiate JVs without approval from Moscow and gives foreign firms the option of owning more than 50 per cent of the JV.[190] Since the enactment of the Joint Fishery Law, over 100 fishery-related JV corporations have been established throughout Russia.

Between Russia and Japan alone there were twenty JVs in the RFE by July 1992, and between thirty and forty by early 1993, covering everything from production and processing to the distribution and marketing of fishery products.[191] A typical JV was the financing by Nichimo of Japan of two new fishing vessels for the Vladivostok Base Trawler and Refrigerator Fleet in 1993. These vessels handle onboard processing of fishmeal for Japanese markets.[192] In return, Japan receives increased fishing quotas. Japan has successfully gained successively larger salmon quotas within Soviet/Russian waters (2,000 tons in 1988, 5,000 tons in 1989, 6,000 tons in 1990, and 8,000 tons in 1991) in return for increased financial and technological participation in bilateral salmon joint ventures.[193] The 1991 Japan-Soviet summit produced a communique stating the two countries' satisfaction with their JVs and emphasizing economic and technological co-operation in fisheries, including co-operation towards rational management of salmon fisheries in the RFE.[194] Other elements of the present Japanese–Russian bilateral fisheries regime include private-level arrangements for Japanese crab fishing off Sakhalin in the Seas of Japan and Okhotsk in exchange for co-operation fees; Japanese sea kelp and sea urchin catches around the Russian-controlled Kaigara Island, north-east of Hokkaido; Japanese purchase at sea of Alaska pollack and herring; and Japanese *madara* fishing in the Russian EEZ.[195]

In spite of this promising start, Russia's economic collapse has seriously affected Russian JVs. Several of the more recently formed JVs have turned out to be little more than fly-by-night operations, due to a clause in the 1987 law stipulating that dissolution of a JV will cause its worth to be split between parties according to the proportion of investment. Even if the foreign enterprise invested all the hard currency and the Russian enterprise simply provided intangibles such as permits, under this clause the Russian partner may be entitled to half the capital in the event of failure.[196] The Japanese government estimated in mid-1993 that less than a fifth of its 916 JVs in Russia were actually operating, and that a majority of those were either operating at a loss or barely breaking even. This was attributed in part to political instability in Russia and a lack of proper legal arrangements there, which caused at least one Japanese company to withdraw from a fisheries JV.[197]

In the case of profitable JVs, local residents and Japanese investors complain that the profit generally either goes to individuals or to the central government rather than into local infrastructure, prompting some Russian JV partners to use part of their revenues to build housing.[198] But the limited scale of such infrastructure investments has caused foreign firms to become much more cautious when entering into JV negotiations with Russia.[199] Moscow is now seeking to form a tripartite trading system co-ordinating Russia's resources, South Korea's capital, and North Korea's labour in seventeen projects, including one for fisheries. Russia is particularly interested in 'taking the helm' of such co-operative ventures.[200]

Russia has much to gain and little to lose from regional co-operation in fisheries management. Russia also has considerable positive experience in fisheries co-operation with its neighbours, through which it has demonstrated the inclination and ability to help form regimes when that suits its interests. Russia co-operates with other nations through twelve international fisheries organizations, in which Russia (and the Soviet Union before it) has always supported all recommendations for the conservation of living marine resources 'based on the best scientific data available' (which in Russian eyes excludes measures adopted by such organizations as the International Whaling Committee).[201] Russia also has numerous bilateral fishing agreements with Japan, South Korea, North Korea, Taiwan, and the United States, and many more are proposed. All of these agreements were constructed within the context of the normalization of bilateral diplomatic relations, and in each case the fishery talks both affected, and were affected by, the progress in negotiations of overall diplomatic relations.[202]

In addition, Russia has recent and direct experience in multilateral agreements for the management of fisheries in specific areas, and for particular stocks. In 1993 Russia and the United States reached an agreement over high seas fishing in the Bering Sea, which led to the withdrawal of the fishing fleets of Japan, China, South Korea, and Poland. This restrictive conservation and management regime approaches a condominium based on the notion of semi-enclosed seas.[203] A multinational convention is being drafted for the management of pollack in this enclave, to be implemented in 1996. The convention is likely to adopt a 'limited entry individual transferable quota (ITQ) system'.[204]

In 1992 Russia concluded a treaty with Japan, Canada, and the United States on the conservation of anadromous stocks in the North Pacific. The convention prohibits all fishing beyond the countries' EEZs, and calls upon the parties to co-operate in collecting and exchanging high seas fishery information. The treaty also established the Anadromous Species Commission, headquartered in Vancouver, Canada, for the purpose of promoting the conservation of North Pacific anadromous stocks.[205]

In 1990 and 1993, private meetings were held in Victoria, British Columbia, Canada, between officials from nine fishing states with interests in the North Pacific. These informal meetings benefit RFE specialists, who must contend with significant constraints to fishery management and normal diplomacy.[206] And in May and June 1993, Russia hosted the first international conference on the living resources of the Sea of Okhotsk, which was attended by representatives from Japan, China, South Korea, and Poland. At this conference it was agreed to establish a scientific committee to prepare a report on the condition of Alaska pollack in the 'Peanut Hole'.[207]

The political controversy surrounding the Southern Kuriles/ Northern Territories dispute has forced Russia, Japan, and South Korea to reach an agreement on fishing around these islands. In early 1993 a group of Russian private companies opened up fishing areas there to South Korean vessels, but Tokyo protested that

South Korea should have negotiated with Japan and not Russia. The resulting agreement stipulates that Japan transfer its fishing rights in an area south-east of the Northern Kuriles (which it obtained from Russia) to South Korea in return for the latter's promise to give up fishing around the Southern Kuriles.[208]

The marine resources of the waters around the Kurile islands are among the most productive and intensively utilized in the world. Competition for these resources exacerbates the tension resulting from the conflicting sovereignty claims of Russia and Japan. Joint development of fisheries in the EEZ attached to the islands would be an interim provisional measure that could reduce tension, enhance confidence, and improve management of the resource.[209]

Environmental Protection

Environmental protection within Russia and co-operation in this area between it and its neighbours is hampered by bureaucratic confusion, ineffectiveness, and economic malaise. Environmental protection duties have been split between the inter-departmental Co-ordinating Council on Ecological Policy and the Ministry of Ecology and Natural Resources (commonly known as the Environmental Ministry, or Minpriroda). The Co-ordinating Council consists of the Ministers of Ecology and Natural Resources; Public Health; Agriculture; Fuel and Electric Power; Science, Higher Education and Technical Policy; and the heads of state committees in the Russian government.[210] Minpriroda has been both stronger and more responsive than its predecessors to the increasing political influence wielded by the growing environmental movement in Russia. On the other hand, the Russian public has hesitated to accept the new agency's powers, since central control in Moscow was to blame for many of the environmental problems in the first place.[211] Moreover, Minpriroda is, in reality, an amalgamation of several previously autonomous agencies, and to some extent struggles which were previously interministerial have simply been replaced by wrangling within the ministry itself.[212] Their mandates largely spelled out in March 1992 by the Russian law entitled 'On the Protection of the Natural Environment', the Co-ordinating Council and Minpriroda nonetheless represent an incremental improvement over the previous Goskompriroda.

In spite of these measures, a stronger umbrella agency and budgetary commitment will be required to co-ordinate research and monitoring of fisheries, marine pollution, and environmental degradation to support national regulatory programs. Although 1.3 per cent of Russia's GNP was allocated for environmental protection in 1992, the 1993 fiscal budget had no allocation for it at all.[213] Meanwhile, Nikolai Vorontsov, head of the former State Committee for the Protection of Nature, estimated that reversing Russia's environmental decline would require around 7 per cent of its GNP.[214]

Aside from setting broad standards and guidelines, and some specific attention to the Arctic, there is little indication that national laws and regulations have had much effect. So much so that it is estimated that Russia is losing 15 to 17 per cent of its GNP to pollution and natural resource damage annually.[215] A 1993 study by the World Bank concluded that the overall health of the Russian population is the worst of all the industrialized countries, due partly to pollution.[216] All of Russia's large cities have unhealthy air and half its people drink unsafe water.[217] Discharges into water bodies have increased since the dissolution of the Soviet Union, while the level of technology available to prevent pollution has declined.[218] Only 25 per cent of Russian sewage discharged into coastal waters is treated to accepted standards, and areas around the Amur River and Sakhalin have pollutants often rising to ten times the official limits.[219] Russia has thus sunk to the environmental protection nadir which the United States and Japan reached in the late 1960s and early 1970s, but unlike those countries it is simultaneously wallowing in an economic depression.

Regulatory agencies are understaffed and chronically short of funds and data. Although pollution fines are occasionally amended to reflect Russia's skyrocketing inflation rate, they have become largely symbolic for polluters. And Moscow, which cannot even collect taxes from the various regional governments, has become quite lax in enforcing them.[220] RFE industries, the very industries on which Russia will probably concentrate to pull itself out of its depression, continue to pollute heavily through inefficient practices ingrained through the decades. Even factories and power plants which were shut down because of public outrage over environmental damage have been quietly reopened.[221]

Such problems are unlikely to be resolved until the depression in Russia is well over. In the meantime, overgeneralized policy and the snail's pace of modifications are likely to cause its environment to deteriorate further.[222] The lack of funds and accountability, and corruption have combined to make it quite easy to buy environmental regulatory approvals and even endangered animals from local authorities. What appear to be scientific permits are often covers for illegal commercial operations that increasingly involve Russian organized crime. In 1993 alone, 4,500 tons of crabs, or fifteen times the allowable quota for the entire country, were taken for 'scientific purposes' by Okhotskuisa, a JV.[223]

Many enterprises in Russia have adopted titles that suggest an environmental orientation in the hope of attracting tax breaks and foreign investment. In reality many of these firms are incapable of performing environmental services.[224] Few enterprises and industries comply with environmental codes. Shutting them down is not only impractical when nearly all of them are violators, but it is also anathema to politicians newly aware of the fact that high unemployment could result in their removal from office.[225] As a result, investment priorities of the MOF and the Environment Ministry tend to take a back seat to the desires of the local government and *nomenklatura* to improve port facilities for trade, and for greater control over local operations. While legislative power is still perhaps overly concentrated in Moscow, enforcement power has devolved to the provinces and become inadequate.[226] This imbalance between central and local goals continues to constrain the development of a national marine environmental policy which would integrate economic and legal concepts, legislative processes, and scientific knowledge.[227] In addition, ministries such as the MOF have been charged with both the development and conservation of resources. 'This was akin to having the fox guard the henhouse, or, to "Russify" the metaphor, the goat guard the cabbage.'[228]

Even if a strong national environmental umbrella institution were in place, there is some doubt that its mandate would include the marine sphere. Though Russia is placing more emphasis on studies of coastal waters, a large portion of its marine research is still concentrated in distant areas such as the Southern Ocean.[229] The current shift of interest to local waters may simply be a result of the crumbling Russian economy, and the lack of fuel and supplies available to research vessels. On the other hand, some of this shift is undoubtedly the result of the thawing of the Cold War and with it, lack of a perceived need for security-related long-distance research.

Most Russians assume that since Russia is still in an early stage of economic development, environmental protection will manifest itself at a later stage when the free market takes hold. This assumption may be misplaced. Vladimir Zhirinovsky's extreme right-wing Liberal Democratic Party now chairs five environmentally related committees in the Russian Duma (lower house of parliament), including the Ecology Committee, the Committee on Natural Resources and their Utilization, and the Committee on Industry, Building, Transport and Energy. Yet neither the LDP nor Boris Yeltsin's Russia's Choice party even mentioned the environment in their 1993 campaign manifestos. Although the Duma is too weak to force dramatic changes anytime soon, LDP control of the environmental sector has put at least a temporary damper on Western enthusiasm for providing aid for the protection of Russia's environment.[230]

One extreme scenario has Russia's environmental problems becoming so overwhelming that the country concentrates on environmental restoration as one of its primary goals, a focus which will ultimately help to shape its domestic and international policy.[231] If this happens, Russia could possibly emerge as a strong player in regional environmental regime building. But by the time it reached this plateau, a regime could very well be in place.

Genuine Russian support for and participation in an environmental regime could build confidence in Russia's willingness to take responsibility for its actions. Soviet international environmental co-operation was traditionally symbolic and driven mainly by political benefits rather than real concern for environmental protection. Although both the USSR and Russia have been active in international regimes like the Montreal Protocol on Substances that Deplete the Ozone Layer,[232] they have been slow to ratify and implement such accords. And unlike the other republics of the CIS, all of whom have had environmental committees since the 1950s, the Russian Federation did not set one up until 1988.[233] Even now, Russia seems to be using its environmental crisis largely to attract increased investment in the country's infrastructure, and for international monitoring of treaties which it should be monitoring itself.[234] There is even suspicion that it hopes to attract 'dirty' industries from Taiwan and South Korea.

Russian participation in a regime might also help to overcome North-East Asian suspicion of Russia's dedication to environmental protection. Up to 1989, the USSR had dumped more than twice as much radioactive waste in the Arctic Ocean alone, as had been dumped in the world's oceans by the rest of the world combined.

The vast majority was dumped by the military, since commercial dumping of radioactive waste had completely stopped by 1984. The military considered itself immune to the stipulations of international treaties, and to many government regulations throughout the Soviet era. Moscow compounded the problem by keeping the military's transgressions secret, and the extent of the USSR's military pollution came to light only in 1993 through Greenpeace reports that were later confirmed by Russia's Environmental Policy Co-ordinating Council.[235] Access to data is still largely controlled by regional authorities, who are loathe to release data which will implicate them in the failure to enforce environmental laws and regulations.[236] They could well have reason for concern because the 1993 revelations of Russia's long-term dumping of radioactive waste in North-East Asian seas may intensify general anti-Russian feeling in the region. These political problems could be overcome by genuine participation in a regional marine environmental protection regime.

The current trend in Russia is towards conformity with international environmental standards and compliance with international laws through a combination of the traditional 'command and control' methods, such as fines and co-ordination of environmental issues, and economic mechanisms, such as environmental charges and tax breaks.[237] The 'Law on the Protection of the Natural Environment' also pledges the Russian Federation to co-operate with other nations to improve the global environment, stating that 'The ecological well-being of one state cannot be secured at the expense of other states without taking their interests into consideration.'[238] In the fall of 1993, Minpriroda undertook to draft new national legislation to strengthen international environmental treaties. Although the dissolution of the Supreme Soviet in October 1993 put this issue on a back burner, the ministry is still pursuing the matter.

Russia also has a specific legislative basis upon which it can build to prevent marine pollution and conserve nature. Soviet legislation passed in the 1980s asserted enforcement rights over pollution by foreign-flagged ships operating in the Soviet EEZ, in order to exercise better control of vessel-source pollution and dumping, although Russian vessels continued to break the laws.[239] By the 1970s, a majority of the fish, amphibians, and birds listed in the USSR Red Book were being conserved and bred in national parks, and gene pools were set up for endangered animals. This system deteriorated drastically throughout the late 1980s and early 1990s.[240] Still, such legislation could provide a basis for a future Russian role in any North-East Asian environmental regime, including maritime environmental regimes.

Russia's considerable oceanographic expertise could also underpin and contribute to a regional marine environmental protection regime. Russian scientists have been monitoring and studying the physical, chemical and biological aspects of oil pollution in the world ocean for over fifty years. Vladivostok's Pacific Institute of Oceanology is the leading marine science research institute on the Sea of Japan, and there are also other large government, university, and Academy of Sciences ocean-related research institutes located there. Although monitoring of pollution in the Sea of Japan has only been carried out since about 1980, the network of stations stretches from Tatar Strait in the north to Tsushima Strait in the south. But there is a need for more modern data. Most of the more than 30,000 Soviet/Russian temperature and salinity observations in the area were made prior to World War II, and little data exist for the northern Sea of Japan and the western and northern Sea of Okhotsk.[241]

Russia has a history of international co-operation in marine environmental protection. In July 1986 Mikhail Gorbachev gave a speech in Vladivostok in which he proposed a Pacific Conference, modelled on the Helsinki Conference, with all countries having a relationship with the ocean participating.[242] Protection of other seas also received Moscow's support. In 1990 the Soviet Union formed a commission with Canada to co-operate in the preservation of the Arctic, and the USSR also supported an initiative by Finland to hold an international conference on Arctic preservation. Moscow has signed international declarations on the protection of the Baltic and Black Seas, and in 1990 proposed a multilateral convention on preserving the marine and biological resources of the Pacific.[243]

In fact, Russia is party to some fifty-six multinational environmental arrangements, although it often fails to comply with them.[244] Some are focused on marine areas. For example, in June 1990, it concluded an accord with the United States for the creation of a Soviet-American international Beringia Park, a complex of specially protected land and sea territories in the Bering Strait.[245] In the Baltic, Moscow has negotiated several maritime boundary delimitation and co-operation agreements with neighbouring coastal states, including the 1974 'Convention on the Protection of the

Marine Environment of the Baltic Sea Area', which was signed by all seven countries bordering the Baltic. Turning to the Arctic, in 1984 the USSR signed a decree for 'Improving Nature Protection in Areas of the Far North and in Sea Regions Lying Adjacent to the Northern Coast of the USSR.'[246] And Russia has agreed to take part in both bilateral and multilateral investigations of the dumping sites in the Sea of Japan, together with Japan, South Korea, and perhaps North Korea.[247]

During the late 1980s, joint Soviet-North Korean investigations of environmental pollution were carried out in the Sea of Japan, and scientists from China, Japan, South Korea, and other countries were invited to participate. There has been discussion of the need to co-ordinate further studies on contact zones (river-sea, water-sediment, water-air), and on automation of environmental data collection, treatment, and analysis.[248] And the 1990 statement of normalization of relations between South Korea and Russia referred to the intention of the two countries to co-ordinate their efforts 'to protect the environment and to that end... co-operate in international and regional organizations.'[249]

Perhaps more important at this juncture, aid organizations have focused on assistance in the environmental sphere. Just as Japan invested in smokestack scrubbers for China in 1993, the threat which Russian pollution poses to both offshore and onshore areas of neighbouring countries should provide an impetus for neighbours to invest in its prevention. Production of cleaner technology might take the form of JVs. Co-operation in research and production would reduce loss, increase efficiency, increase the quality of produced goods, keep environmental degradation to a minimum, and provide a good basis for improved relations between the nations involved.

Turning to aid organizations, USAID has proposed assisting the CIS in pollution and environmental cleanup. Its far-reaching plan calls for technical assistance in identifying cleanup priorities, as well as policy-level 'institution building' to strengthen regulatory controls.[250] A World Bank technical assistance loan to Russia is being directed at environmental education and the strengthening of environmental institutions at federal, republic, and local levels. Although its impact will be constrained by the fact that 80 per cent of industrial facilities in Russia are unable to meet the new standards, it is nevertheless a strong start.[251] At the request of the Ministry of Environmental Protection and Russian environmental groups, the World Wide Fund for Nature announced a $17 million plan in 1994 for conserving Russia's biodiversity and natural resources, including provisions for the establishment of new wildlife preserves and financing for endangered species' recovery programs.[252] And Japan is assisting Russia to construct a floating-type nuclear waste processing plant in waters off Vladivostok.[253]

The rise of a Green movement in Russia could stimulate and reinforce Russian participation in a regional marine environmental protection regime. The far-reaching law 'On the Protection of the Natural Environment' of March 1992 gave unprecedented rights to citizens, allowing them to form environmental associations, obtain complete information about the environment, organize meetings and demonstrations regarding proposed projects, demand that state environmental impact assessments be undertaken, and file lawsuits demanding the termination of environmentally damaging activities. The law strengthened citizen movements by providing for civil, criminal, and administrative liability for the violation of environmental regulations, including full compensation for damages to health, property, and the environment in general.[254] Most elements of Russia's environmental movement are demanding greater environmental protection and more local autonomy in environmental policy formulation.[255] Citizen environmental protests have become increasingly common. Local officials are now often called to task for weak enforcement of environmental regulations. And Russian fishery and environmental data have been declassified and even published in English-language journals, contributing to greater public knowledge.

The Green movement actually started in the 1960s, when citizens began protesting about pollution in Lake Baikal. Since then 'Greens' have organized a 'Save the Amur' movement, blocked geothermal drilling on Kamchatka, publicized the location of atomic waste dumps, and prevented nuclear-powered ships from calling at ports in the region.[256] By the early years of *perestroika*, government officials viewed environmental protests not as a political threat, but as a catalyst for their reforms. Many of the emerging national leaders fought their first political battles as environmental advocates, and in doing so, gained both valuable organizational skills and political exposure.[257] Environmental groups even held 15 per cent of the seats in the last Soviet Congress of People's Deputies.[258] These non-governmental organizations (NGOs) have clearly emerged as significant players in the political arena. Pressure from them and the general public has already prompted Moscow to establish tougher environmental standards on foreign enterprises in Russia, tougher even than those imposed on purely domestic enterprises. Such standards include comprehensive environmental impact assessments and environmental protection programs.[259]

In Russia's currently strained circumstances, there is a danger that environmental protection might increasingly be seen as coming at the cost of economic well-being. Polls conducted in 1989 and 1990 put environmental issues at or near the top of the list of the USSR's most serious problems, but by 1991, both public concern and environmental activism had decreased significantly.[260] Some now see NGOs as obstructing economic progress. And a united voice is lacking on methods of implementing environmental reform.[261] Still, there is a burgeoning epistemic environment-oriented community of academics, officials, and scientists from within and outside Russia. These groups provide informed editorials on environmental and fishery issues that have an impact on international relations, and their opinions are likely to carry some weight, especially in the RFE.[262]

Projections

In addition to the central government, four distinct groups may emerge to control fisheries and environmental issues in Russia:[263] republic-level government leaders (both legislative and administrative), local government organizations, business firms, and NGOs. Moscow's authority will continue to deteriorate for the foreseeable future, so control over fisheries may eventually reside in the fishery enterprises themselves. There are even now serious doubts regarding the MOF's ability even to represent fishery interests in international discussions, and provincial and local agencies and enterprises are likely to assume much of this role. The diffusion of authority will confuse and frustrate attempts to negotiate bilateral and multilateral arrangements. Furthermore, Moscow is increasingly ambivalent about its role in UNCLOS and the UN Conference on High Seas Fishing.[264]

Four factors will determine the future of fisheries co-operation between Russia and North-East Asia:[265]
1. the progress of Russian market reform;
2. the commitment of the region's leaders to the development of a modern, market-oriented fishing industry in the RFE;
3. the commercial interests of North-East Asian nations, and

4. political relations between Russia and North-East Asia, with the dispute over the Southern Kuriles/Northern Territories being the major issue.[266]

The same four factors can be cited for marine environmental issues, with the impact of incipient NGOs constituting a potential fifth factor. Consideration of these factors shows that the market will play a dominant role in the Russian environmental agenda. Environmental protection will continue to be constrained by the country's deep economic depression and the inertia of its Soviet past. As an example, the Russian Law on Environmental Protection includes a clause which allows a firm to waive pollution fees if they are likely to be an undue burden. The power to impose or waive fees rests firmly at the local level, with no role or even a standard of appropriate behaviour codified by the central government in Moscow. This extreme decentralization will reinforce the current abuse of the regulatory system. And although enterprises have the option of investing in their own pollution control rather than paying the pollution fees, enforcement of compliance will place an additional burden on the meagre manpower of regulatory agencies.[267]

Russia is gradually increasing its domestic management of, and international co-operation in, marine fisheries and environmental protection. Strong incentives for improvement in marine fisheries policy include an influx of hard cash and technology in return for fish exports to North-East Asia; possible leverage in negotiations with Japan over the Southern Kuriles issue;[268] and access to information on fish stocks currently controlled by Japan. Regarding marine environmental protection, Russia could receive additional funding and technology, and partially satisfy the increasingly strident calls for environmental protection within the Russian Federation.

Domestic factors limiting Russia's roles in North-East Asian multilateral maritime interactions are its relatively small percentage of temperate seacoast, its perception of itself as a continental power with vast resources, its lack of infrastructure, and the tensions between its national and local marine policy goals.

Although the RFE was not considered part of North-East Asia until the late 1980s, it is now being inexorably drawn into the region's quickening development agenda. The sooner Russia determines how it wishes to proceed, the better. Reasons for greater North-East Asian interest in the RFE include lingering concern over its nuclear weapons, and the 'environmental security' threat that faulty nuclear plants, submarines, and waste disposal in North-East Asian seas constitutes to these nations. Furthermore, the potential cornucopia of raw materials, including fishery resources, has long fascinated North-East Asia. With its current troubles, it is highly unlikely that Russia at this point in its history could play a pivotal role in multilateral fisheries and/or marine environmental management in North-East Asia—at least without significant international assistance. But Russia's role in such management is likely to grow proportionately with its transition to a market economy and the increased standard of living and the confidence that will bring.

Notes

1. The section is based on a draft manuscript by Zha Daojiong and Mark J. Valencia, 'The politics of Chinese national reunification and maritime regime building in Northeast Asia'.
2. For a recent statement of China's policy on the Taiwan issue, see China's White Paper entitled 'The Taiwan Question and Chinese Reunification,' *People's Daily*, (overseas edition), 1 September 1994, 1, 5.
3. For an update, see George Yu and David J. Longenecker, 'The Beijing–Taipei struggle for international recognition', *Asian Survey*, May 1994, 475–88.
4. Frank Ching, 'Jiang Zemin goes fishing', *Far Eastern Economic Review*, 2 March 1995, p. 40; Julian Baum, 'Pressure cooker', *Far Eastern Economic Review*, 24 August 1995, pp. 16–17.
5. Jun Zhan, *Ending the Chinese Civil War*, New York: St. Martin's Press, 1994; 'Major breakthrough' in Taiwan–China talks, *Honolulu Advertiser*, 8 August 1994; 'Taipei–Beijing achieve major breakthroughs', Agence France Presse English Wire, 4 August 1994.
6. Ma Zhongshi, China Institute of Contemporary International Relations, personal communication dated 1993.
7. Lin Chang-Pin, 'Beijing and Taipei dialects in post-Tiananmen interactions', *China Quarterly*, (December 1993): 780.
8. Frank Ching, 'Taiwan looks for loophole', *Far Eastern Economic Review*, 2 February 1995, 29.
9. Lin, 'Beijing and Taipei dialects', p. 785.
10. 'Mainland vessels to be confiscated if detained twice', *FBIS-CHI-94*, 29 March 1994, 50.
11. 'Chinese people' here means Chinese living on the mainland as well as those on Taiwan.
12. 'Further progress in Taipei–Beijing talks', Agence France Press English Wire, 5 August 1994; *China Information Bulletin*, September 1994, 94.
13. Julian Baum, 'Jiang talks strait', *Far Eastern Economic Review*, 16 February 1995, pp. 14–15.

14. Interview by Zhou Daojiong with Chinese maritime safety researchers in Beijing, January 1994.
15. *Far Eastern Economic Review*, 8 July 1993, p. 22.
16. Interview by Zhou Daojiong with Chinese maritime affairs researchers in Beijing, January 1994.
17. Robert Scalapino, 'Back to the future', *Far Eastern Economic Review*, 26 May 1994, p. 38.
18. *People's Daily*, overseas edition, 9 June 1992, p. 1.
19. According to one account, Liaoning, Shandong, Jiangsu, Hebei, Zhejiang, and Fujian provinces have all drafted ambitious plans to extract more wealth from the ocean. See Yang Jiuyan and Wu Lunkai, 'Marine-related economy and a proposition for an "Oceanic Guangdong"', *Xueshu Yanjiu*, 6 (1993): 33.
20. 'The two sides of the Taiwan Straits hold secret talks in Singapore on offshore oil exploration and exploitation', *World Journal*, 25 April 1994, p. 6; 'China and Taiwan hold secret maritime exploration talks in Singapore', *World Journal*, 25 April 1994, p. 6; *World Journal*, 12 September 1994, p. 7.
21. Ralph N. Clough, *Reaching Across the Taiwan Strait: People-to-People Diplomacy*, Boulder: Westview Press, 1993, p. 206.
22. Arthur A. Stein, *Why Nations Cooperate*, Ithaca: Cornell University Press, 1990, p. 172.
23. A case could be made that there is a tacit agreement between Taiwan and the other three countries. What the present *status quo* does not offer is the binding nature of that consensus.
24. Mark J. Valencia and John Klarquist, 'National marine environmental policies and transnational issues' in Joseph R. Morgan and Mark J. Valencia, *Atlas for Marine Policy in East Asian Seas*, Berkley: University of California Press, 1993, pp. 136–7.
25. For the mainland's assessment of cross-straits development since April 1993, see 'One year since the "Wang-Koo" talks', *People's Daily* (overseas edition), 28 April 1994.
26. This section is an elaborated version of Noel A. Ludwig and Mark J. Valencia, 'Building Northeast Asian maritime regimes: Japan's potential role', *Marine Policy*, 19 (2): pp. 83–96.
27. Gregory W. Noble, 'Japan in 1992: just another aging superpower?', *Asian Survey*, 33, 1 (1993): 10.
28. 'Tokyo: US meet on flag-of-convenience issue', *FBIS*, 20 November 1992, p. 1.
29. Ivan Zrajevskij, 'The Northwest Pacific Region Action Plan: progress, problems and lessons learned', in Hyung Tack Huh, Chang Ik Zhang and Mark J. Valencia, (eds.), *Proceedings of the International Conference on East Asian Seas: Cooperative Solutions to Transnational Issues, Seoul, 21–23 September 1992*, Korean Ocean Research and Development Institute and East–West Center 1992, pp. 35–42.
30. *Daily Yomiuri*, 20 November 1993, p. 3; *Financial Times*, 9 March 1994, p. 10; *FBIS-EAS*, 9 February 1993, p. 1.
31. Mark J. Valencia, 'Sea of Japan: transnational marine resource issues and possible cooperative responses', *Marine Policy*, 14, 6 (1990): 511.
32. Anthony Bergin and Marcus Haward, 'The last jewel in a disintegrating crown: the case of Japanese distant water tuna fisheries', *Ocean Development and International Law Journal*, 25 (1994): 187–215.
33. Tsuneo Akaha, 'Japanese–Russian fishery joint ventures: opportunities and problems', *Marine Policy*, 17, 3 (1993): 199–212; Tsuneo Akaha, 'Muddling through successfully: Japan's postwar ocean policy development', Paper presented at the International Conference on Ocean Governance: The Making of a National Ocean Policy, Kuala Lumpur, 27–8 April 1994, p. 4.

34. Tsuneo Akaha, 'Balancing developmental needs and environmental concerns, domestic and international interests: Japan's ocean policy in the post-Cold War era', Paper prepared for the 27th Annual Conference of the Law of the Sea Institute, University of Hawaii, 13–16 July 1993, 28–9.
35. 'Fishery "White Paper" calls for resource management', *FBIS-EAS*, 22 April 1993, p. 7.
36. Akaha, 1994, 'Japanese–Russian fishery joint ventures: opportunities and problems', p. 4.
37. Tsuneo Akaha, *Japan in Global Ocean Politics*, Honolulu: University of Hawaii Press, 1985: 158; *Far Eastern Economic Review*, 16 January 1992, p. 40.
38. Hideo Takabayashi, 'Regional cooperation in marine resource management in the North Pacific', in Douglas M. Johnston (ed.), *Regionalization of the Law of the Sea*, Cambridge, Mass: Ballinger Publishing Co., 1978, pp. 146–8; Tsuyoshi Kawasaki, 'Fisheries problems in the Yellow and East China Seas', *Occasional Paper WP-88-2*, Honolulu: Environment and Policy Institute, East–West Center, p. 19; Bergin and Haward, 'The last jewel in a disintegrating crown', pp. 202–3.
39. Olav S. Stokke, 'Transnational fishing: Japan's changing strategy', *Marine Policy*, 15, 3 (1991): 238.
40. Jin Hyun Paik, 'Strengthening maritime security in Northeast Asia', Paper presented at the 8th Pacific Roundtable, Kuala Lumpur 5–8 June 1994; Kawasaki, 'Regional cooperation in marine resource management in the North Pacific', pp. 25–6.
41. Tsuneo Akaha, 'Bilateral fisheries relations in the Seas of Japan and Okhotsk: a catalyst for cooperation or seed of conflict?', Paper presented at the Joint Meeting of Asian Studies on the Pacific Coast and the International Studies Association, Monterey, California 29–30 October 1993, p. 12; 'Assemblies urge application of 200 nm zone', *FBIS-EAS-94-097*, 19 May 1994, p. 10.
42. Stokke, 'Transnational fishing: Japan's changing strategy', pp. 237–8.
43. Hans J. Buchholz, *Law of the Sea Zones in the Pacific Ocean*, Germany: Institute of Asian Affairs, 1989, p. 66; Valencia, 'Sea of Japan: transnational marine resource issues and possible co-operative responses', p. 512.
44. Bergin and Haward, 'The last jewel in a disintegrating crown: the case of Japanese distant water tuna fisheries', pp. 200–1.
45. Akaha, 'Japanese–Russian fishery joint ventures: opportunities and problems', p. 13.
46. Takabayashi, 'Regional cooperation in resource management in the North Pacific,' pp. 146–7.
47. Sachiko Sakamaki, 'The eyes have it', *Far Eastern Economic Review*, 26 May 1994, pp. 66–8.
48. Tsuneo Akaha, 'Bilateral fisheries relations in the Seas of Japan and Okhotsk: a catalyst for co-operation or seed of conflict?' in Norton Ginsburg, Elisabeth Mann Borgese, and Joseph R. Morgan, *Ocean Yearbook* 11, Chicago: University of Chicago Press, 1994, p. 405.
49. *Japan Aquaculture Economic News*, 1 January 1994; 'Nations to join panel on Sea of Japan fisheries', *FBIS-EAS*, 9 September 1992, p. 10.
50. Takeshi Umezu, 'Squid watch', *Look Japan*, 39, 453 (1993): 31.
51. Hideo Sato, 'Japan's role in the post-Cold War world', *Current History*, 90, 555 (1991): 148; Valencia and Klarquist, 'National marine environmental policies and transnational issues', pp. 136–41.
52. Akaha, 'Japanese–Russian fishery joint ventures: opportunities and problems', p. 15; *Christian Science Monitor*, 27 September 1990, p. 8.
53. Valencia and Klarquist, 'National marine environmental policies and transnational issues', p. 140.

54. Ibid., pp. 138.
55. Noburo Kumamoto, 'Japanese environmental law and ocean resources', *Ecology Law Quarterly*, 16, 267 (1989): 274; 'Pain in Japan' *The Economist*, 11 December 1993, pp. 77–8; Hidefumi Imura, 'Japan's environmental balancing act: accommodating sustained development', *Asian Survey*, 34, 4 (1994): 357.
56. Akaha, 'Japanese–Russian fishery joint ventures: opportunities and problems', p. 1.
57. Akaha, 'Balancing developmental needs and environmental concerns, domestic and international interests', p. 3.
58. Saburo Kato, 'Points of departure', *Look Japan*, 38, 435 (1992): 14–15; *Japan Times*, 13 October 1993, p. 2.
59. Kunio Nishimura, 'On the ball', *Look Japan*, 38, 435 (1992): p. 27.
60. 'Tokyo pledges money, efforts to conserve wetlands', *FBIS-EAS*, 15 June 1993, pp. 1–2.
61. 'OECD reviews green policy' *Japan Times*, 14 October 1993, p. 8.
62. Fred Pearce, 'Can Toyota save the world?', *Audubon*, (November–December 1992): 132.
63. *Asahi Evening News*, 18 February 1993, p. 3.
64. 'Japan called top donor of non-military assistance' *Japan Times*, 1 July 1993, p. 3; 'Asia seen as leader on the environment', p. 14.
65. 'Environment Agency proposal', *FBIS-EAS-94-171*, 2 September 1994.
66. Richard A. Forrest, 'Japanese aid and the environment', *The Ecologist*, 21, 1 (1991): 30.
67. Hanns W. Maull, 'Japan's global environmental policies', *Pacific Review*, 4 (1991): pp. 259–60; Ian Austin and Mary J. Koontz, 'Japan's 100-year environmental plan unveiled', *Pollution Prevention*, 1, 5 (1993): 20.
68. *Daily Yomiuri*, 24 August 1993, p. 1.
69. *Japan Times*, 28 October 1993, p. 4.
70. 'Hata: Tokyo should take international initiative', *FBIS-EAS*, 27 October 1993, p. 2.
71. 'Hopes high as green talks end', *Japan Times*, 27 October 1994, p. 3.
72. Masayuki Takeyama, 'Japan's foreign negotiations over offshore petroleum development: an analysis of decision-making in the Japan-Korea Continental Shelf Joint Development Program', in Robert L. Freidheim (ed.), *Japan and the New Ocean Regime*, Boulder: Westview Press: 300–2.
73. 'Japan agrees to widen Seoul participation in ODA', *FBIS-EAS*, 5 May 1993, p. 22.
74. 'Environmental agreement reached with ROK', *FBIS-EAS*, 30 June 1993, p. 3; Asahi Evening News, 22 March 1994, p. 5.
75. *FBIS*, Ibid.; 'Tokyo, Beijing agree on environment co-operation', *FBIS-EAS*, 1 December 1993, p. 4.
76. 'Seoul, Tokyo sign wastewater treatment pact', *FBIS-EAS*, 30 August 1993, pp. 25–6.
77. 'Russia seek co-operation on 12 economic schemes', *FBIS-EAS*, 8 September 1992, p. 23.
78. 'Tokyo to ofter $8.1 billion in aid to Russia', *FBIS-EAS*, 13 April 1993, p. 5; *Richmond Times-Dispatch*, 4 June 1994, p. 6.
79. 'ROK, Japan, Russia to test for waste dumping', *FBIS-EAS*, 14 March 1994, p. 38; *FBIS-EAS*, 22 March 1994, p. 4.
80. 'Tokyo to ofter $8.1 billion in aid to Russia', *FBIS-EAS*, 13 April 1993, p. 5.
81. 'Joint Economic Declaration', *FBIS-EAS*, 13 October 1993, pp. 4–5.
82. 'Northeast Asia drafting, environmental plan', *FBIS-EAS*, 9 February 1993, p. 1.

83. *Mainichi Daily News*, 4 December 1993, p. 17; 'Seoul, Tokyo discuss reduction of PRC pollutants', *FBIS-EAS*, 12 January 1994, p. 23; *Asahi Evening News*, 22 March 1994, p. 1.
84. 'Miyazaura hopes to work with US on pollution', *FBIS-EAS*, 9 April 1993, p. 2.
85. Kishore Mahbubani, 'Japan adrift', *Foreign Policy*, 88 (Fall 1992): 126–7; Karel van Wolferen, 'Japan's non-revolution', *Foreign Affairs*, 72, 4 (1993): 64.
86. Sato, 'Japan's role in the post-cold war world', p. 146; Charles Smith, 'War and remembrance', *Far Eastern Economic Review*, 25 August 1994, pp. 22–4; Robert Scalapino, 'Historical perceptions and current realities regarding Northeast Asian regional cooperation', North Pacific Cooperative Security Dialogue, Working Paper No. 20, October 1992; Tsuneo Akaha, 'International relations of Northeast Asia: imperatives of regional economic co-operation against pre-War and Cold War leagues.' Paper presented at the Annual Meeting of Asian Studies on the Pacific Coast (ASPAC), 16–18 June, 1995.
87. 'Environment chief resigns over World War II remark', *Japan Times*, 16 August 1994, p. 1; Teresa Watanabe, 'Another Japanese official retracts remark on War', *Honolulu Advertiser*, 13 August 1994, p. A-5; 'Foreign Minister writes on ROK–Japan summit', *FBIS-EAS*, 10 November 1993, pp. 31–2.
88. Amy Borrus and Robert Neff, 'Why shirking the burden isn't in Japan's best interests', *Business Week*, 28 January, p. 34.
89. Mahbubani, 'Japan adrift', p. 128.
90. 'Japan increasingly aware of international role', *Journal of Commerce and Commercial*, 28 March 1991, p. 4A.
91. Jessica Shin, 'Irreconcilable differences? Nationalism stalls resolution of island dispute', *Harvard International Review*, 16, 2 (1994): 48; Mark J. Valencia and Noel A. Ludwig, 'Southern Kurile Islands/Northern Territories Resource Potential', *Geojournal*, 24 (1991): 227–31.
92. Shinichi Kitaoka, 'Ghost of Japan's pan-Asianism', *Journal of Commerce and Commercial*, 26 May 1992, p. 8A.
93. Michael Blaker, 'Japan in 1994', *Asian Survey*, 35, 1 (January 1995): 1–12.
94. 'Fishery White Paper calls for resource management', *FBIS-EAS*, 22 April 1993, p. 7.
95. Akaha, *Japan in Global Ocean Politics*, p. 40.
96. 'Russian military shoots over sea-poaching', *Honolulu Advertiser*, 7 February 1994, p. A-8.
97. See, for example, Aurelia George, 'Japan as America's global partner: problems and prospects', *Journal of Northeast Asian Studies*, 11, 4 (1992): 11.
98. Mahbubani, 'Japan Adrift', pp. 136–7.
99. Maull, 'Japan's global environment policies', p. 261.
100. Akaha, Japanese–Russian fishery joint ventures: opportunities and problems, pp. 3–7, 14.
101. Yoichi Funabashi, 'Japan and the new world order', *Foreign Affairs*, 70, 5 (1991): 60.
102. Mark J. Valencia, 'Preparing for the best: involving North Korea in the New Pacific Community', *Journal of Northeast Asian Studies*, 13, 1 (Spring): 1.
103. Frank Ching, 'Securing Northeast Asia', *Far Eastern Economic Review*, 11 November 1993, p. 42.
104. John Merrill, 'North Korea in 1992', *Asian Survey*, 33, 1 (January 1993): 42–9; John Merrill, 'North Korea in 1993', *Asian Survey*, 34, 1 (January 1994): 15; Sungwoo Kim, 'Recent economic policies of North Korea', *Asian Survey*, 33, 9 (September 1993): 864–78; Nayan Chanda and Shim Jae Hoon, 'North Korea: poor and desperate', *Far Eastern Economic Review*, 9 September 1993, p. 17.
105. Merrill, 'North Korea in 1993'.

106. 'Jilin Province, DPRK to develop Chonjin port', *FBIS-CHI-94-018*, 27 January 1994, p. 13.

107. Mark J. Valencia, 'The Pyongyang International Conference and Field Trip', East–West Center, August 1992.

108. Mark J. Valencia, 'The proposed Tumen River scheme', in Hong Yung Lee and Chung Chongwook, eds., *Korean Options in a Changing International Order*, Institute of East Asian Studies and Center for Korean Studies, Berkeley: University of California Press, 1993, pp. 187–202; Mark J. Valencia, 'Work on Tumen River delta will benefit states', *The Straits Times*, 24 November 1990; John Whalen, 'Status of the Tumen River Area Development Programme: Progress, accomplishments and remaining tasks', Paper presented to the Conference on Regional Economic Cooperation in North-east Asia, Yongpyeong, South Korea, 26–8 September 1993.

109. 'DPR Korea's Fisheries Industry', *INFOFISH International*, 4/80 (1988): 39–42.

110. 'Bilateral fisheries relations in the Seas of Japan and Okhotsk: a catalyst for cooperation or seed of conflict?' p. 397.

111. Chung Young-Hoon, 'South and North Korea cooperation in fisheries: toward establishing a joint fishing zone', Master's thesis, College of Marine Studies, University of Delaware, 1993, p. 46.

112. 'North Korean crab fleet laps up capitalism in Japanese ports', *New Sunday Times*, 12 June 1994, p. 34.

113. Ibid.

114. Chung, 'South and North Korea co-operation in fisheries: toward establishing a joint fishing zone', pp. 46–7.

115. For background and analysis of North Korea's jurisdictional claims, see Choon-ho Park, 'The 50-mile military boundary zone of North Korea', *American Journal of International Law*, 72, 4, (1978): 866–75; 'DPRK gunboats fired on PRC fishing boats', *Yonhap*, 27 January 1993.

116. Chua Thia Eng, Robert Cordover, Miles Hayes, Celso Roque, David Shirley, Gurpreet Singhota and Philip Tortell, 'Prevention and Management of Marine Pollution in East Asian Seas', Formulation Mission Report Prepared for the United Nations Development Programme Division, Regional Bureau for Asia and the Pacific, April 1993, pp. 2.11–2.27.

117. Peter Hayes, 'Economic dimensions of restoring North Korea's environment', Paper presented to the Korea Development Institute/*Korea Economic Daily* Seminar on the North Korean Economy, Seoul, 17 October 1994.

118. Chua et al., 'Prevention and Management of Marine Pollution in East Asian Seas'.

119. Valencia and Klarquist, 'National marine environmental policies and transnational issues', p. 139.

120. Economic and Social Commission for Asia and the Pacific, *Report of Expert Group Meeting on Environmental Cooperation in North-East Asia, Beijing, 24–6 November 1994*.

121. Mark J. Valencia, 'East Sea: sea of co-operation', *The SISA Journal*, 23 December 1993; 'DPRK wants joint nuclear waste probe with Russia', *FBIS-EAS-93-127*, 6 July 1993, p. 35.

122. 'Pyongyang protests Russians radioactive pollution', Ibid., *FBIS-EAS-93-082*, 30 April 1993, p. 10; Statement by the DPRK, Plenary Meeting, Maritime Safety Committee, IMO, 61st session, 8 December 1992.

123. International Maritime Organization, Marine Environment Protection Committee 33/WP.7/Add.1, para. 13.4, p. 21.

124. Statement by the DPRK, Plenary Meeting, Maritime Safety Committee, IMO, 61st session, 8 December 1992.

125. United Nations Environment Programme, 'Draft Action Plan for the Protection and Development of the Marine and Coastal Environment of the North-West Pacific Region', September 1993; United Nations Environment Programme, 'Draft Action Plan for the Protection, Management and Development of the Marine and Coastal Environment of the North-West Pacific Region', 1994; United Nations Environment Programme, *Report of the Third Meeting of Experts and National Focal Points on the Development of the North-West Pacific Action Plan, Bangkok, 10–12 November 1993*, UNEP(OCA)/NOWP.WG3/6.

126. Peter Hayes and Lynda Zarsky, 'Regional cooperation and environmental issues in Northeast Asia', *Nautilus: Pacific Research*, 1 October 1993, p. 16.

127. Ibid., p. 13.

128. Ibid., p. 14.

129. 'China, Seoul agree to involve DPRK in project', *FBIS-EAS-94-067*, 7 April 1994, p. 33.

130. 'Miyazawa hopes to work with US on pollution', Kyodo News Service, 30 June 1993.

131. Seoung-Yong Hong, 'A framework for emerging new marine policy: the Korean experience', Paper presented at the International Conference on Ocean Governance: The Making of a National Ocean Policy, Malaysia Institute of Marine Affairs, Kuala Lumpar 27–8 April 1994, pp. 1–5.

132. Paik, 'Strengthening maritime security in Northeast Asia'.

133. Hong, 'A framework for emerging new marine policy: the Korean experience', pp. 1–10.

134. Ibid.

135. Ibid.

136. Shim Jae Hoon, 'Destination: Pyongyang', *Far Eastern Economic Review*, 14 July 1994, p. 28.

137. Park Seong-Kwae, 'The status of fisheries in Korea with emphasis on distant-water operations', Paper presented at the 27th Annual Conference of the Law of the Sea Institute, 13–16 July 1993, p. 6–3.

138. Ibid., p. 6.

139. Ibid.

140. Mark J. Valencia, 'East–West Environment and Policy Institute Occasional Paper No. 1987', International Conference on the Yellow Sea, pp. 33, 37, 38, 41.

141. Sun Yun Hong, 'Remarks', in Hyung Tack Huh, Chang Ik Zhang and Mark J. Valencia (eds.), *Proceedings of the International Conference on East Asian Seas: Cooperative Solutions to Transnational Issues*, Seoul: Korea Ocean Research and Development Institute, East–West Center, 1992, pp. 69–70.

142. 'Talks held with Japan on fishing dispute', *Yonhap*, 23 February 1994.

143. Hong, 'Remarks', p. 69.

144. Seoung-Yong Hong, 'Marine Policy in the Republic of Korea', Paper presented at the XXVII Law of the Sea Institute Conference.

145. Park, 'The status of fisheries in Korea with emphasis on distant water operations'.

146. Pak In-chal, 'Seoul to protest to PRC on illegal fishing', *The Korea Herald*, 24 August 1993, p. 3; 'South Korea asks China to curb fishing off its coast', *The Korean Herald*, 4 August 1995, p. 2.

147. *Korea Herald*, 18 March 1993.

148. Joh Jung-Jay, 'Measures to prevent marine pollution urgently needed.' *The Korea Times*, 15 June 1995, p. 3; *The Korea Herald*, 1 August 1995, p. 3.

149. Y.H. Seung and Y.C. Park, 'Physical and environmental character of the Yellow Sea', in Choon-Ho Park, Dalchong Kim and Seo-Hang Lee (eds.),

The Regime of the Yellow Sea: Issues and Policy Options for Cooperation in the Changing Environment, Seoul: Institute of East and West Studies, Yonsei University, 1990, p. 31; Mark J. Valencia, Chen Lisheng and Chen Zhisong, 'Yellow Sea marine and air pollution: status, projections, transnational dimensions and possibilities of cooperation', in Won Bae Kim (compiler), *Report on Regional Development in the Yellow Sea Rim*, Honolulu: East–West Center, p. 382.

150. 'Further on Han's remarks', *FBIS-EAS*, 27 October 1993, p. 14; 'Foreign minister describes diplomatic goals', *FBIS-EAS*, 29 December 1993, p. 34; 'Washington's ire: the US takes aim at China and Japan', *Far Eastern Economic Review*, 20 January 1994, pp. 56–7.

151. 'Gong promotes regional security dialogue in N-E Asia', *The Korea Times*, 4 August, 1995, p. 2.

152. Sang Don Lee, 'Remarks', in Hyung Tack Huh, Chang Ik Zhang and Mark J. Valencia (eds.), *Proceedings of the International Conference on East Asian Seas: Cooperative Solutions to Transnational Issues*, Seoul: Korea Ocean Research and Development Institute and East–West Center, 1992, pp. 44–5.

153. 'Seoul urges Asia to work on pollution control', *FBIS-EAS*, 8 February, 1993.

154. Shim Jae Hoon, 'Not on my island', *Far Eastern Economic Review*, 20 July, 1995, p. 68.

155. 'Spokesman denounces Russia's nuclear waste disposal', *FBIS-EAS*, 12 April 1993, p. 16; 'DPRK wants joint nuclear waste probe with Russia', *FBIS-EAS*, 21 October 1993, p. 5; 'Further reporting on Russian nuclear waste', *FBIS-EAS*, 21 October 1993, pp. 24–5.

156. *FBIS*, 6 July 1993 p. 35.

157. *The Korea Times*, 29 October 1993, p. 3.

158. Dalchoong Kim, 'The Yellow Sea in the Northeast Asian setting: need for cooperation', in Choon-ho Park, Dalchoong Kim and Seo-Hang Lee (eds.), *The Regime of the Yellow Sea: Issues and Policy Options for Cooperation in the Changing Environment*, Seoul: Institute of East and West Studies, Yonsei University, 1990, p. 3.

159. Hong Liu, 'The Sino-South Korean normalization: a triangular explanation', *Asian Survey*, 33, 11 (November 1993): 1085; Robert A. Scalapino, 'Northeast Asia—prospects for cooperation', *The Pacific Review*, 5, 2 (1992): 108.

160. This section is based on a draft 1994 manuscript by Noel Ludwig and Mark J. Valencia, 'Fisheries and environmental protection regimes in Northeast Asian seas: Russia's potential role'.

161. Scalapino, 'Northeast Asia: prospects for cooperation', p. 102; Burnham O. Campbell, 'Trade and regional cooperation in developing Northeast Asia: the general picture and the role of the Russian Far East', Paper presented at the Conference on the Russian Far East and the North Pacific: Prospects for Cooperation, Honolulu, Hawaii, 19–21 August 1993.

162. 'Joint Economic Declaration', *FBIS-EAS*, 13 October 1993, pp. 4–5.

163. Leszek Buszynski, 'Russia and the Asia-Pacific region', *Pacific Affairs*, 65, 4 (1992): 497–8; *International Herald Tribune*, 20 October 1993.

164. Elena N. Nikitina and Peter H. Pearse, 'Conservation of marine resources in the Former Soviet Union: an environmental perspective', *Ocean Development and International Law*, 23 (1992): 374.

165. Douglas M. Johnston and Mark J. Valencia, 'The Russian Far East and the North Pacific region: prospects for cooperation in fisheries' in Mark J. Valencia, (ed.), *The Russian Far East and the North Pacific Region*, Boulder: Westview Press, 1995: 155.

166. Tsuneo Akaha, 'Japanese–Russian fishery joint ventures and operations:

Opportunities and problems', *Marine Policy*, 17, 3 (1993): 210; Peter H. Christiansen, '*Bezvlastia*: adrift at the edge of Russia', Letter to Peter B. Martin, Institute of Current World Affairs, Hanover, New Hampshire, 10 January 1994, p. 3.

167. John J. Stephan, 'The Russian Far East', *Current History*, 92, 576 (1993): 333; Nobuo Arai, 'Fishery development in the Russian Far East', Paper presented at the Conference on US–Japan Cooperation in the Development of Siberia and the Russian Far East, 24–6 July 1993, p. 5.

168. Sophia Drewnowski, 'Former Soviet states initiate environmental reform', *Engineering News-Record*, 231, 15 (1993): E10.

169. Arai, 'Fishery development in the Russian Far East'.

170. *Russian Far East Update*, August 1993, p. 3; Christiansen, '*Bezylastia*: adrift at the edge of Russia', p. 19.

171. Collected Russian abstracts from the International Conference on the Sea of Japan, Nakhodka, Russia, Summer 1989; Arai, 'Fishery Development in the Russian Far East', p. 7.

172. Christiansen, '*Bezylastia*: adrift at the edge of Russia', p. 11.

173. Ibid.

174. Jeffrey L. Canfield, 'Recent developments in Bering Sea fisheries conservation and management', in *Ocean Development and International Law*, 24 (1993): 267.

175. Pavel A. Minakir, 'The Far East economy: the crisis', in Mark J. Valencia (ed.), *The Russian Far East and the North Pacific Region: Emerging Issues in International Relations*, Honolulu: East–West Center, 1992, p. 14.

176. Nikitina and Pearse, 'Conservation of marine resources in the former Soviet Union', p. 376; Arai, 'Fishery development in the Russian Far East', p. 1.

177. Vladimir Kotov and Elena Nikitina, 'Russia in transition: obstacles to environmental protection', *Environment*, 35, 10 (1993): 15; Arai, 'Fishery Development in the Russian Far East', p. 12.

178. *New York Times*, 20 May 1994, p. A11.

179. Kotov and Nikitina, 'Russia in transition: obstacles to environmental protection', p. 14.

180. *Russia in Asia Report No. 16*, University of Hawaii, Honolulu, January 1994, p. 67; Robert H. Boyle, 'The cost of caviar', *Amicus Journal* (Spring 1994): 23.

181. 'Russia suggests supervisors aboard fishing boats', *FBIS-EAS*, 29 October 1992, p. 10; *FBIS-USR*, 11 November 1993, pp. 86–8; *FBIS-USR*, 1 December 1993, pp. 91–2; *Honolulu Advertiser*, 7 February 1994, p. A-8; Nikolai Burbyga, 'Warships take civilian vessels under protection', *Current Digest of the Post-Soviet Press*, 45, 28 (1993): 23–4.

182. *Russian Far East Update*, August 1993, p. 3.

183. *Russia in Asia Report No. 16*, University of Hawaii, January 1994, pp. 71–2.

184. Stephan, 'The Russian Far East', p. 333.

185. *Russia in Asia Report No. 16*, University of Hawaii, January 1994, p. 88; Christiansen, '*Bezylastia*: adrift at the edge of Russia', p. 3.

186. Arai, 'Fishery development in the Russian Far East', p. 12.

187. Collected Russian abstracts, pp. 42–3.

188. Ibid.; Chitin, the main constituent of the shells of such crustaceans as crab and krill, is non-toxic and biodegradable, and for these reasons its uses are receiving increasing attention. Russia alone is thought to be capable of producing up to 2 million kilograms of chitin per year, and shellfish waste from around the region could be shipped to a single large processing plant on the Sea of Japan.

189. Pavel A. Minakir, 'Economic reform in Russia and the Russian Far East',

Paper presented at the Conference on the Russian Far East and the North Pacific Region: Prospects for Cooperation, Honolulu, Hawaii, 19–21 August 1993, pp. 10, 16.
190. Ibid., p. 40.
191. Akaha, 'Japanese–Russian fishery joint ventures and operations', p. 203; Arai, 'Fishery development in the Russia Far East', p. 9.
192. *Russian Far East Update*, August 1993, p. 3.
193. Akaha, 'Japanese–Russian fishery joint ventures and operations', p. 201.
194. Ibid., p. 202.
195. Ibid., pp. 205–6.
196. Jeff Wise, 'Russian capitalism plays by its own rules', *The Asian Wall Street Journal Weekly*, 13 December 1993, p. 13.
197. '88% of Russian Far East joint ventures inactive', *FBIS-EAS*, 30 August 1993, p. 3.
198. Akaha, 'Japanese–Russian fishery joint ventures and operations', p. 210.
199. Arai, 'Fishery development in the Russian Far East', p. 11.
200. 'Russian studying co-operation with DPRK, ROK', *FBIS-EAS*, 5 January 1994, pp. 20–1.
201. Viatcheslav K. Zilanov, 'Living marine resources: their conservation, rational utilization, and management in the USSR and the world', *Oceanus*, 34, 2 (1991), pp. 31–2.
202. Akaha, 'Japanese–Russian fishery joint ventures and operations', pp. 199–200 and 207; Johnston and Valencia, 'The Russian Far East and the North Pacific Region', pp. 5–7; Buszynski, 'Russia and Asia-Pacific region', p. 498; Canfield, 'Recent developments in Bering sea fisheries conservation and management', p. 268.
203. Canfield, 'Recent Developments in Bering sea fisheries conservation and management', p. 257.
204. Ibid., p. 280.
205. Akaha, 'Japanese–Russian fishery joint ventures and operations', p. 201.
206. Johnston and Valencia, 'The Russian Far East and the North Pacific Region', p. 176.
207. Alex O. Elferink, 'Fishing in the Sea of Okhotsk high seas enclave: towards a special legal regime?' Paper presented at the Conference Toward the Peaceful Management of Transboundary Resources, Durham, UK, 14–17 April 1994, pp. 4–5.
208. Johnston and Valencia, 'The Russian Far East and the North Pacific Region', p. 169.
209. Mark J. Valencia and Noel A. Ludwig, 'Southern Kurile Islands/Northern Territories resource potential', *GeoJournal*, 24, 2 (1991): 231.
210. Andrew R. Bond and Matthew J. Sagers, 'Some observations on the Russian Federation Environmental Protection Law', *Post-Soviet Geography*, 33, 7 (1992): 463.
211. Steven J. Marcus, 'After the USSR: environmental management, market-style', *Technology Review*, 95, 1 (1992): 66.
212. William Zimmerman, Elena Nikitina and James Clem, 'International law and global environmental change: Soviet Union/Russia case study', Paper presented at the Workshop on International Law and Global Environmental Change: A Systematic Study of International Environmental Accords, Washington, DC, 2–5 February 1994, p. 21.
213. Kotov and Nikitina, 'Russia in transition: obstacles to environmental protection', pp. 16–17.
214. Gail Fondahl, 'Book review', *International Environmental Affairs*, 5, 3 (1994): 278.

215. Randolph M. Lyon, 'Environmental management in the Former Soviet Union: a comment', *Comparative Economic Studies*, 34, 2 (1992): 64.
216. *Honolulu Advertiser*, 7 March 1994, p. A-8.
217. Pollution choking Russia, ministry report concludes. *Honolulu Advertiser*, 4 June 1995, p. A-16.
218. Kotov and Nikitina, 'Russia in transition: obstacles to environmental protection', p. 16.
219. Baruch Boxer, 'Marine environmental protection in the Seas of Japan and Okhotsk', *Ocean Development and International Law*, 20 (1987): 203.
220. Kotov and Nikitina, 'Russia in transition: obstacles to environmental protection', p. 14.
221. Natalia Mirovitskaya and Marvin S. Sooros, 'Socialism and the tragedy of the commons: reflections on environmental practice in Soviet Union/Russia', Paper presented at the annual meeting of the International Studies Association, Washington, DC, 29 March–1 April 1994, pp. 2–3.
222. Kotov and Nikitina, 'Russia in transition: obstacles to environmental protection', p. 12.
223. Zimmerman et al., 'International law and global environmental change: Soviet Union/Russia case study', pp. 27, 40.
224. Kotov and Nikitina, 'Russia in transition: obstacles to environmental protection', p. 17.
225. Ibid., p. 19.
226. Ibid., p. 13.
227. Boxer, 'Marine environmental protection in the Seas of Japan and Okhotsk', p. 201.
228. Zimmerman et al., 'International law and global environmental change: Soviet Union/Russia case study', p. 8.
229. Boxer, 'Marine environmental protection in the Seas of Japan and Okhotsk', p. 204.
230. *Greenwire*, 20 January 1994, pp. 11–12; and 4 February 1994, pp. 11–13; Mirovitskaya and Sooros, 'Socialism and the tragedy of the commons', p. 11.
231. Maria Chersakova, 'Sounding the alarm', *UNESCO Courier*, 1 March 1992, p. 22.
232. Zimmerman et al., 'International law and global environmental change: Soviet Union/Russia case study', p. 16.
233. Ibid., pp. 2–6.
234. Ibid., p. 41.
235. Ibid., pp. 29–32.
236. Rowland T. Maddock, 'Perestroika, glasnost, and environmental regeneration in the Soviet Union', *International Environmental Affairs*, 3, 3 (1991): 185–7.
237. Drewnowski, 'Former Soviet states initiate environmental reform', p. E10.
238. Bond and Sagers, 'Some Observations on the Russian Federation Environmental Protection Law', p. 472.
239. Boxer, 'Marine environmental protection in the Seas of Japan and Okhotsk', p. 204.
240. Zimmerman et al., 'International law and global environment change: Soviet Union/Russia case study', p. 24.
241. Collected Russian abstracts, p. 38.
242. Mikhail Gorbachev, 'USSR is an Asian and Pacific power', Speech given in Vladivostok in July 1986 and quoted in Gordon Livermore (ed.), *Soviet Foreign Policy Today: Reports and Commentaries from the Soviet Press*, Columbus, Ohio: The Current Digest of the Soviet Press, 1986, pp. 180–3.

243. USSR Ministry of Foreign Affairs, 'The foreign policy and diplomatic activity in the USSR (November 1989 to December 1990)', *International Affairs* (Moscow), (April 1991): 90–1.
244. Kotov and Nikitina, 'Russia in transition: obstactes to environmental protection', p. 19.
245. USSR Ministry of Foreign Affairs, 'The foreign policy and diplomatic activity in the USSR (November 1989 to December 1990)', p. 90.
246. Boxer, 'Marine environmental protection in the Seas of Japan and Okhotsk', p. 203.
247. *Russia in Asia Report No. 15*, University of Hawaii, July 1993, p. 152.
248. Collected Russian abstracts, pp. 42–3.
249. This document is printed in full in *The Sejong Review*, 1, 1, pp. 255–7.
250. *Engineering News Report*, 6 September 1993, p. 3.
251. Drewnowski, 'Former Soviet states initiate environmental reform', p. E10.
252. *Greenwire*, 14 January 1994, p. 13.
253. 'Russians offered aid to build floating nuclear waste plant', *Japan Times*, 17 May 1994, p. A-1.
254. Drewnowski, 'Former Soviet states initiate environmental reform', p. E10; Zimmerman et al., 'International law and global environmental change: Soviet Union/Russia case study', pp. 13–14.
255. Maddock, 'Perestroika, glasnost, and environmental regeneration in the Soviet Union', pp. 182–3.
256. Stephan, 'The Russian Far East', p. 334.
257. Fondahl, 'Book Review', p. 277.
258. Maddock, 'Perestroika, glasnost, and environmental regeneration in the Soviet Union', p. 182.
259. 'CIS going green', *Marine Pollution Bulletin*, 26, 11 (1993): 590.
260. Fondahl, 'Book Review', p. 278.
261. Marcus, 'After the USSR: environmental management, market-style', p. 68.
262. Some Soviet scholars were discussing the possibility of ozone depletion as early as the early 1970s. But Soviet public opinion played a very limited role in engendering action in the USSR, even as both academic and public understanding of this problem broadened during the 1970s and 1980s. Zimmerman et al., 'International law and global environment change: Soviet Union/Russia case study', p. 16.
263. John Palmisano and Brent Haddad, 'The USSR's experience with economic incentive approaches to pollution control', *Comparative Economic Studies*, 34, 2 (1992): 53.
264. Johnston and Valencia, 'The Russian Far East and the North Pacific Region', p. 167; 'UN amends seabed laws', *Japan Times*, 30 July 1994, p. 3.
265. Akaha, 'Japanese–Russian fishery joint ventures and operations: opportunities and problems', p. 212.
266. Japan continues to insist on the return of the islands before freeing up aid and investment, while Russia insists that expanded economic relations will make a solution easier. 'Russian deputy premier arrives for five-day Japan visit', *Kyodo*, 27 November 1994.
267. Palmisano and Haddad, 'The USSR's experience with economic incentive approaches to pollution control', p. 60.
268. Scalapino, 'Northeast Asia-prospects for cooperation', p. 48.

5 Pollution and Environmental Protection

This chapter describes the status of marine pollution, delineates the problems and issues, reviews the existing initiatives and regimes and their inadequacies, focuses on the example of the Yellow Sea, and concludes with an outline of an ideal environmental protection regime for North-East Asian seas.

The idea that critical environmental and resource management issues carry the seeds of conflict is actually quite old.[1] Broadly defined, 'environmental security in the ocean' includes all marine resource management issues, as well as issues in pollution control, nature conservation, and coastal community protection. In virtually all regions of the world there is a long history of contending interstate claims in such contexts. They usually begin as mere *differences* of policy and practice, but frequently deteriorate into *issues*. These issues then assume an official character as they are placed on the agenda of diplomatic discussions and negotiations. If no abatement or resolution results from normal negotiation, frustration sometimes endows the issue with a symbolic significance, and other irritants may contribute to further deterioration until the contending parties are forced to acknowledge the existence of a *dispute*. The incidence of such disputes seems to be rising globally as well as in North-East Asia.

Environmental Problems and Issues

Status of Pollution
The Yellow Sea/East China Sea

The Yellow Sea has been publicly designated as one of the seven 'dying' seas of the world and in the second worst condition after the Black Sea.[2] The Yellow River carries some 750 tons of heavy metals into the Sea each year, producing concentrations of 1,000 times the permissible level of some pollutants. Other pollutant sources in the Yellow Sea include industrial wastewater and domestic sewage from the coastal urban and rural areas. Every year

181 million tons of domestic sewage and industrial wastes and 0.5 million tons of Chemical Oxygen Demand (COD) are discharged into the nearshore areas of the Eastern Yellow Sea. The concentration of these pollutants in nearshore waters has adversely affected fish breeding. Worse still, the number of fish species has supposedly declined from 141 to 24 over the past thirty years. And heavy metals such as mercury and cadmium pose serious health hazards to coastal inhabitants.

Oil spills from oil and gas exploration, tankers and ports, and waste dumping in the offshore areas cause more damage. The main sources of oil are: vessels (85%), land based (9%), oceanic platforms (1%), and unknown sources (5%).[3] In the eastern Yellow Sea, an estimated total of 550 tons of oil was spilled from 1979 to 1985 despite intensive efforts to reduce accidents by the Korea National Maritime Police, the body responsible for the monitoring of marine oil pollution in South Korea. The average concentration of oil in the surface waters of the eastern Yellow Sea coastal waters increased from 1.9 microlitres per litre in 1985 to 2.2 in 1987. Between 1983 and 1987, total economic loss of South Korean resources, including aquaculture areas, due to oil spilled was estimated at 17 million *won*.

Among the most polluted areas in the region are the southern Korean Peninsula, which has a high nearshore concentration of hydrocarbons, moderate Chemical Oxygen Demand (COD), and frequent red tides; the west coast of South Korea near Inchon, which also suffers from oil pollution, and frequent red tides; the Bo Hai, which has both wide-spread low-level hydrocarbon pollution and scattered high concentrations of oil pollution, moderate COD, industrial waste, and frequent red tides; and the Yangtze and Yellow river mouths and deltas, which have oil and industrial pollution, high COD, and frequent red tides. Major pollutants entering the western Yellow Sea include COD, sewage, oil, mercury, cadmium, lead, and inorganic nitrogen.

In addition to major fisheries resources (see Chapter 6), this region contains many ecological resources which are vulnerable to pollution. The Ryukyu archipelago includes whale calving grounds, turtle breeding areas, coral reefs, mangroves, protected areas, and marine mammals. There is also a remarkable concentration of turtle breeding areas, mangrove stands, coral reefs, and protected areas in the Sakishima Gunto. The adjacent and heavily used Korea Straits are spawning grounds for both demersal and pelagic fish, as well as whales. The island of Taiwan is enveloped by mangroves and coral reefs, some of which are protected. The south-west coast of South Korea also harbours many aquaculture sites and protected areas.

Although some shallow coastal water areas near river mouths, ports, and coastal cities on the west and south coasts of the Korean Peninsula can be considered polluted, offshore environmental quality is still normal. So the most obvious conflict areas are close to shore where aquaculture and various pollutants occur together. The inner shores of the Bo Hai are clear examples, as are numerous places on the coasts of Japan, where aquaculture activities and shore-based pollution are both widespread. In the more open reaches of the Yellow and East China Seas, tar balls on the surface demonstrate the potential conflict between whale calving areas, spawning grounds for both pelagic and demersal species, and oil pollution.

Both South Korea and China have ambitious plans for developing their coastal areas bordering the Yellow Sea as part of the concept of a Yellow Sea Economic Circle. In the Yellow Sea, as elsewhere, coastal zone development has implications for ocean management. In October 1989 South Korea began to implement its West Coast Development Plan, which consists of 126 diverse projects to be completed by 2001.[4] Two-thirds of a planned total of twenty-three thermal and nuclear power stations will be located on the west coast. These plants will produce thermal effluent, and the nuclear power plants may discharge trace amounts of radioactive substances as well. The construction of more thermal power stations also means that petroleum-poor South Korea will have to import more oil and coal, particularly from China. Between 1991 and 1995, China exported about 10 million tons of crude oil to Japan.[5] Much of this crude moved through the Yellow Sea, increasing tanker traffic there.

There are eleven sewage treatment plants on the west coast, but, apart from Seoul, only six west coast cities have sewage treatment plants, so that only 28 per cent of domestic sewage is currently treated. The South Korean government plans to construct 33 more sewage treatment plants and by 2001, 258 cities and towns are expected to have them. That will allow 75 per cent of all domestic sewage to be treated before entering the rivers and the sea.[6]

The concept of China's Bo Hai-Yellow Sea Economic Region was formulated in 1984 as a part of its open door policy for regional development.[7] It includes three provinces, Hebei, Liaoning, and Shandong, and two independent municipalities, Beijing and Tianjin. In 1988 the major ports of this region handled about 182 million tons of goods or about 41 per cent of the total handled by China's major coastal ports.[8] Principal coastal cities in the region include Tianjin, Dalian, Qinhuangdao, Yantai, Qingdao, Shijiushuo, and Lianyungang. Dalian, Tianjin, and Lianyungang are three of the four terminals linked to the major Eurasian landbridges across China and, as such, they will probably be expanded and upgraded. Dalian, Tianjin, Qinhuangdao, Qingdao, and Lianyungang are also connected to large inland oilfields and coal mines in North and/or North-East China by railroads and pipelines.[9] These port cities and the Yellow Sea are clearly important for China's exports of coal and petroleum to South Korea, Japan, and other countries, and for China's north-south interregional transportation.

The Sea of Japan

The coastal rim of the Sea of Japan is relatively free of pollution although there are significant exceptions: mercury off Niigata, and oil and heavy metals in Peter the Great Bay. Recent revelations of dumping of nuclear materials and chemical munitions in the Sea by the former Soviet Union are especially worrying.[10] If tourism and a marketable fish catch are to be sustained, the Sea must be kept free of both land- and sea-based sources of pollution. But the relatively pristine nature of the Sea of Japan, an ironic beneficial by-product of the ideological Cold War, may be short-lived. As visionaries and industry eye the possibility of a Sea of Japan Natural Economic Territory,[11] it is becoming increasingly urgent to put a process or infrastructure in place to manage the Sea's environment.

Institutional Issues

In few other semi-enclosed seas are multilateral measures for marine pollution control as deficient as those in North-East Asia. Indeed beyond coastal waters, most North-East Asian seas are a 'mare nullius' in terms of marine environmental protection.[12] According to regime theory, international anarchy prevails.[13] And sensitive political relations and uncertain boundaries have not been conducive to information-sharing and co-operation on many matters, let alone the environment. This situation has made it difficult to evaluate the nature and extent of support for international environmental activities or even national positions thereon.

There is a general dearth of capacity and will to co-operatively monitor marine pollution. Nor is there any formal infrastructure to bring about the critical mass of international collaboration and co-operation in monitoring and research activities that would delineate the spatial distribution of a contaminant and its subsequent effects, and, in particular, whether it would cross national boundaries. The lack of such a structure prevents the development of well-co-ordinated co-operative baseline studies, and co-ordination in emergencies, such as a spill of oil or other toxic and hazardous materials.

Existing monitoring and research programs are ineffective because they stop at artificial, politically determined borders, rather than at a physical or chemical border. Among the countries bordering the Yellow and East China Seas, there are wide discrepancies in the level and effectiveness of marine pollution monitoring and research in support of regulation. Japan is clearly far superior to the others in terms of marine environmental knowledge and technology. Russia has the capability but neither the will nor the means to fully utilize it. China has carried out extensive surveys and research in its 'off-shore' areas since the early 1970s. But comprehensive marine research programs in South Korea and Taiwan have begun more recently. It is unknown, but doubtful, whether North Korea has undertaken such activities.

Except in response to occasional tanker accidents that have destroyed coastal fisheries and aquaculture areas, and to the severe effects of untreated industrial effluents on public health, there has been only minimal overt recognition by the North-East Asian coastal states in recent years of the long-term effects of land-source, vessel, and other pollution on people and the marine environment. Scientific questions on factors affecting the health of marine species and ecosystems are poorly articulated, and the relevance of national laws and policies to regional environmental protection has not been seriously considered. A review of national legislation shows little evidence of laws and regulations being developed with specific relevance to natural features or processes that may affect pollutant transport, circulation, transformation, and dispersion. Laws and policies are couched in terms that separate legal justification and intent from the reality of people, ecosystems, and place. This problem exists in other regions as well but it is more important in North-East Asia, because the apparent failure to relate law more directly to nature through improved scientific understanding supports a general impression of regional disinterest in marine environmental issues.[14]

Impressions aside, the degree of concern with marine pollution is quite varied, and practical policy towards it is even more diverse. Japan is clearly the leader in marine pollution policy and prevention in the North-East Asian region, but even it is now backsliding in policy and enforcement. Marine pollution awareness and prevention are much more recent phenomena in China, South Korea, Taiwan, and Russia. Their laws and regulations are sufficiently strict, but a wide gap exists between the law and its implementation and enforcement. Although marine pollution is becoming a critical problem in these countries, industrial and economic growth remains the dominant national ethos.

Efforts to implement unfamiliar and untried regulations have had mixed results. In some cases, planners and policymakers have made *ad hoc* modifications of the traditional criteria for selecting development priorities and allocating funds because of the possibility, if not the reality, of environmental regulation. Consequently, organizational and institutional arrangements for managing marine environmental protection have become *ad hoc* and variable. This variability has, in turn, reduced governments' willingness and ability to understand and regulate marine pollution and resource conservation.

Uncertain jurisdiction due to the lack of agreed EEZ boundaries in North-East Asian seas complicates any effort towards co-operation, and perhaps necessitates the involvement of an international organization as an intermediary or co-ordinator. Only North Korea, Russia, and Taiwan have declared 200 nautical mile EEZs. China's unwillingness to specify the limits of its claims leads to uncertainty and tension. And this uncertainty is reinforced by Beijing's continuing stridency regarding coastal state enforcement rights. Some analysts of China's Marine Environment Protection Law suggest that the law's intent, under Article 41 and other sections, is to treat violations of it as a crime, punishable in criminal terms, although this may not be in keeping with international practice.[15]

Modelling of hypothetical oil spills from point sources in areas of active exploration in North-East Asia shows that spills could easily cross claimed maritime boundaries, and eventually impact on valuable and vulnerable marine resources: fisheries, coastal aquaculture, fragile wetlands, fish spawning grounds, and endangered species such as seabirds, whales, seals, and porpoises.[16] If jurisdiction is uncertain, the responsibility for cleanup and compensation may be as well. So a major transnational environmental disaster, such as an oil spill, could well bring into focus questions of jurisdiction and responsibility. Co-operation and a co-ordinated effort to clean up such a spill may be hampered. For example, the movement of personnel and equipment across a hypothetical median line without prior permission, even for the express purpose of combating marine pollution, could be constrained or even considered provocative.

South Korea and Japan have both established and used dump sites in areas of uncertain or overlapping jurisdiction in the Sea of Japan. Similarly, South Korea and China dump waste in the Yellow Sea. Furthermore, China is rapidly expanding its requirement for energy, with coal as a major source of fuel. Increased combustion of coal produces greater quantities of fly ash, which will require disposal. China is considering disposing of this fly ash, as well as calcium carbonate residue from the production of fertilizer, in the Yellow Sea. In these circumstances, agreement on dump sites and the co-operative monitoring thereof is advisable.

Increasing intra-regional trade presents new issues for regional environmental regulation.[17] On the one hand, co-operation tends to accelerate economic growth. But without environmental controls, economic growth increases resource depletion rates and generates ever more toxic industrial pollution, which enters the marine environment. But nations may be reluctant to raise environmental standards because they consider that raising standards will increase short-term production or resource extraction costs, thereby undermining their international competitiveness. Governments may even try to gain competitive advantage by seeking foreign investment through low or lax environmental regulations, creating so-called 'pollution' or 'resource extraction havens.' In North-East Asia, such a strategy may be especially attractive to nations seeking to lure Japanese or South Korean companies facing increasingly stringent domestic environmental regulations at home, or nations which need foreign investment to harvest their timber, mineral, and marine resources. Ironically, a patchwork of different national environmental standards and regulations will impede regional economic co-operation by increasing transaction costs.

The 'pollution/resource extraction haven' strategy has several other negative implications. First, if it is pursued by all the developing countries of North-East Asia, a 'vicious cycle' of standards-lowering competition could result in regional environmental degradation—including of the marine environment. As well as high long-term social and health costs, rapid resource depletion and ecological decline are likely to carry high opportunity costs. Second, companies and industries attracted by 'pollution havens' are likely to be low-growth 'sunset' industries which face a limited future. A development strategy based on non-dynamic companies is unlikely to bring the technology transfer and knowledge spillovers which are crucial to sustainable, self-generating economic growth. Third, products manufactured or extracted from 'pollution/resource extraction havens' may face import barriers in the increasingly environmentally and health conscious markets of the Organization for Economic Co-operation and Development (OECD). North-East Asian timber resources or polluted fish may be especially vulnerable to global campaigns by environmentalist groups such as Greenpeace.

In the region, there was, and still is, except in some polluted coastal areas, little public awareness of the importance of marine environmental protection. Governments still tend to see environmental problems as peripheral issues to be acknowledged but effectively ignored. Their attempts to draft regulations have invariably been hindered by the need to balance the interests of competing national and provincial level sectors, such as coastal and offshore shipping interests, fishing and fish processing enterprises, coastal inland development construction and water conservancy bureaucracies, port and harbour administrations, and agriculture and industrial ministries.

The main constraints to regional co-operation in marine environmental protection remain poor political relationships and environmental apathy. Transboundary pollution, co-ordination of regulations and their implementation, and a 'tragedy of the commons'[18] mentality are the main pressing issues. Two trends are apparent: increasing marine pollution with concomitant damage to living resources in semi-enclosed seas, especially in the Yellow Sea, and a growing environmental consciousness, which may spill over into the marine sphere. What is not clear is whether improving intra-regional relations and environmental consciousness will overtake and mitigate an environmentally damaging ethos before irreversible damage is done.

The Nuclear Waste Dumping Controversy[19]

The revelations of Russian and Japanese nuclear waste dumping in the Sea of Japan may prove to be the necessary trigger for North-East Asian regional co-operation on marine environmental protection. The news that the former Soviet navy has dumped eighteen decommissioned nuclear reactors and 13,150 containers of radioactive waste since 1978, most of it in the Sea of Japan, created an uproar in the world environmental community. It jolted nuclear-sensitive Japan and South Korea, in particular, and even drew a rare comment from North Korea. Adding fuel to the fire, a Russian naval vessel dumped nearly a thousand tons of low-level waste in the Sea of Japan shortly after Russian President Boris Yeltsin's visit to Japan in October 1993.

Despite protests by Japan, and both North and South Korea, Russia subsequently announced that it would have to continue dumping radioactive waste because it has no place to store it on land. Then, in a stunning case of the 'pot calling the kettle black', Japanese Science and Technology Agency Chief Satsuki Eda admitted that Tokyo Electric Power Co. dumps annually ten times more radioactive waste into the Sea of Japan than the 900 tons dumped by the Russian navy. This revelation prompted demonstrations in South Korea. Although most scientists agree that the dumped waste poses no immediate threat to the environment or humans, the longer term effects are unknown, particularly after the Russian containers corrode.

The initial report of Russian dumping prompted co-operation to deal with it at hastily arranged bilateral meetings of the relevant Russian and Japanese ministers and experts, proposals for joint South Korean, Japanese, and Russian surveys at specific dump sites, and a call by Japan for an international co-operative fund to help Russia treat its nuclear waste. North Korea even offered to host an international seminar on regimes for pollution control.[20] In March 1994 a joint expedition, involving Japan, South Korea, Russia, and the International Atomic Energy Agency (IAEA), began a search for signs of radioactive waste contamination in the Sea of Japan. The participants shared the costs of the expedition equally, and the scientists used a Russian vessel.[21] More recently, it has been revealed that Russian chemical munitions were also dumped in the Seas of Japan and Okhotsk until the mid-1980s.[22]

The interaction between Japan and Russia regarding Russia's dumping of radioactive waste is indicative of the general problems of co-ordination between the two nations. On 2 April 1993, when it was discovered that Russia was dumping nuclear waste, Japan officially protested to Moscow, and on 15 April, Foreign Minister Muto demanded of Russian Foreign Minister Kozyrev that it be stopped. Kozyrev suggested that the two countries launch a joint expedition to determine the extent of the pollution, and establish a joint working group to monitor the situation. But the details of his proposal were unclear. At the first meeting of the working group on 11 and 12 May 1993, Japan and Russia agreed to send an expedition in the summer or fall. It was delayed due to lack of preparation by Russia. On 13 October 1993 President Yeltsin visited Japan and agreed to a joint expedition by the end of 1993 or early 1994. The expedition was duly launched on 18 March 1994 and included South Korea and the International Atomic Energy Agency as well. Their participation accorded with Japan's view that the issue was a regional question and part of a wider North-West Pacific problem.

In October 1993 Russia's dumping of more nuclear waste shortly after Yeltsin's visit was publicized by Greenpeace. Japan's Environment Agency was severely criticized for failing to know about the dumping in advance. Its immediate objective, to stop the next dumping, was achieved when the new Japanese Foreign Minister, Hata, called his Russian counterpart, Kozyrev, who suspended the dumping. More than twenty meetings have occurred since.

Japan's view is that Russia has primary responsibility for controlling its radioactive waste. But Tokyo does recognize Russia's current economic difficulties and is willing to extend assistance, *provided* it does not enhance the war fighting capability of Russia's Pacific fleet. Indeed, Japan has already contributed $77 million. But progress has not been as rapid as Tokyo hoped. The original agreement was that Russia, Japan, and South Korea should be equal partners in the investigation and monitoring of the dumped material. Japan would bear two-thirds of the cost and South Korea and Russia one-third. In the end, Russia supplied the ship and staff at the expense of Japan and South Korea.

Besides the immediate stopping of the ocean dumping of nuclear waste, Japan's objectives have expanded to conducting research experiments on the waste, and to finding a way to permanently resolve the dumping issue. Tokyo asked Russia to make a concrete proposal to meet these objectives, and if it did, promised to contribute financial and technical assistance.

Several problems arose. Russia was apparently not willing or able to formulate a concrete proposal. And it kept changing its position. Moscow initially wanted a 10,000-ton chemical tanker within four weeks to store the waste. With great difficulty, Japan found one. But Russia turned it down, arguing that the vessel was not made to store radioactive waste or to withstand ice conditions (although Vladivostok harbour does not freeze). Moscow asked instead for a land-based waste processing plant. Japan agreed to supply the technology. But when Japan tried to clarify the project and co-ordinate it with the local government concerned, Moscow said that it, *and it alone*, was mandated to negotiate with Japan.

When Japan attempted to sign the agreement on 6 April 1994, Moscow said, incredibly, that it had to co-ordinate with the local government after all! By then it was clear that the central government could not deal with local government opposition, and that any successful co-operation agreement would have to be with both the local government and the central government. But this was not the end. Russia changed its position yet again, arguing that its citizens opposed a land-based facility. Bringing the negotiations full circle, Moscow asked to be supplied with a floating nuclear waste processing facility instead. A conceptual design for a floating plant was eventually completed and the two countries established a joint secretariat to implement the project.

Japan's present priorities are to: support the Monaco Institute which provides technical backup to the IAEA; monitor dumped waste; and prevent dumping of waste in the Sea of Japan and the North-West Pacific, including off Sakhalin and Kamchatka, because currents there may bring the pollutants to Japan. Tokyo believes the co-operative effort should not be limited to the Sea of Japan but extend to the North-West Pacific and involve the United States and Canada, both of which could eventually be affected by the pollution. Japan's overall policy is to stimulate and use the weight of world opinion to persuade Russia to stop the dumping. So it has tried to internationalize the issue, including a pertinent clause in the G7+2 economic summit (discussions among the industrialized economies plus the European Union and the Russian Federation). Even so, negotiations have been bilateral so far. South Korea has been encouraged to participate, and did participate in the waste dumping expedition. North Korea has not shown much concrete interest in joint co-operation but has asked for compensation from Russia. Moscow refused, saying there was no proof of damage.

Japan also believes resolution of the waste dumping problem is the first step towards including the North-West Pacific, and ultimately, the global ocean in a ban on the dumping of radioactive waste. It would like to persuade Russia to remove its opposition to the amendment to the London Dumping Convention (LDC). Japan has concluded that Russia can be moved by adverse publicity, and that it does need help and will use it if appropriately offered.

In summary, Japan considers that the dumping, while catastrophic, has had some benefits. It has led to the first ever government to government co-operation among Russia, Japan, and South Korea, and it could form the basis for broader co-operation in marine environmental protection and more. By contributing financially and keeping its word, Japan has increased the confidence of the local Russian government in its neighbours, an important result since the central government contributed very little to the process. One development which will not help this process, though, is the Russian Pacific fleet's plan to jettison 600 tons of decommissioned ammunition in the Sea of Japan. And in July 1995, Russia announced it may be forced to renew dumping of liquid radioactive waste into the Sea of Japan.[23]

Existing Regimes and their Inadequacies

Regional law drafting and policy development in North-East Asia is quite limited, and is mainly a response to the International Maritime Organization (IMO), of which all North-East Asian governments except Taiwan are members, and to initiatives related to the Law of the Sea. Even then, the extent and level of participation and implementation reflects countries' varied domestic interests and priorities.

Thirteen IMO treaties focus specifically on pollution prevention from ships (Table 5.1). Russia and Japan have subscribed to the

Table 5.1 IMO Pollution Convention Signatures

Convention	China	DPRK	Japan	ROK	Russia
Convention for the Prevention of Pollution from Ships, Annex 1/2	X	X	X	X	X*
Convention for the Prevention of Pollution from Ships, Annex 3		X	X		X
Convention for the Prevention of Pollution from Ships, Annex 4		X	X		X
Convention for the Prevention of Pollution from Ships, Annex 5		X	X		X
Convention on the Prevention of Marine Pollution by Dumping	X		X		X
International Convention Relating to Intervention on the High Seas in Cases of Oil Pollution Casualties, 1969			X		X
International Convention Relating to Intervention on the High Seas in Cases of Oil Pollution Casualties, 1973 Protocol					
International Convention on Civil Liability for Oil Pollution Damage, 1969	X		X	X	X
International Convention on Civil Liability for Oil Pollution Damage, 1976 Protocol	X				X
International Convention on Civil Liability for Oil Pollution Damage, 1984 Protocol	X				
Convention on the Establishment of an International Fund for Compensation for Oil Pollution Damage, 1971			X		X
Convention on the Establishment of an					

Table 5.1 *Continued*

International Fund for Compensation for Oil Pollution Damage, 1976 Protocol
Convention on the Establishment of an International Fund for Compensation for Oil Pollution Damage, 1984 Protocol
Convention Relating to Civil Liability in the Field of Maritime Carriage of Nuclear Material, 1971

* with exception of III, IV, V.

most pollution treaties, ten and eight, respectively. China has ratified five. South Korea has ratified only two treaties, the original Civil Liability Convention and the Convention for the Prevention of Pollution from Ships. North Korea has acceded to Annexes 3, 4, and 5 of the Prevention Convention, and all but South Korea have signed the Civil Liability Convention. Only Russia has joined the 1973 Intervention Convention. All six have acceded to the International Convention for the Prevention of Pollution from Ships, 1973, as modified by the Protocol of 1978 (MARPOL 73/78). For the two Koreas, accession perhaps reflected the dominance of shipping and shipbuilding interests, whereas for China, it signalled a desire to identify with international environmental and shipping interests.

China, Japan, and Russia are parties to the 1972 LDC, but in the latest vote banning at sea disposal of all nuclear wastes, both China and Russia abstained.[24] China subsequently declared it would adhere to the ban.[25] North Korea, South Korea, and Taiwan are not members. LDC members have an obligation to report ocean-dumping permit activity so that all concerned countries are aware of the nature and quantity of wastes that enter their shared waters. The signatories also approved permanent total bans on toxic industrial waste disposal at sea beginning in 1996. The ban includes the export of waste to non-signatory countries for ultimate disposal at sea, and the burning of waste at sea. Japan has said it will continue to dump industrial waste at sea[26] because a ban would severely impact upon Japanese industry. It is the largest marine dumper in the world, dumping 4.5 million tons per year in the Pacific Ocean and the Sea of Japan.

Prospects for improved transnational co-operation may depend on better understanding by both policymakers and the public of the impacts of pollution. Monitoring and research would provide this critical information. And the effective study of transboundary contamination requires excellent co-operation and synoptic sampling to enable integration of data across a region. All North-East Asian countries are members of the 10-nation Working Group for the Western Pacific (WESTPAC), which was established by UNESCO at its Tenth Assembly in 1977 to plan and co-ordinate multilateral ocean science programs. WESTPAC has focused on intercalibration exercises, with the collaboration of the Global Investigation of Pollution in the Marine Environment (GIPME) and the Intergovernmental Oceanographic Commission (IOC) Group of Experts on Methods, Standards, and Intercalibration (GEMSI).

The WESTPAC III meeting in September 1983 concluded that a major emphasis of initial WESTPAC program activities should address the overall need for training in bioindicator sampling techniques and contaminant analysis, particularly for organochlorine and hydrocarbons, and that national centres should be identified. The Program Group also recommended that the WESTPAC Task Team on Marine Pollution Research and Monitoring Using Commercially Exploited Shellfish as Determinants should participate actively in the IOC's Marine Pollution Monitoring Programs (MARPOLMON). But progress has been slow and sporadic. The organization involves scientists only, and some mutual suspicion still hinders data sharing. No real multilateral co-operative effort which involves all North-East Asian countries has yet been undertaken.

The UNCLOS creates an international umbrella framework for developing coherent national marine pollution policies.[27] It is generally accepted that Part XII of the UNCLOS—Protection and Preservation of the Marine Environment—is the strongest and most comprehensive global agreement ever negotiated on the marine environment. It is of major significance for at least two reasons. First, it states that all countries have a general obligation to protect and preserve the marine environment, as well as a duty to enforce international regulations to protect the marine environment from all sources of pollution. Second, it places signatories under an obligation to enforce generally accepted international rules and standards established in other maritime conventions, such as those negotiated under the auspices of the IMO, even if they are not parties to them.

Section One of Part XII of the UNCLOS obligates nations to take all necessary measures consistent with the Convention to prevent, reduce, and control pollution of the marine environment from any source. This obligation is complemented by Section Five's call for the enactment of national legislation and regulations controlling specific sources: land-based, dumping, vessels, and seabed activity. Enforcement of marine pollution laws is dealt with in Section Six, which stipulates that coastal states are to enforce their land-based pollution laws against their own polluters. Coastal states are also made responsible for the protection of the marine environment out to the boundary of their EEZs, which can extend up to 200 nautical miles from baselines.

UNEP experts subsequently fashioned the Montreal Guidelines on land-based pollution (LBMP) in 1985 to help integrate the regional harmony called for in UNCLOS with the responsibility for preventing transnational pollution.[28] Its purpose is to serve as a checklist for regional conventions and national legislation. When viewed as a checklist, the core of the Guidelines lies in Guideline No. 13 on the development of control strategies, and No. 16 on the adoption of national laws and procedures. Although the body of the Guidelines appears to be softened by compromise, the scientific recommendations contained in the annexes bolster its credibility. Despite its weaknesses, the Montreal Guidelines could be helpful to the nations of North-East Asia, which have yet to reach a regional agreement, and which continue to look for guidance in refining their LBMP laws and regulations.

The UNCLOS also provides a framework for countries wishing to build co-operative marine environmental protection regimes. It calls on states bordering semi-enclosed seas to harmonize their policies regarding protection of the marine environment. The countries in the North-East Asian region have similar wastes and, other than Japan, a similar level of technology for disposing of them. Theoretically, they might adopt similar or uniform standards. The fact that they do not reflects both a lack of communication and real differences in national priorities for environmental protection in general, and for specific pollutants and pollutant sources in particular (see Tables 5.2 and 5.3). Although they all have standards for effluent and water quality, the strictness of those standards differs widely. Russia's water quality and effluent standards are, on paper, generally much stricter than those of its neighbours. Taiwan's effluent standards are considerably lower than those of Japan, and China's water quality standards are the most relaxed of all. Of course, enforcement of these standards is another matter and, other than in Japan, is generally quite lax.

Moreover, countries regulate different substances and use different policy instruments. Such differences are consistent with the Convention, since it provides that states 'shall use the best practicable means at their disposal and within their capabilities to prevent, reduce and control pollution'.[29] But it is expected that the countries with strong environmental standards will try to encourage other countries to adopt similar standards, and thus reduce any competitive disadvantage to their own industries.[30]

Now that UNCLOS is in force, the region's countries must decide how to adjust their national initiatives so that they are compatible with emerging international legal and technical obligations or, conversely, the extent to which each of them wishes to ignore or deviate from international practice. The basic need to draft national regulations that reflect and incorporate the vaguely defined intent of UNCLOS Articles 192 and 194 will require that choice to be made. These articles charge states with the 'duty to protect and preserve the environment' and obligate them 'to take all measures necessary to prevent, reduce, and control marine pollution and to ensure that activities under their jurisdiction or control do not cause pollution damage to other states or otherwise spread beyond the seas where they exercise sovereign rights'.[31] Yet, there are no agreed scientific criteria to determine the precise meaning of such terms as 'prevent, reduce, and control'. It is also difficult to determine how to justify and enforce legal prescriptions, given the limitations of scientific and technical knowledge. And there is a large gap between acceptance of a vaguely defined legal framework, which moves from 'obligations of responsibility' to 'obligations of regulation and control', and the willingness and ability of states to establish and enforce standards and rules.

Potential controversy for the region is also hidden in some UNCLOS articles. Article 65 allows a coastal state or a competent international organization to regulate and limit exploitation of marine mammals more strictly than provided for within Part V of the Convention. States are to co-operate with a view to conserva-

Table 5.2 Water Quality Standards in Northeast Asia

	Japan A[1], B[2], C[3]	Taiwan A[5], B[6], C[7]	China A[8], B[9], C[10]	South Korea A[11], B[12], C[13]	Russia A[14], B[15]
T°C			<4°C above ambient		<5°C above ambient
pH	7.8–8.3	7.5–8.5; 7.5–8.5; 7.0–8.5	7.5–8.4; 7.3–8.8; 6.5–9.0	7.3–8.3; 6.5–8.5; 6.5–8.5	6.5, 8.5
COD	<1mg/l 2, 3, 8	(BOD) 2, 3, 6	<1mg/l 3, 4, 5	1.0, 2.0, 4.0	2.0
DO	>1mg/l 7.5, 5, 2	>1mg/l 5.0, 5.0, 2.0	5.0, 4.0, 3.0	>95, >85, >80	6.0, 4.0
Coliform	<1,000 MPN[4]/100ml	(A only) <1,000 MPN/100ml	10,000/1 (700 for shellfish culture)	<200 <1,000	
N-hexane extracts	0				
Phenols		0.01	0.005, 0.01, 0.05		0.001[15]
Mineral oil and fat		0.2, 2, —	0.05, 0.01, 0.05		
Heavy metals		(mg/l)			
Arsenic		0.05	0.05, 0.1, 0.1	0.005	0.05
Cadmium		0.01	0.005, 0.01, 0.01	0.01	0.005
Chromium		0.05	0.1, 0.5, 0.5	0.05	0.01
Copper		0.02	0.01, 0.1, 0.1		0.05

Parameter						
Cyanide	0.02	0.02	0.1	0.5		0.1
Lead	0.1	0.05	0.01	0.1		
Mercury	0.002	0.0005	0.001	0.001		0.0001
Selenium	0.05				0.01	
Silica		0.01	0.02	0.03		
Sulfide						
Zinc	0.04	0.1	1.0	1.0		0.01
Agricultural chemicals						
Organic phosphorous	0.1					0.0
Inorganic phosphorous		0.015	0.03	0.045		
Organic chloride		0.001	0.02	0.04		
Inorganic nitrogen		0.01	0.2	0.3		
DDT, DDD, DDE	0.001					0.1
Heptachlor and heptachlor epoxide	0.001					0.1
Suspended sediment					<10; <25; —	0.05
Floating material		No surface oil, froth, other				No surface oil, froth, other
Colour, odour, smell		Normal				

Notes.

1. Fishery, class 1: bathing, conservation of natural environment, and uses listed in B and C. Fishery, class 1: for aquatic life such as red sea bream, yellowtail, seaweed, and those of fishery, class 2.

Table 5.2 *Continued*

2. Fishery, class 2: industrial water, and uses listed in C. Fishery, class 1: for aquatic life such as grey mullet and laver.
3. Conservation of the environment: up to the limits at which no unpleasantness is caused to people in their daily life (including a walk by the shore).
4. With regard to the quality of fishery, class 1, for planting oysters, the number of coliform groups shall be less than 70 MPN/100 ml.
5. First class water for aquaculture and swimming.
6. Second class water for aquaculture, industry, and environmental protection.
7. Water which has the lowest standards for environmental protection.
8. First-class waters (>10mg/l from artificial sources).
9. Second-class waters (>50mg/l from artificial sources).
10. Third-class waters (>150mg/l from artificial sources).
11. Fisheries habitat.
12. Beaches.
13. Industrial area.
14. Sanitary protection zone I (for sewage release; 2–7 miles).
15. Sanitary protection zone II (for pollution from vessels and offshore installations; 2–12 miles).

Sources: Marine Environment Protection Policy Analysis, Bureau of Environmental Protection, Taiwan, October 1988; Japan Law No. 136 of 1970, amended by Law No. 137 of 1970; *Marine Protection Law of the People's Republic of China*, adopted at the 24th Meeting of the Standing Committee of the 5th National People's Congress on 23 August 1982, effective 1 March 1983; Republic of Korea, Law No. 3078, 31 December 1977; Hong Seong-Yong, *Marine policies toward the 21st century: world trends and Korean perspectives*, Occasional Paper No. 1, Ocean Industry and Policy Division, Korea Ocean Research and Development Institute p. 232; USSR Ministry of Water Resources, *Regulations Protecting Littoral Sea Waters from Pollution*, Moscow, 1984.

Table 5.3 Effluent Standards in Northeast Asia (mg/l)

	China	Japan	Taiwan A, B, C,	South Korea	Russia
pH		5–9	5–9		—
BOD	<50[1]	av. 160 / 120	400–600–1,200	50–300	—
COD	10	av. 160 / 120	600–900–1,800		—
Organic phosphorus suspended solids	<150[1]	1.0 / av. 200 / 150	400–600–1,200	50–300	—
Coliform bacteria	<250/ml[1] / <1,000/ml[2]	3,000 mps/100 ml	200,000–300,000		
N-hexane extracts		5 mineral oil / 30 animal and vegetable oil			
Total oil and fat			100	70 ppm	—
Phenols	0.5	5	5		P
Arsenic		0.5	5		$As^3 + 0.01$
Cadmium	0.1	0.1	1		—
Chromium (total)		2	5		—
Chromium (hexavalent)		0.5			—

Table 5.3 Continued

Copper	0.03		3	2-2-6	0.001
Cyanide			1.0	5-10-10	P, 0.05
Fluorine			15		—
Dissolved iron			10		$F^3 + 0.05$
Lead	0.1		1.0	10	$Fb^2 + 0.01$
Dissolved manganese			10		—
Mercury (total)	0.001		0.005	0.2	$Pmg^2 +$ and substances containing it
Mercury (alkyl)			0		—
Nickel			3		$Ni^2 + 0.01$
Zinc	1.0		5	4	$Zn^2 + 0.01$
PCB			0.003		—
Change to C				1, 1, —	
Oil and grease					

1. Sewage released from ships within four miles of the coast.
2. Between 4 and 12 miles from the coast.

Sources: Japan Law No. 136 of 1970, amended by Law No. 137 of 1970; *Marine Environment Protection Policy Analysis*, Bureau of Environmental Protection, Taiwan, October 1988; Republic of Korea, Law No. 3078, 31 December 1977; USSR Ministry of Water Resources, *Regulations Protecting Littoral Sea Waters from Pollutants*, Moscow, 1984; People's Republic of China, 'Regulations on Protection of Marine Environment from Damage by Land-Based Pollutants', in Shi E-hou et al., 'Review of China's marine environmental protection in the past 15 years and future prospects', *Marine Environmental Science*, 6, 2 (1987): 1–7.

tion, and 'in the case of cetaceans shall in particular work through the appropriate international organizations for their conservation, management and study,' that is, the International Whaling Commission. Japan, however, believes that small cetaceans are not under the jurisdiction of the IWC.

Unfortunately, there is a long list of institutional problems regarding protection of the environment in the region. These include[32] lack of lateral co-ordination within and between environment and resource management agencies; lack of integration between economic and environmental policies and planning processes; inadequate authority and budgets for environment agencies to ensure sufficient weight in decision making processes; lack of environmental information; undeveloped managerial and administrative resources at all levels of government; low levels of public awareness and underfunded and understaffed environmental education programs; short-term horizons which drive political and organizational decisions; outdated provisions in existing environmental laws and gaps on key issues such as solid hazardous wastes and radioactive pollution; slow enactment of domestic legislation to implement international treaty commitments; and inefficient markets, prices that do not fully reflect environmental costs and benefits, and uncertain ownership or control of natural resources.

On the other hand, regional co-operation can bring economic benefits from: knowledge spillovers and accelerated learning curves; economies of scale in data collection and information management, including storage and dissemination; economies of scale in scientific, managerial and administrative training; better and cheaper enforcement mechanisms; economies of agglomeration (the creation of one or more centres or fora for regional environmental management) including knowledge spillovers, reduced transport costs, and cheaper inputs; reduced transactions costs of trade and investment stemming from a common environmental regulatory framework; resource pooling; and elimination of regional standards-lowering competition.

Current Initiatives

In the past, environmental quality has been balanced or traded off against economic growth.[33] New thinking holds that environmental and developmental goals are compatible, and should be integrated whenever possible. This concept is called 'economically sound and sustainable development', and it underlies the fundamental consensus achieved at the 1992 Earth Summit, especially in the summit's Agenda 21 and Rio Declarations. Current North-East Asian initiatives in environmental management are summarized in Table 5.4.

General Environmental Protection Initiatives

The first North-East Asian Conference on the Environment was organized jointly by Japan's Environment Agency and its Ministry of Foreign Affairs, and held in Niigata, Japan in October 1992.[34] Delegations from China, Russia, Mongolia, and South Korea attended, along with UNEP, UNDP, and ESCAP. The conference was primarily a meeting of the representatives of environmental ministries and aimed at developing co-operation among them. It sought to promote a frank policy dialogue on environmental problems 'of common concern to the region as a whole.' Reports were received on the state of the environment and the status of measures to protect it. The goal was to convene the meeting annually, and eventually to upgrade it to ministerial-level gatherings similar to Europe's environmental ministers' conference.

The Niigata Conference reached the following specific conclusions:
1. Co-operation in environmental protection should be strengthened and broadened to include the public and private sectors as well as NGOs and grass-roots organizations.
2. There should be regular exchange of information, experience, and expertise leading eventually to a policy dialogue on environmental problems of common concern to the region as a whole.
3. The North-East Asian Conference on Environmental Co-operation should be held on an annual basis, preferably hosted each time by a different country of the region.
4. Active participation of local governments should be sought in the process of planning for and implementation of a co-operative program.
5. Regional co-operation might initially be focused on selected priority areas in which activities have already been initiated, or those which require urgent action.

Table 5.4 Current North-East Asian Regional Initiatives in Environmental Management

Project	Objectives	Participating Countries	Milestones
NOWPAP, UNEP*	Preservation of the ocean environment in North-East Asia	North Korea, South Korea, Japan, China, Russia	First experts meeting (Vladivostok, October 1991) Second experts meeting (Beijing, October 1992) Third experts meeting (Bangkok, November 1993) Agreement on principles of the projects First intergovernmental meeting (Seoul, September 1994)
Senior Officials Meeting on Environmental Co-operation in North-East Asia*	Discuss intergovernmental co-operation on environmental protection	China, Japan, South Korea, Russia	First meeting (Seoul, February 1993) Second meeting (Beijing, November 1994)
North-East Asian Conference on Environmental Co-operation*	Assess each country's environmental situation and promote regional co-operation	South Korea, Japan, China, Russia, Mongolia	First meeting (Niigata, October 1992) Second meeting (Seoul, September 1993) Third meeting (Kinosaki, September 1994) Fourth meeting (Pusan, Fall 1995)

Table 5.4 Continued

UNDP Northeast Asia Environmental Project	Development of technology for prevention of air pollution and for renewable energy	South Korea, North Korea, China, Mongolia	UNDP North-East Asia Environmental Meeting (Ulan Bator, July 1991) Project suspended
UNCED** and Regional Environmental Co-operation	Implementation of UNCED Discuss regional environmental problems and co-operation	South Korea, China, Japan, Russia, Mongolia	Seoul, September 1992 Established informal North-East Asia network (Irkutsk, August 1993) Kyoto, November 1994
UNDP Tumen River Area Development Program — Environmental Component	Assess environmental impact of Tumen River Project	China, Mongolia, North Korea, South Korea, Russia	Resource, Industry and Environment Working Group meeting (Helsinki, March 1993) Preliminary Environmental Assessment Workshop (Beijing, April 1994)

* intergovernmental
** non-governmental

6. Suggested modalities of regional co-operation included an information sharing and exchange network, and collaborative research and training, including joint surveys and monitoring activities on acid rain, coastal and inland water pollution, and biodiversity.
7. A particularly keen interest was expressed in case studies on the use of economic instruments for environmental management.
8. Specific financial and institutional mechanisms for promoting these activities were to worked out at the next North-East Asian Conference on Environmental Co-operation.
9. The United Nations agencies should play an active role in co-ordinating the efforts and activities of the participating countries as well as of the United Nations system as a whole.

A second meeting was held in September 1993 in Seoul, and a third in September 1994 in Kinosaki, Japan. The latter meeting called for: integrated regional and national strategies for achieving sustainable development; a comprehensive and integrated approach to the development of sustainable cities; emphasis on action at the regional level and increased public awareness of the importance of biological diversity; promotion of research as well as exchange of information and co-ordination of research programs to maintain biological diversity; and co-operation between all levels of society to ensure the effectiveness of efforts to achieve sustainable development.[35] A fourth meeting was held in Pusan in September 1995.

A Meeting of Senior Officials on Environment Co-operation in North-East Asia, organized by ESCAP in conjunction with UNEP and UNDP, took place in Seoul in February 1993, and was attended by representatives of the foreign and environmental ministries of China, Japan, Mongolia, South Korea, and Russia, along with representatives of UNEP, UNDP, ESCAP and ADB.[36] Its objective was to try to co-ordinate the evolving separate environmental initiatives in the region. The participants considered a consultant's report which listed possible areas of collaboration, and emphasized energy-related air pollution and capacity building as important cross sectoral themes. They also suggested concentrating on only one or two substantive issues at the outset in order to demonstrate the utility of co-operation, and then expanding these activities incrementally. Although they cautioned against an overly ambitious program, the participants recognized that identifying priority areas also necessitated the adoption of an overall strategy for regional environmental co-operation, and a support arrangement.

The following priority areas were adopted within which specific projects for regional co-operation could be developed: energy and air pollution; capacity building; ecosystem management, in particular deforestation and desertification; and intercalibration of pollution measurement equipment. Although Japan wanted to add marine pollution to the priority list, the meeting concluded that coastal and marine pollution issues should be addressed within the UNEP/NOWPAP framework.

The Seoul meeting also requested administrative, technological, and financial assistance from ESCAP, UNDP, UNEP, and ADB. ESCAP agreed to function as a temporary co-ordinating office with financial assistance from UNDP. The next steps are to agree on the contents of the projects for each country, and to obtain financial assistance from international organizations.

This 'North-East Asia Regional Environment Programme' has several areas of emphasis:

1. Energy and air pollution, specifically:
   - the development of clean coal technology and clean fuel,
   - joint research on air pollution,
   - joint monitoring (although China objects), and
   - a joint plan for reduction of air pollution.
2. Preservation of the ecosystem:
   - organization of a North-East Asian network for nature-preservation, including exchange of information and materials and a joint conservation plan,
   - co-operation in research on migratory birds, and
   - joint research on ecosystems.
3. Technology advancement:
   - standardization of measurements and methods, and
   - training and development of programs through UNDP's Environmental Technology Centres in Osaka and Shiga, Japan.

The Seoul meeting concluded that intergovernmental initiatives were superior to those of international organizations because the latter take too long and are less efficient.

ESCAP's future plans include:

1. Preparation of a list of projects for priority co-operation, and the submission of materials and reports on current situations for the next ESCAP meeting. The document is to include the

detailed content of co-operative projects and a list of projects in the three agreed priority areas: energy and air, preservation of ecosystems, and technological advancement;
2. Co-operation among environmental authorities, preparation for the North-East Asian environmental officials meeting, including exchange of information and manpower, and a dialogue regarding policies; and preparation for the meeting through organizing a preliminary committee. The issues to be discussed will include: economic means for executing environmental policies; comparison and transferability of measurement and methodology to prevent pollution; development of training and education programs for environmental specialists; and standardization of each country's definitions and perceptions on various poisonous chemicals.
3. Consolidation of bilateral co-operation.

The Seoul meeting also produced a plan for consolidating co-operation in environmental protection in North-East Asia. It agreed that elements of a North-East Asia Regional Environment Plan must include recognition of each country's capacity for exchange of information and materials, as well as for providing and receiving technology assistance; recognition of the transfer of air pollutants; recognition of the problem of land-based ocean pollutants; and the willingness of international organizations such as UNDP and ESCAP to help overcome political impediments among some countries in the region and to provide the necessary financial assistance. The objectives should be to develop and increase the exchange of information, policies, and personnel in the short term, and in the long term, to develop a standardized system of environmental measurement and joint research and, eventually, an organization for protection of the North-East Asian environment.

At a second meeting of senior environment officials in Beijing in November 1994, government officials from North Korea participated for the first time.[37] This meeting recommended the following project proposals: regional biodiversity management, regional seed research and information bases for forests and grasslands, and environmental pollution data collection, intercalibration, standardization, and analysis.

ECO ASIA 94, an environment congress for the Asia-Pacific region, was hosted by Japan's Environment Agency in June 1994 as a follow up to the ESCAP initiative. The meeting focused on future co-operation strategies; regional initiatives for sustainable development, and the role of the Asia-Pacific region in implementing agreements adopted at UNCED; measures to promote regional co-operation; and a long-term environment plan initiated by Japan's Environment Agency at a previous conference. This Long-Term Perspective on Environment and Development in the Asia-Pacific Region requires countries to draw up policy options according to simulations of economic growth to the year 2025 and the resulting impact on the environment.[38]

The Asia Foundation/NGO North-East Asian Environment Programme arose out of a symposium held in Seoul in September 1992 that supported the development of an informal environmental network. It was preceded by a joint memorandum of understanding between Russia and South Korea, which called for the creation of a regional environmental forum, and stimulated by the growth of NGOs in North-East Asia.[39]

The meeting of the Second International Symposium on Environmental Co-operation was held near Irkutsk between 17–20 August 1993. The agenda included environmental co-operation in areas such as transfer of air-borne pollutants; efficiency of economic inducements in environmental policies and experience in executing such policies; policies regarding prevention of ocean pollution; environmental conditions, rules, and policies of each country; and future plans on how to proceed. The Irkutsk meeting further hoped to activate a 'North-East Asia Environment Co-operation Informal Network'; consolidate co-operation with the private sector; and support activities of the temporary ESCAP Committee on Environmental Science and Research.

In addition, the Irkutsk meeting created the North-East Asia and North Pacific Environmental Forum, which concentrates on developing co-operation among NGOs. The Forum provides a mechanism whereby *people* in the region can exchange ideas and information, enhance the public's awareness of environmental issues, promote dialogue and co-operation among governments and NGOs, support surveys and joint projects, and develop a method for fostering the Forum's work. The Forum last met in Kyoto, Japan, in November 1994 to focus on grassroots activities to stop environmental destruction.[40]

The environmental component of the UNDP's Tumen River Area Development Programme is perhaps the most advanced of the several regional environmental activities, and may establish important legal and political precedents for other regional environmental agreements. This mammoth undertaking will involve industries that pollute heavily, such as the preprocessing of minerals and timber using coal-fired energy. If the project is to receive seed financing from the ADB or the World Bank, it must undertake extensive environmental impact assessments and be so designed as to mitigate significant impacts.

In October 1992 a preliminary environmental assessment of the proposed Tumen River scheme was presented to the Program Management Committee's second meeting. The report stated that the hinterland, deltaic, and adjacent coastal areas were ecologically fragile, and noted the paucity of environmental and resource data for the area.

In May 1993 the third meeting of the Program Management Committee reviewed a draft set of 'Environmental Principles' with the following objectives in mind:

- the achievement of 'environmentally sound and sustainable development' in accordance with UNCED, international environmental law and agreements, and multilateral donor requirements;
- co-operation and co-ordination of the relevant governments on environmental concerns and on their preparation of impact assessments of projects on national territory; co-ordination of environmental protection projects developed within the zone by the Tumen River Development Corporation will be the responsibility of institutions responsible for implementing the scheme; and
- permission by member states for non-governmental organizations to participate in environmental assessment procedures.

The Global Legislators' Organization for a Balanced Environment is a group of legislators from Russia, Japan, the United States, and Europe who meet twice-yearly to co-ordinate their countries' environmental legislation. The last meeting was in Tokyo in August 1993. It is conceivable that participation in this Group could be expanded to include other North-East Asian countries or be a model for a separate initiative that does so.[41]

The APEC environment ministers met in Vancouver on 23 March 1994 to discuss a regional environmental strategy, and marine pollution was among the topics discussed.[42] The meeting issued two documents. The first was the APEC Environmental Vision

Statement, which urged APEC Senior Officials to develop multi-sectoral exchanges at the regional level, including the exploration of an Asia Pacific Round Table on the Environment and the Economy. The second document was a Framework of Principles for Integrating Economy and Environment in APEC, including sustainable development, internationalization of environmental costs, fostering of science and research, technology transfer, application of the precautionary approach, mutually supportive trade and environmental policies, environmental education and information to the public, financing for sustainable development, and the role of APEC itself. Environmental initiatives and objectives for APEC's working groups include technical exchange on red tide/toxic algae; compilation of national reports on land-based sources of pollution, coastal zone management policies, and responsible national agencies; and co-ordination and interaction with other organizations in the APEC region to pursue the UNCED Oceans chapter recommendations.

Initiatives for Marine Environmental Protection

Despite the relatively poor record of the region's entities in joining or adhering to international conventions protecting the marine environment, the new wave of environmentalism, combined with the muting of the Cold War in North-East Asia, has stimulated a proliferation of multilateral discussions and program proposals for environmental protection. But the motives and rationale for these new initiatives may be broader than concern for the environment. By calling attention to politically benign but mutually threatening environmental issues, states sometimes can achieve broader objectives. North Korea specifically sees co-operation on environmental issues as one way of reducing tension in the region. And the suggested emphasis of an Indian Ocean regional grouping is on environment and maritime issues as a precursor to more difficult issues like trade.[43] Although marine environmental protection is a minor peripheral issue in relations among the East Asian coastal states, negotiations or provisional agreement on environmental questions may permit parties to avoid more controversial issues such as boundary delimitation.

It can be argued that because the prevention of pollution of an international space such as the ocean cannot be adequately managed by the individual ocean users, herein lies an incentive for states to establish a regime of joint decision making to co-ordinate their restraint in the common use of the ocean.[44] New norms have been created and need to be operationalized:[45] the obligation to protect and preserve the marine environment, the duty not to transfer or transform pollution, and the human right to a quality marine environment. Other norms are evolving, such as the obligation of monitoring and environmental assessment, and of providing technical assistance to developing countries for the protection and preservation of the marine environment.

Recent Developments

The North-West Pacific Region Action Plan (NOWPAP)

Of the several ongoing multilateral co-operative efforts pertaining to marine environmental protection, the most advanced is the United Nations Environment Programme's (UNEP) NOWPAP, which has been undertaken as part of UNEP's Regional Seas Programme.[46] However, implementation is lagging well behind expectations. Globally, the UNEP has almost two decades of experience. Its programme encompasses thirteen regional seas and involves the participation of some 140 coastal countries and island states and territories. Nine 'action plans' are currently operational. Nine conventions and twenty-eight protocols have been signed, and seven conventions are in force.

The substantive aspect of any regional programme is outlined in an 'action plan' that is formally adopted by an intergovernmental meeting of the governments of a particular region before the programme becomes operational. In the preparatory phase leading to adoption, governments are consulted through a series of meetings and missions about the scope and substance of an action plan suitable for the region. In addition, reviews on the specific environmental problems of the region are prepared with the co-operation of appropriate global and regional organizations to assist governments in identifying the most urgent problems in the region and to set priorities for the activities outlined in the action plan. UNEP co-ordinates the process leading to the adoption of the action plan directly, or in some regions, indirectly through existing regional organizations.

All action plans have similar structures, although the specific activities for a particular region are dependent upon the needs and priorities of that region. An action plan usually includes the following components:

1. *Environmental assessment.* Environmental assessment is undertaken to assist national policy-makers to manage their natural resources in a more effective and sustainable manner, and to provide information on the effectiveness of legal/administrative measures taken to improve the quality of the environment. It involves the assessment and evaluation of the causes of environmental problems, as well as their magnitude and impact on the region. Emphasis is given to such activities as baseline studies; research and monitoring of the sources, levels, and effects of marine pollutants; ecosystem studies; studies of coastal and marine activities and social and economic factors that may influence, or may be influenced by, environmental degradation; and the survey of national environmental legislation.

2. *Environmental management.* Each regional programme embraces a wide range of environmental management activities. They include: co-operative regional projects on training in environmental impact assessment; management of coastal lagoons, estuaries, and mangrove ecosystems; control of industrial, agricultural, and domestic wastes; and formulation of contingency plans for dealing with pollution emergencies.

3. *Environmental legislation.* An umbrella regional convention, elaborated by specific technical protocols, most often provides the legal framework for co-operative regional and national actions. Legislation is particularly important because the legal commitment of governments clearly expresses their political will to manage individually and jointly their common environmental problems.

4. *Institutional arrangements.* When adopting an action plan, governments agree on an organization to act as its permanent or interim secretariat. Governments are also expected to decide on the frequency of intergovernmental meetings to review the progress of the agreed workplan, and to approve new activities and necessary budgetary support.

5. *Financial arrangements.* UNEP, together with selected United Nations' and other organizations, provides 'seed money', or catalytic financing, in the early stages of regional programmes. However, as a programme develops, the governments of the region are expected progressively to assume

full financial responsibility. Government financing is usually channelled through special regional trust funds, to which governments make annual contributions. These funds are administered by the organization responsible for the secretariat functions of the action plan. In addition, governments contribute by supporting their national institutions participating in the programme, or by financing specific project activities.

China, South Korea, and Russia would like to see some form of regional co-operation in combating marine pollution problems. Japan originally supported the idea of NOWPAP in principle but felt that there was not a consensus on the issue, and that the lack of diplomatic relations between some of the countries would prevent a consensus emerging. It was probably also leery of having to provide the bulk of the financial backing. But Japan did support an initial meeting of experts in the region to discuss common problems, as a first step toward attaining a consensus.[47] Apparently UNEP itself was reluctant to act until all the relevant countries asked it to initiate a program. Moreover, it seemed that the UNEP Oceans and Coastal Areas Programme (OCA/PAC) was not fully supportive of the concept either, because of continuing uncertainty over the stability of Russia and its effects on political relations in the region, as well as the fact that Japan and South Korea were developed countries and should not need UNEP's help.

Nevertheless, at the Fifteenth Session of the UNEP Governing Council, held in Nairobi between 15–26 May 1989, the states of the North-West Pacific officially indicated that UNEP assistance would be welcome in resolving regional marine pollution problems.[48] Consequently, in its Decision 15/1, the UNEP Governing Council called for the preparation of new action plans for the seas not yet covered by the Regional Seas Programme. The North-West Pacific region was specifically mentioned as the region for which an action plan should be prepared. As a follow-up, a letter from UNEP's assistant executive director was sent to UNEP Focal Points (national, individual, or institutional co-ordinators) in China, Japan, North Korea, South Korea, and the then USSR to explore their governments' interest in the development of a regional action plan for the protection and development of the marine and coastal environment of the North-West Pacific. Reactions to UNEP's involvement in the development of NOWPAP were unanimously positive.

Next, the Oceans and Coastal Areas Programme Activity Centre of UNEP convened an informal meeting on NOWPAP in Nairobi on 29 May 1991. Representatives of China, North Korea, Japan, South Korea, and the then USSR participated. Different views on the geographical coverage and scope of the Action Plan indicated that further consultation among countries was necessary at this early stage. So a consultative meeting of Focal Points was convened in Vladivostok in October 1991 to consider the geographical coverage of the Action Plan, and to formulate a workplan and timetable for its development. The participants agreed on a workplan and timetable, as well as on the contents of national reports which should be submitted to UNEP.

The geographical coverage issue remained controversial. The majority considered that the Action Plan should initially cover the marine environment and the coastal areas of the Sea of Japan and the Yellow Sea without prejudice to its possible future extension to the East China Sea. On the other hand, the view was also expressed that geographical coverage should be considered and defined on the basis of further deliberations, including the possible content of the Action Plan. Although North Korea did not participate in the Vladivostok meeting, as a result of UNEP's entreaties, the General Bureau of Environment Protection and Land Administration of the DPRK was nominated as the National Focal Point for the development of NOWPAP.

The establishment of the following structures was suggested in the national reports submitted to UNEP:
- a regional co-ordinating centre (China);
- a regional centre on the monitoring and assessment of the state of marine environment (Russia);
- a regional centre for information and data exchange (South Korea); and
- a permanent task-force or group of experts from the riparian countries (China and South Korea).

The participants agreed that the national reports prepared by the National Focal Points for future meetings would cover the status of the marine environment and coastal areas; national policies and measures to deal with marine pollution; and proposals for steps to be taken in a regional Action Plan. They noted that regional co-operation in response to a pollution emergency would be appropriate for joint activities in the future.

A second meeting of National Focal Points was convened by UNEP in Beijing in October 1992 to consider the regional overview

and a draft Action Plan prepared by UNEP. The following information was to be included in the national reports prepared for this meeting:
1. the present state of the marine and coastal environment;
2. national policies, measures, and relevant activities dealing with marine pollution problems; and
3. proposals on ways and means to resolve the differences regarding the contents of the Action Plan.

Not all objectives were met. China, South Korea, and Russia submitted their reports to UNEP. Although North Korea was represented for the first time, it did not submit a useful report. The National Focal Point of Japan submitted only the first two parts of its national report, but not the last relating to proposals on ways and means to resolve the problems. Japan also insisted that the section of the Action Plan on Biodiversity and Ecological Resources be deleted except for the material on wetland reserves and genetic resources. And IMO declined an invitation to prepare a document on 'Legal implications of international conventions and a review of possible mechanisms for regional co-operation in combating marine pollution especially in the case of pollution emergencies'. The Third Meeting of Experts and National Focal Points on Development of the NOWPAP was held in Bangkok in October 1993.[49]

The overall goal of the NOWPAP is 'the wise use, development and management of the coastal and marine environment so as to obtain the utmost long-term benefits for the human populations of the region, while protecting human health, ecological integrity and the region's sustainability for future generations.' Subsidiary and complementary goals are:
- the control, halting, and prevention of any further degradation and deterioration of the coastal and marine environment and its resources;
- the recovery and rehabilitation of coastal and marine environments that have been degraded and which still have the potential for recovery; and
- the long-term sustainability of coastal and marine environmental quality and resources as assets for the present and future human populations of the region.

NOWPAP's general objectives include monitoring and assessment of the state of the regional marine environment, creation of an efficient and effective information base, integrated coastal area planning, integrated coastal area management, and establishment of a collaborative and co-operative legal framework.

In more specific terms, NOWPAP seeks to:
1. assess regional marine environmental conditions by coordinating and integrating monitoring and data gathering systems on a regional basis, making the best use of the expertise and facilities available within the region on a consistent and collective basis;
2. collate and record environmental data and information in order to form a comprehensive database and information management system which will serve as a repository of all relevant, available data, act as a sound basis for decisionmaking, and serve as a source of information and education for specialists, administrators, and others;
3. develop and adopt a harmonious approach towards coastal and marine environmental planning on an integrated basis, and in a pre-emptive, predictive, and precautionary manner;
4. develop and adopt a harmonious approach towards the integrated management of the coastal and marine environment and its resources, in a manner which combines protection, restoration, conservation, and sustainable use;
5. develop and adopt a regional framework of legislative and other agreements for mutual support in emergencies, collaboration in the management of contiguous bodies of water, and co-operation in the protection of common resources as well as in the prevention of coastal and marine pollution.

UNEP provided US$417,000 to cover some of the costs of the secretariat and the implementation of the Action Plan in 1994–5. The first projects have been approved. They include, in order of priority: (1) establishment of a comprehensive data base and information management system; (2) a survey of national environmental legislation, objectives, strategies, and policies; (3) establishment of a collaborative, regional monitoring program; (4) development of effective measures for regional co-operation in marine pollution preparedness and response; and (5) commencing the establishment of regional activity centres and their networks.

A NOWPAP Intergovernmental Forum is to be established to provide policy guidance and decisionmaking for the Action Plan, and will include representatives of relevant regional and international organizations. The NOWPAP states will continue to work towards the development of a regional convention for the protection and management of the coastal and marine environment and resources. A Regional Coordinating Unit (RCU) is to be established with the assistance of UNEP to ensure the integrated and managed execution from within the region of Action Plan projects. Until the RCU is set up, UNEP will co-ordinate projects and prepare a program based on regional government priorities. And most important to the implementation of the Plan, the regional governments must adopt this program and agree to establish a NOWPAP Trust Fund to finance its implementation.

The first NOWPAP Inter-Governmental meeting was held in Seoul in September 1994. It was considered critical both for continued progress on the Action Plan as well as regional co-operation in general. The fact that North Korea did not attend was a disappointment for UNEP. In addition to setting project priorities for 1994 to 1996, the meeting selected Japan to host the second NOWPAP Intergovernmental Meeting in 1996 and decided to implement a trust fund as soon as the amount available exceeds US$50,000.[50]

NOWPAP faces many problems. Some parties remain uncomfortable with the geographic definition of the region for co-operation. The present compromise definition is the Yellow Sea and the Sea of Japan, with provision for including adjacent areas in the future. This excludes Taiwan and hence the China/Taiwan problem. But the Sea of Japan and the Yellow Sea are quite different areas and have different problems. Moreover, South Korea and China prefer to focus initially on the Yellow Sea where they have already initiated a co-operative project. But Japan is more interested in the Sea of Japan where some co-operation is underway because of the Russian dumping of nuclear waste there. Moreover, different countries use different names for the seas, underscoring the political difficulties in the way of co-operative work.

Disagreement continues over the priorities of projects for co-operation. The present compromise is regional assessment, establishment of a data base, monitoring, and co-operation in emergencies. The differences in country priorities reflect different levels of economic development. China, North Korea, and Russia all welcome monitoring, but Japan is not as interested because it is already well established there.

Prevention of disposal of radioactive waste at sea was added as a NOWPAP task although Japan expressed reservations, presumably because it wishes to retain this option. There are also differences regarding the means of implementation of the projects. In one instance, Russia proposed the use of a Russian ship and North Korea agreed, but Japan and China were not interested in financing it. China and Japan want to use work already done or being done by the coastal states to develop a standard method. But South Korea is more interested in a contingency plan for emergencies, particularly for the Yellow Sea.

Harmonization of legislation and the ultimate goal of a regional convention also present technical and political problems. The coastal nations have different levels of economic development and different discharge standards. China emphasizes its industrial development, rather than pollution control. North Korea is reluctant to participate and thereby reveal the state of its environment. Moreover, the industrial structure and technology differ from country to country, as does the preferred use of maritime areas. And even if the water quality and discharge standards were to be made similar, enforcement standards would differ widely. In China, for example, the concentration is strictly controlled but not the total amount discharged. Japan would have difficulty signing a formal convention, presumably because it has poor relations with Russia and North Korea, and because it feels that adopting the Plan may obligate it to ratify all relevant international agreements. So Japan wants to separate the international legal agreement from the rest, and to make the regime *ad hoc*. For its part North Korea strongly suggested that such excuses should not be accepted as a means to avoid full participation in NOWPAP.

There is also disagreement over the all-important allocation of costs. Japan's Environment Agency does not have the resources to fund the program, which means that funds would have to come from the Ministries of Finance (MoF) and Foreign Affairs (MFA). Japan could provide funding as part of its ODA but Tokyo does not consider Russia a developing country, and it does not have political relations with North Korea. On top of all this, the MoF and MFA do not want to send representatives to UNEP meetings where they will be embarrassed by requests for money. Moreover from Japan's own point of view, Japanese have little expertise in foreign countries, do not generally speak foreign languages, and are familiar only with the problems of an advanced country.

Alternative models for allocation of costs might include an equal division, but this is not feasible for China, Russia, and North Korea. Contributions could be proportional to each country's UN dues, or most of the costs might be borne by Japan and South Korea. The latter option is complicated by rivalry between the two over leadership and the financial support of the regime. Japan wants to lead the initiative but South Korea feels it should be led by North and South Korea because of their geographic location bordering both seas.

UNEP's entire Regional Seas Programme approach has been criticized both generally and specifically. Some argue that the Programme needs a new perspective and restructuring.[51] It is linked to only one sector: prevention and control of marine pollution. But some critics believe that protection of the marine environment should be addressed in holistic terms, which would require an intersectoral approach and new leadership at the national and local levels. Another criticism is that the Programme has only infrequently yielded results of immediate practical value to decisionmakers. But action plans have been less successful where the scientific capability was not present or scientists lacked access to their government's decision-making processes. Indeed, after withdrawal of UNEP support, a region's momentum tends to persist only in areas with a strong indigenous scientific capability and network. The Mediterranean is a typical precedent in this regard. There, discrete networks of scientists interacted independently of one another, neither exchanging information nor co-ordinating their actions. As a result, environmental assessments were conducted in a vacuum.

Moreover, UNEP's approach generally caters only to environmental elites: the environmental ministries/agencies, academics, and NGOs. It needs to involve the private sector, unions, and the general public, and to address the ministries responsible for activities which generate pollution: energy, tourism, port authorities, public works, and agriculture. UNEP's regional conventions have rarely led to the desired national legislation and decision making. Instead a general malaise follows the withdrawal of UNEP support. Governments take several years to regularize their allocations to the trust funds, and support fades in terms of attendance at meetings, financial contributions, and level of activity. Institutional arrangements, then, face enormous problems. Nor is this all. Traditionally, there has been great competition and duplication of effort among the UN specialized agencies dealing with oceans. And financial support has not reached the expected level or degree of autonomy necessary to meet present and projected needs. Finally, the Regional Seas Programme was structured during the Cold War and it needs to be reconfigured according to new political and economic realities.

Nevertheless, the regional seas programmes have made some advances. They have educated regional elites on the need for a more comprehensive regional environmental policy. They have created or extended domestic constituencies for environmental protection. They have contributed to the body of international environmental law. They have enhanced knowledge of the quality of regional seas. And they have transferred marine science technology and knowledge to many developing countries, thereby increasing their capacity and confidence to participate in regional marine environmental protection regimes. It is this latter achievement that is likely to be most significant and long lasting.

The UNDP/GEF Program on Prevention and Management of Marine Pollution in East Asian Seas[52]

In response to a number of requests from East Asian nations regarding management of the marine environment, the Regional Programme Division of the United Nations Development Programme's Regional Bureau for Asia and Pacific is formulating a program entitled 'Prevention and Management of Marine Pollution in East Asian Seas', with support provided from the pilot phase of the Global Environment Facility. This Asia-Pacific Regional Environment Facility would be administered by the Asian Development Bank, with Japan as a major contributor. The participating countries are the ASEAN nations (the Philippines, Malaysia, Indonesia, Brunei Darussalam, Singapore, Thailand), China, North Korea, Vietnam, and Cambodia. The approved budget is US$8 million, with Australia contributing an additional A$5 million.

The long-term objective of the program is to support the efforts of the participating governments in the prevention, control, and management of marine pollution, at both the national and regional levels, on a long-term and self-reliant basis. The program concept currently consists of four main project areas, which are defined by the following objectives:

- to assist in the prevention, control, and management of marine pollution problems through proper assessment of the state of marine pollution, including the effects of marine,

coastal, and other land-based activities on biodiversity and environmental quality;
- to assist with the development of policies, plans, and programs on prevention, control, and management of marine pollution, including measures for their support and implementation at both the national and subregional levels;
- to strengthen national and subregional institutional infrastructures and implementing mechanisms, and upgrade technical skills and management capabilities on prevention/control of pollution, as well as management and enhancement of the marine environment; and
- to establish appropriate financial arrangements and/or mechanisms for the long-term sustainability and self-reliance of national and subregional efforts to protect marine environments.

North Korea subscribes to the objectives of the East Asian Seas Marine Pollution Program and intends to take part. It is particularly interested in participating in the proposed network of information management and marine pollution monitoring centres, and wants assistance to upgrade the equipment and facilities of the West Oceanographic Research Institute to enable it to do so.

Marine Scientific Research Initiatives

The Intergovernmental Oceanographic Commission (IOC) was established in 1960 as a functionally autonomous body within UNESCO with a mandate to organize basic oceanographic research. The IOC's Subcommission for the Western Pacific (WESTPAC) was set up in 1989.[53] Its Secretariat is to be established in Bangkok, which hosted the second session of the Commission in January 1993. The next session, planned for 1996, will probably be in Tokyo.

The goals of an IOC regional subcommission are to:
- define regional problems and develop marine scientific research programs;
- implement IOC global marine scientific research programs at a regional level;
- facilitate the regional exchange of scientific data, especially to developing countries; and
- identify training, education, and mutual assistance needs.

At its first meeting in Hangzhou, China in February 1990, WESTPAC identified nine projects as necessary to achieve these general objectives and incorporated them in a Medium Term Plan for the period between 1991 and 1995. The nine projects are:

- Ocean Science in Relation to Living Resources
  Toxic and anoxic phenomena associated with algal blooms (red tides)
  Recruitment of Penaeid Prawns in the Indo-Western Pacific
- Marine Pollution Research and Monitoring:
  1. Monitoring heavy metals and organochlorine pesticides using the Musselwatch program
  2. Assessment of river inputs to seas in the WESTPAC Region
- Ocean Dynamics and Climate
  Banding in *Porites* coral as a component of ocean climate studies
  Ocean dynamics in the north-west Pacific
  Continental shelf circulation in the western Pacific
- Ocean Science in Relation to Non-living Resources
  WESTPAC palaeogeographic map
  Margins of active plates

The North Pacific Marine Science Organization (PICES) has goals, objectives, interests, and scientific projects similar to some of those of NOWPAP.[54] These include data and information exchange, common assessment methodology for marine pollution, marine pollution monitoring techniques (such as mussel watch and sediment monitoring), land-based sources of pollution, fluxes and their impacts on the marine environment, cross-boundary transportation of contaminants from the North-West Pacific region to the open ocean, intercalibration exercises, and development of environmental criteria and standards. The PICES/Marine Environment Quality Scientific Committee will focus on algal blooms and priority chemical and biological contaminants in the whole North Pacific region. The PICES Third Annual Meeting was held in Japan in October 1994.

The Advisory Committee on Protection of the Sea (ACOPS) program in the North-West Pacific is an extension of its Arctic program, which was the subject of a conference held in Arkhangelsk, the Russian Federation, in July 1993.[55] Conclusions and recommendations adopted there were endorsed by the Ministerial meeting of the eight circumpolar countries (Russia, Canada,

United States, Iceland, Denmark, Norway, Sweden, and Finland) held in Nuuk, Greenland, in September 1993. The ACOPS 1994 program concentrated on the overlap between the Russian Far East and the North-west Pacific. The ACOPS conference to discuss these issues was held in August 1994 in Petropavlovsk, Kamchatka. Land-based sources of marine pollution were a priority topic.

With the assistance of the World Bank, both China and South Korea have drafted an Action Plan for Monitoring and Protection of the Yellow Sea Large Marine Ecosystem (YSLME). A conference on the protection of this was held at Qingdao, China, in June 1993. The large marine ecosystem is a new concept linking exploitation, development, and administrative efforts. It sets aside national rights and interests to consider the whole ecosystem of the Yellow Sea. The goal is to establish bilateral co-operation to protect the ecosystem and make possible sustainable utilization of the biologic resources of the Yellow Sea.

Problems and Inadequacies of Existing Regimes

There is considerable redundancy of activities in the programs conducted under the auspices of WESTPAC, UNDP/GEF, PICES, and NOWPAP. WESTPAC anticipates conducting training in the modelling of coastal circulation in order to predict and control accidental oil spills. It is also developing a WESTPAC Action Plan as a follow-up to UNCED. Both activities appear to be similar to ones contemplated by NOWPAP. But WESTPAC activities can also complement the national marine scientific and technological capabilities of North-East Asian states.

WESTPAC's SEAWATCH program which monitors the marine environment may be helpful in the implementation of NOWPAP. Work by North-East Asian WESTPAC members (which includes all six states that participate in NOWPAP) on continental shelf circulation, ocean dynamics, paleogeographic mapping, tectonics and coastal zones, and on musselwatch and harmful algal blooms, is either more active in that region than in East or Southeast Asia, or is implemented on a western Pacific-wide basis without subregional focus. The objectives of the UNDP/GEF Program also seem greatly to overlap those of the NOWPAP, and the Program includes North Korea and China in its terms of reference too. A mechanism similar to the Co-ordinating Body on the Seas of East

Asia (COBSEA) operative in South-East Asia may be needed to co-ordinate WESTPAC and UNDP/GEF activities with NOWPAP's.

Despite the numerous initiatives, it will clearly be far easier to implement environmental assessment, legislation, and institutional arrangements than a regional management and financial structure. The concept of the EEZ is not yet ingrained in the psyche of policymakers. Besides, the more obvious problems and the initial effects of new ones are most likely to arise in waters close to land, and national attention is therefore concentrated on protecting the health of the coastal waters rather than the offshore, especially in enclosed and semi-enclosed seas. And countries generally resist the involvement of other nations in their coastal waters, no matter how well-intentioned.

Despite efforts at national, regional, and international levels, the current sectoral and monodisciplinary approach to the multiple use of marine and coastal resources will not provide an effective framework for achieving sustainability. Aside from physical and ecological degradation of the coastal and near-shore zones, and of course, nuclear waste dumping, pollution from land-based sources is at present the single most important threat to the North-East Asian marine environment, contributing some 70 per cent of the pollution load of the oceans. Intensified human activities in the coastal zone there cannot be supported if the marine environment is considered as an 'infinite sink' or receptacle for wastes and an endless supply of free resources.

Prospects for improved transnational co-operation in resource development and use may depend, therefore, upon better understanding of the causes and consequences of marine pollution in both coastal and open-sea areas. Indeed, increased knowledge is extremely important to the creation of regimes, and accounts for the expansion and strengthening of marine pollution regimes worldwide.[56] The most successful efforts to deal with marine environmental problems appear to have been carefully nurtured, with simultaneous institution-building, scientific, and treaty-drafting activities at the regional level, but this can come about only with strong and sustained littoral state support and state or international organizational leadership. Environmental consciousness in the region must be raised further, new institutional arrangements developed, and new economic theory applied, incorporating environmental benefits and pollution costs. Laws must be harmonized and co-operative monitoring achieved, especially regarding future industrial development. Particular emphasis should be placed on ocean dumping, red tides, and the environmental hazards of nuclear power and dumped nuclear waste.

The North-East Asian countries have fundamental differences in their approaches to regional co-operation in environmental protection.[57] China believes co-operation in North-East Asia should focus on urgent issues: industrial pollution, soil erosion, desertification, decrease in agricultural output, marine pollution, and depletion of marine resources. It prefers an informal mechanism to facilitate periodic meetings and exchange of relevant information and personnel in: environmental management, legislation, pollution control, monitoring and data collection; resource accounting, pricing policy, and economic incentives; joint research on hazardous waste, acid rain, and environmental management; pilot projects on desulphurization in power plants, toxic and hazardous waste treatment facilities, the prevention and control of lake eutrophication; and the prevention of marine pollution. Beijing also believes that the developed countries in the region and international institutions should contribute technical and financial assistance to projects in these issue areas. Its position may reflect in part the fact that China is more an exporter than an importer of pollutants.

For this reason, Beijing is also quite sensitive about transboundary environmental issues. In fact, China opposes joint research on the monitoring of air pollution because WHO's GEMS initiative already has a program for that purpose. It also supports North Korea's participation in regional environmental programs.

Regarding the establishment of a forum for regional environmental co-operation, Japan prefers to start with an exchange of information and knowledge and then gradually to move to policy-oriented dialogue on common environmental concerns. It supports the establishment of a central secretariat to organize meetings, publish a newsletter, and administer subcommittees which would handle concrete issues. But Japan considers that the establishment of a framework for implementation of multilateral co-operation will take several years. Hence it feels that discussions on a new institutional mechanism are premature, and that it is preferable to implement joint programs. Japan has suggested four priority areas for programs: regional marine conservation, acid rain, and air and water pollution.

Tokyo opposes a regional forum composed solely of officials of environmental agencies, because it feels that economic ministries must also be involved. Furthermore, Japan is concerned that any regional environmental co-operative body must not become just another channel of assistance but rather, it should spawn concrete projects involving sharing of domestic experiences, monitoring of the regional state of the environment, and transfrontier pollution. Tokyo will seek to ensure that such projects do not duplicate Japan's existing bilateral and multilateral assistance projects, including the existing UNEP Environmental Technology Centers in Osaka and Shiga.

Russia clearly requires financial assistance to protect its environment. It prefers ecosystem management, and more practical and action-oriented co-operation programs.

South Korea emphasizes the necessity of regional co-operation for environmental protection. For Seoul, co-operation should include technical projects, as preferred by China, and environmental management projects, such as a joint survey of the state of the environment, as preferred by Japan. South Korea attempted to arbitrate between China and Japan by proposing a top priority project on energy and air pollution. It also supports a co-ordinating mechanism for environmental assessment and management which would channel financial assistance from UNDP and ADB, and institute regional projects. Seoul's priorities include transboundary air pollution, marine pollution, capacity building, technical co-operation, and waste management. It wants the many initiatives on environmental co-operation in North-East Asia (see Table 5.4) to be harmonized.

The most formal forum, the Senior Officials Meeting, revealed the sharpest differences. China opposes a focus on transboundary air pollution. Japan opposes new institutionalization of co-operation, and providing financial assistance to it. South Korea supports both approaches. These disparate positions may be explained in part by the perceptions that China is more an exporter than an importer of marine pollutants; Japan is neither except when oil is spilled from its tankers; and South Korea is perhaps more an importer than an exporter of pollutants. Russia certainly exports more marine pollutants than it receives, and North Korea probably does also. At this stage, then, less formal fora appear more efficient since they do not produce highly binding mechanisms. Although inconsistencies and overlaps exist, different fora may actually play complementary and reinforcing roles, and thus support a trend towards the establishment of an efficient co-operative regional mechanism.

Transnational Environmental Protection Approaches: The Example of the Yellow Sea

There are several important transnational environmental problems in the Yellow Sea that are not being adequately addressed by current regimes.
- Oil spills from drilling or tankers are a serious potential transnational marine pollution problem. By the year 2000, there will be over 10,000 tons of oil spilled annually on the Chinese side of the Yellow Sea.[58] In summer, an oil spill originating in Korean waters would spread to Chinese waters, while in winter an oil spill originating in Chinese waters would spread to Korean waters (Figures 5.1 and 5.2).[59] Despite this inevitability, the necessity of cleaning up oil spills and how much it will cost are still under discussion in both China and South Korea.

Protection of living resources from marine pollution is a transnational problem because spawning, breeding, and wintering grounds are distributed on both sides of the Yellow Sea, and fish migrate throughout the Sea. Marine pollution at the mouth of the Yalu River is a transnational problem because it is an international boundary and the River is already polluted by both oil and mercury.

China, South Korea, and North Korea could co-operate to control marine pollution in the Yellow Sea using the following approaches:

1. Setting uniform standards for seawater quality classification and concentrations of pollutants in effluent (Tables 5.2 and 5.3).

The seawater quality standards set by South Korea are slightly different from those of China. South Korean standards do not allow mercury at all in any class of water, while Chinese standards set limits for mercury concentrations in waters of different classes.

A positive basis for co-operation is the fact that China and South Korea, and indeed, all North-East Asian entities except possibly North Korea, use both effluent and water quality standards, thus avoiding the argument of whether pollution is the

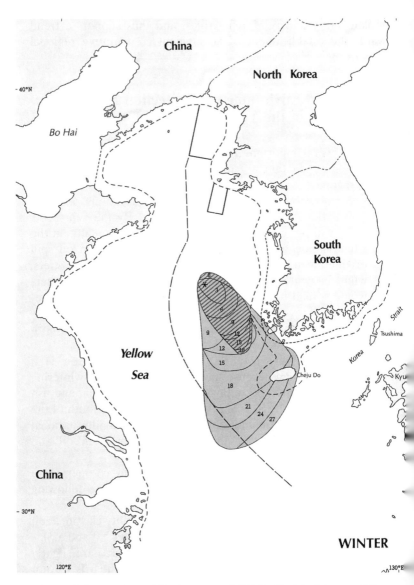

Figure 5.1 Projected Trajectory of Oil Spilled in Korean Waters in Winter
Source: Dong Soo Lee and Mark J. Valencia, 'Marine Pollution' in Joseph R. Morgan and Mark J. Valencia (eds.), *Atlas for Marine Policy in Northeast Asian Seas*, Berkeley: University of California Press, 1993, p. 132.

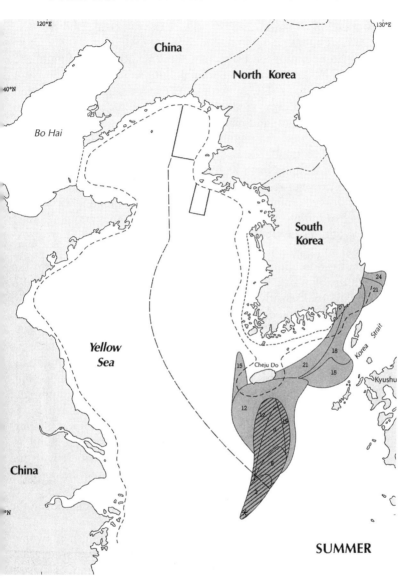

Figure 5.2 Projected Trajectory of Oil Spilled in Disputed Waters in Summer
Source: Dong Soo Lee and Mark J. Valencia, 'Marine Pollution' in Joseph R. Morgan and Mark J. Valencia (eds.), *Atlas for Marine Policy in Northeast Asian Seas*, Berkeley: University of California Press, 1993, p. 133.

introduction of harmful substances or the effects caused by their introduction.

2. Co-operation on protecting the Yalu River mouth and the northern Yellow Sea from marine pollution.

According to UNCLOS Article 207 'states shall adopt laws and regulations to prevent, reduce and control pollution of the marine environment from land-based sources, including rivers, estuaries, pipelines and outfall structures, taking into account internationally agreed rules, standards and recommended practices and procedures'. China and North Korea should establish a bilateral agreement to prevent, reduce, and control land–based pollution.

3. Co-ordination of marine environmental monitoring through the use of remote sensing and patrol vessels. It may be appropriate to consider establishing traffic separation schemes and tanker exclusion zones.

4. Co-operation in modelling and contingency planning for the clean-up of transnational oil spills. A regional team could also estimate the sensitivity of Yellow Sea ecological resources to spilled oil and the economic losses due the spillage.

5. Establishment of a unified regional marine dumping policy.

Marine dumping can be controlled by: assigning particular areas for particular kinds of industrial wastes and dredged materials; establishing standards for different materials being dumped; establishing the maximum amount and the timing of marine dumping; and establishing a permit system to certify dumping activities. At least one special marine dumping zone for each important port in the Yellow Sea should be specified. The first step could be exchange of information on marine dumping activities, with the eventual goal being the establishment of a unified marine dumping policy for the whole Yellow Sea. The location of marine dump sites should avoid damage to fishery resources, and the circulation at the chosen site should favour short residence times.

6. Co-operation on the protection of valuable fish spawning, breeding, and wintering grounds for sustainable fishery production.

Yellow Sea fishery resources have been declining over the last thirty years. Although overfishing is a major factor, the impacts of increasing marine pollution in the fish spawning grounds cannot be ignored. But research on the distribution of fish spawning, breeding, and wintering grounds by scientists of the littoral countries has produced different results. To protect fishery resources from ma-

rine pollution, it is necessary to properly define protected spawning zones and juvenile fish zones.
7. Co-operation on research and information exchange.

Co-operation among the Yellow Sea bordering states in marine pollution research could be the first step towards a broader program of co-operation, and would also save money by reducing or avoiding overlapping studies. Without similar data sets from China and South Korea, it is difficult to make valid comparisons or achieve a reliable overall synthesis.

Sea Use Planning for the Yellow Sea?

Sea use planning is a comprehensive strategy for minimizing conflicts among uses and users of specific sea areas over a defined time period. It integrates the goals of safety, efficiency, health, sustainability, and harmony. Its overall goal is to develop and maintain efficient and sustainable uses of the sea, while preventing irreversible change to living marine resources and the ecosystem. Five factors need to be considered in sea use planning: economic, environmental, social/cultural, time, and uncertainty (see Figure 5.3 and Appendix I).

There are numerous constraints on sea uses which drive decisions in sea use planning: oceanographic conditions, availability of resources, existing pollution, present sea uses, future economic and/or social needs, financial conditions, ocean technology, and potential conflicts between or among sea uses and users. Consequently, sea use planning is an integration of, as well as a balance between, these constraints. And decision-making is more a result of negotiation and balance between the various sea users than a unilateral process.

In the Yellow Sea, regional co-operation on sea use planning is necessary because:
1. fishery resources have been seriously depleted by competition among the fishermen from the littoral countries and Japan;
2. the impact of marine environmental pollution is increasing, and the importance of establishing a framework for international co-operation in the control of transboundary marine pollutants is increasingly recognized;
3. marine transportation is rapidly increasing, due to new direct shipping links between China and South Korea, and the

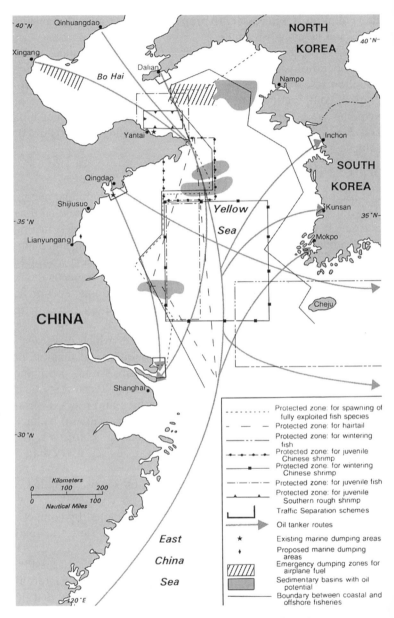

Figure 5.3 Proposed Ocean Zoning in the Yellow Sea
Source: Huang Yunlin, 'Sea Use Planning: The Case of the Yellow Sea', Ph.D. dissertation, Department of Geography, University of Hawaii, 1993, pp. 262–3.

POLLUTION AND ENVIRONMENTAL PROTECTION   229

   safety of shipping, especially of oil tankers, is of growing concern;
4. there are potential petroleum resources and exploration is ongoing, all of which affects other uses and users; and
5. individual countries bordering the Yellow Sea lack sufficient information for decision-making on resource management there.

Objectives of sea use planning in the Yellow Sea would include protection of fishery resources, prevention, control and clean-up of oil spills, guaranteeing the safety of maritime transportation, and protecting the marine environment from marine dumping and oil exploitation.

The introduction of 200 nautical mile EEZs by China and South Korea in the Yellow Sea will complicate boundary issues. But the bordering countries should not wait for an agreement on boundaries to begin co-operation on managing marine resources and protecting the marine environment. First, it is not necessary to define the exact area of co-operation. And second, even if all the boundaries were agreed, international co-operation would still be required because of the transnational nature of pollutants, fish stocks, and shipping. Finally, since none of the littoral countries has the capability to understand the entire marine ecosystem, and to monitor and enforce their own marine laws and regulations in such a large area, regional co-operation would be cost effective. Given the sensitivities over national jurisdiction and the lack of extended jurisdiction in the Yellow Sea, the area for international co-operation should be that beyond territorial seas and contiguous waters, that is beyond 24 miles from the coastline. A specific use plan for the Yellow Sea is outlined in Appendix I and illustrated in Figure 5.3.

The Ideal Regional Marine Environmental Protection Regime

The ideal regime must satisfy many theoretical needs as well as national interests. Above all it should rectify existing inadequacies and disparities in capacity. It should rationalize the redundancy of the existing and proposed international programs. It should provide the consultative channels, or infrastructure for co-operation, needed for synchronic monitoring, co-ordinated baseline studies, and prevention and clean-up of transnational pollution. It should co-ordinate policies and regulations for national zones and tailor them to fit natural features and processes, such as current systems and ecological zones, whether nearshore, offshore, temperate, or boreal. It should foster co-ordination and sharing of the results of research in individual zones. It should serve to educate the public and policymakers as to the causes and consequences of marine pollution, thereby evening up the degree of knowledge and concern among countries, particularly for offshore living resources and ecosystems.

Perhaps most important, it must provide opportunities to upgrade the capacities of North Korea and others to assess, monitor, prevent, control, and combat marine pollution. Without this assistance, North Korea and perhaps China may not be able to participate effectively in negotiations and ensure their concerns are reflected in policy formulation and would thus be less likely to comply with resultant agreements. Indeed such assistance may be the major incentive for North Korea, China, and perhaps Russia to participate in the regime. And Japan and South Korea may be motivated by the opportunity to establish a diplomatic relationship with North Korea. Moreover, the increased scientific knowledge that would accrue could reduce ecological and economic uncertainty regarding the distribution, effects, and costs of pollution. This could, as in the Regional Seas Programme for the Mediterranean, mitigate developing countries' fear that developed countries are trying to make them pay for pollution the developed countries caused, and to make them less competitive by diverting resources from economic development.[60]

The geographic scope of the regime is a critical feature. The Sea of Japan, the Yellow Sea, and the East China Sea are both physically and biologically connected by the movement of water and biota, particularly migratory fish (see Figure 5.4). Only South Korea currently borders all three seas, although Japan has interests and activities in all of them as well. Russia is not a littoral state of the Yellow or East Seas. China may argue that its access rights through the Tumen River and the North Korean port of Chongjin, entitle it to an interest in Sea of Japan affairs.[61] Although North Korea does not border the East China Sea, the biota and ecosystems of the Yellow and East China Seas are sufficiently linked to transfer pollutants and their impacts. North Korea would have a

POLLUTION AND ENVIRONMENTAL PROTECTION 231

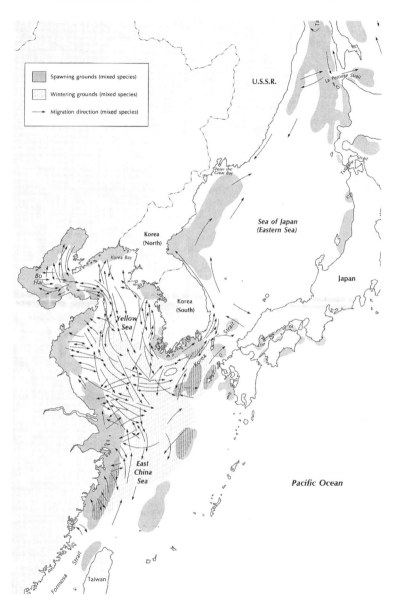

Figure 5.4 Yellow Sea: Fish Migrations and Jurisdictional Claims
Source: Joseph R. Morgan and Mark J. Valencia, 'Integrations' in Joseph R. Morgan and Mark J. Valencia (eds.), *Atlas for Marine Policy in Northeast Asian Seas*, Berkeley: University of California Press, 1993, p. 129.

genuine interest in East China Sea affairs upon reunification with South Korea.

There should be an overarching consultative mechanism which includes all North-East Asian entities, but is comprised of two working committees: one for the Sea of Japan which would include Japan, North Korea, South Korea, Russia, and perhaps China, and one for the Yellow Sea/East China Sea which would include China, Japan, North Korea, South Korea, and Taiwan. China would probably insist on excluding the Bo Hai from the regime because Beijing considers the Bohai to be internal Chinese waters. Since the China/Taiwan relationship makes policy co-ordination in the East China Sea particularly troublesome, the initial focus of the Yellow Sea/East China Sea working committee would be on the Yellow Sea, and perhaps include Taiwan only as an observer.

Although the East China Sea presents formidable jurisdictional problems, these could be finessed by *ad hoc* arrangements, which do not specify who has jurisdiction over which area, but only that the partners collectively have jurisdiction over the entire area. The outer boundary of the geographic scope of the working committee does not have to be precisely defined, but the core area must be agreed upon by the participants. Japan should be included at this early stage because of its fishing, oil development, and shipping interests in the Yellow Sea, because of the links between the waters and ecosystems of the Yellow Sea with the Sea of Japan, and because of its ability to provide money, technology, and knowledge for the regime. The two working committees would be related by annual reports to plenary sessions and via their common members: South Korea, North Korea, and Japan.

There is a convergence of factors making the formation of a marine environmental regime in the Sea of Japan increasingly likely. Only four states border the Sea, thereby reducing the complexity of the bargaining process. There is a growing recognition that successful efforts in environment regimes could have spillover effects in other spheres closer to the core of international relations, such as security and trade. The Sea of Japan's unsettled maritime frontiers may offer an opportunity for innovative management approaches. The interconnectivity of the waters and their biota, including fisheries resources, is increasingly apparent. Furthermore, the Sea of Japan is considered relatively unspoiled, and therefore a prime candidate for preservation. The nuclear waste disposal controversy has provided the exogenous shock to the system necessary to enhance regime formation and speed it along. It has raised public awareness and placed the issue of marine pollution on national policy agendas. And it is creating an epistemic community of scientists and environmentalists in the bordering countries who are pressing for policy action.

The stimulants include fear and uncertainty about the long-term effects of radioactive pollutants, and the high cost of gathering information, as well as the necessity for co-operation when undertaking research beyond national jurisdiction. Finally, the obvious need for the continuing management and enforcement of anti-dumping regulations and broader pollution monitoring and control measures argues for a mechanism to co-ordinate policies and approaches. Pressure to formulate sound environmental policy has increased since the UNCLOS came into force in November 1994, obliging nations to protect the marine environment and providing further ammunition to NGOs to persuade their governments to do so.

A different but equally strong argument can be made for the establishment of a regime for the Yellow Sea. This Sea is showing signs of severe industrial and oil pollution, even though it contains significant fisheries and ecological resources. It is increasingly clear to policymakers, particularly in China and South Korea, as well as the regional public, that with the quickening pace of economic development, pollution will continue to increase, and damage those resources. Only three countries border the Sea, although a fourth, Japan, has activities and interests there. Discussions of jurisdictional issues are ongoing between China and South Korea and settlement, or provisional arrangements in lieu of a settlement, could include an environmental component. The incentive to conclude some sort of agreement may increase when the UNCLOS comes into force. That treaty urges countries to enter into provisional arrangements pending a settlement of boundary disputes.[62] Of course the North/South Korea dynamic is an obstacle, but, if the nuclear tensions between the two countries continue to subside, co-operation on marine environmental matters would be an opportunity for each to feel the other out and to build confidence for co-operation on a broader front.

Because of the tentative relations among the littoral countries their sensitivity regarding national jurisdiction, and intra-regional competition and mutual suspicions, the Yellow Sea regime should initially be consultative, and 'self-help' in nature. Each country must manage its own waters. But a loose consultative mechanism can discuss common policies, co-operative research, education, and training. The lead of a capable medium power, South Korea, might be acceptable to all parties. As it has historically, South Korea can serve as a buffer between the major powers and mitigate their reluctance to follow each other's lead. To decrease sensitivities, areas of overlapping claims around Tok Do/Takeshima and the Kuriles might be initially excluded from the purview of the consultative mechanism.

The consultative mechanism could also become the focal point for rationalizing NOWPAP, WESTPAC, UNDP/GEF and other international organization initiatives with a marine environmental focus, and for co-ordinating the implementation of Law of the Sea environmental responsibilities. As confidence builds and a habit of consultative and co-operative behaviour gains a foothold, an organization could be added. It could involve conferences of environmental ministers, which would make the delegates directly responsible to their populaces. Laggards would then find it more difficult to retard progress.[63]

The process of environmental regime creation for both the Sea of Japan and the Yellow Sea should be more or less the same. The suggestion is to recognize the natural course of events and to allow the regime to form in stages; in other words, to evolve. It should begin, as it is showing signs of beginning, with a limited and temporary focus on monitoring and possible clean-up of dumped radioactive materials in the Sea of Japan and serious pollution in the Yellow Sea. But policymakers should be prepared to move rather quickly beyond this limited *ad hoc* arrangement to a broader co-ordination regime, which would agree on rules and procedures while leaving each member free to implement them in its own way at its own pace. This more advanced arrangement would focus on service functions: information exchange, data gathering and analysis, consultation, co-ordination of research programs, and planning for joint action in emergencies. Gradually and incrementally addressing ever more competing uses of the seas can produce a more coherent, comprehensive, balanced set of arrangements. The trend from a use-oriented to a resource-oriented approach can move successively from pollution protection to species conservation, to collective management, and then to more refined monitoring and research. Eventually the parties could agree on pollution reduction targets as well as on reporting on implementation, and on improved public access to information.[64]

What specifically needs to be done? The most appropriate means for co-operation in the region are marine environmental monitoring and assessment; the development, upgrading, and harmonization of marine environmental legislation;[65] the transfer of marine pollution control technology; and combatting marine pollution, especially in the case of pollution emergencies stemming from incidents involving vessels or offshore drilling. Education and training should be an integral part of all areas of co-operation. Environmental monitoring and assessment should be decision-oriented and receive high priority. Research priorities might include a synthesis of information on the state of marine pollution and dumping, perhaps resulting in a dynamic computer-based atlas of ecology, sea use, and pollution in North-East Asian seas.

Ultimately, national legislation would need to be harmonized, and regime members would need to prepare and adopt an umbrella convention on the protection of the marine environment. Supporting efforts might include joint assessment of priorities for marine resource management and areas most at risk. Recommendations for integrated coastal zone and marine environment management might be developed at the regional level. Co-operative projects on training in environmental impact assessment, the co-ordinated creation of marine parks, management of wetlands, and the control of industrial, agricultural, and domestic wastes are also priorities.

Co-operation on transnational issues should include studies of transboundary pollution, including intercalibration, baseline studies, co-ordination in emergencies, and enforcement of environmental regulations near or across disputed boundaries as well as the transnational effects of ocean disposal; and adjustment of national initiatives to be compatible with emerging international legal and technical obligations, for example, the UNCLOS, the LDC, and UNEP's Regional Seas Programme, and with one another.

Decisionmakers should give consideration to a number of possible measures to enhance overall maritime safety and environmental protection beyond the IMO-co-ordinated international conventions. These mitigating or precautionary actions might include the establishment of tanker exclusion zones to protect coastal environments, or mobile safety zones, with escorts, around tankers. Another could be the formation of regional pollution response teams, multinational in composition and authorized to act immediately, regardless of the national jurisdiction of the waters affected. These teams might initially be established for the Sea of Japan and for the Yellow Sea. Emergency response vessels, powerful tugs with pollution control equipment, tanks, towing, and other emergency gear, would have to be readily available. Existing search-and-rescue organizations and facilities could provide the nucleus around which these pollution response teams could be built. The countries bordering particular seas may also consider establishing a regional compensation fund for damage caused by polluting accidents.

Although the environmental scope of the regime should be expanded to include all major sources of pollutants—atmospheric, shipborne, dumping, drilling, and land-based—it should not become multisectoral. Multisectoral approaches are expensive, and unfamiliar to bureaucrats. They lack a public constituency and suffer from the inherent conflicts that create their apparent need. But an expanded scope will allow for side-payments and for each participant to play multiple roles. With an expanded scope, Japan could be satisfied that its concerns with the dumping of radioactive material will be addressed, while Russia and North Korea would be assured of assistance in training and research. In the Sea of Japan, Japan and Russia can claim to be both victims and perpetrators of the dumping of radioactive material. Similarly, in the Yellow Sea, China, North Korea, and South Korea are all polluters and all suffer from its effects.

The regime must be simple but not too loose. The first step is to define the problem accurately by acknowledging the varied capacity and will for co-operation in a structured manner to monitor and control marine pollution, particularly pollution beyond the narrow coastal zone. This requires intensive self-examination, and specification of needs and intents. Rights and rules must be defined and agreed. In general, compliance with a marine pollution regime[66] largely depends on such variables as the source of the pollution regulated by the regime; the geographical scope of the regime (global or regional); the inclusiveness of the regime's membership; the extent to which the regime exhibits attributes of collective goods (that is the extent to which non-members or non-parties enjoy the advantages provided by the arrangement and know that they can use it even if they do not contribute to its maintenance); the speed at which the regime moves towards becoming legally binding; the degree of economic and political homogeneity of the participants; the extent to which they perceive their environmental challenge and, more generally, the extent of their socialization to the basic principles of the regime; the strength of national environmental lobbies; the state of national antipollution legislation and compliance mechanisms, including technical personnel and the scientific and technological infrastructure; the efficiency and effectiveness of national enforcement; and the level of investment necessary to implement the substantive provisions of the regime.

Initially each government would manage its own jurisdictional areas according to agreed standards, perhaps using those of centrally located South Korea as a base, but with the monitoring capability of Japan, the most developed country in the region. Decisions should be by consensus and implemented by voluntary acquiescence to the rules. There is little point at this early stage in attempting to coerce compliance, and voting in international fora is generally an anathema to Asian countries. Compliance would be achieved through detection, publicity, and persuasion, as well as the costs, including loss of face, of purposefully and continually 'defecting' from the regime, once having joined. Differences would be discussed openly in the working committee meetings and, if necessary, in the plenary.

Over time, a higher functional level may be reached at which laws and policies are harmonized, and common standards and regulations are set, monitored, and enforced by national teams in their own waters. Eventually an organization might be envisaged with a secretariat comprised of technical and policy representatives from each party, charged with developing recommendations for *regional* policies, laws, standards, procedures, training, research and environmental assessment, and management. A regional self-help arrangement is perhaps preferable to an international organization-led and managed effort because it is more efficient, more flexible, and easier to control and use. But an international organization could be the catalyst to start the process, as UNEP has attempted to do.

Leadership is a critical need at both the state and individual level. South Korea should take the regional lead because of its geographic location bordering all three seas, because of its middle power status and thus its inability to exercise hegemony, and because of its ability to play a mediating role between the major powers: China, Japan, and Russia. That South Korea may be ready and willing to play such a leadership role is evidenced by its web of bilateral arrangements in this sector, and by its environmental initiatives on the world and regional stage.

Since Prime Minister Won Shik Chung's major environmental

speech at UNCED declaring support for a North-East Asia Environment Co-operation Organization,[67] South Korea has emerged as a leader in regional environmental initiatives. It has signed an agreement with China[68] which covers co-operation on air and marine pollution and the disposal of hazardous waste. The agreement also provides for exchange of information and experts, co-operative research, the establishment of a joint commission, and the eventual conclusion of a convention on environmental co-operation. Present joint efforts between South Korea and Japan include an attempt to conclude a convention on environmental co-operation (a first draft has already been written), and to continue joint projects and Japan International Co-operation Agency (JICA) special training. South Korea and Russia, and South Korea and Mongolia are also exploring co-operation in environmental protection. Although the stage is set, it awaits the appearance of a leader of sufficient energy and stature to convince and entice its colleagues in other countries to join in moving a regime onto the policy agenda.

When and if a formal organization and a secretariat is required, it should be headquartered in Seoul. The costs could be allocated equally, although China, Russia, and North Korea could contribute in kind. In other words, Japan and South Korea might bear the bulk of the monetary costs, especially for overheads, salaries, ship operations, technical equipment, and training, while China, Russia, and North Korea might contribute the bulk of the scientists and vessels necessary for the research.

The benefits of a regime will be positive but varied for each participant. Although all participants will lose the ability to treat the sea as a free waste dump, all will clearly benefit from cleaner seas. Though considerable, these benefits are also unquantifiable because of the long-term nature of the impacts of an environment which is less polluted than it might have been, and the uncertainty regarding the causal relationships between pollution and ecosystem damage.

Perhaps the most important benefit will be the evening up of the levels of marine environmental technology and expertise throughout the region. The overall objective of the arrangement is to manage the marine environment of North-East Asian seas. The most important spin off may be the provision of greater equity: equity in the sense of increased national responsibility to control pollution with potential transnational effects, and equity in the sense of a transfer of technology and knowledge from the rich for the benefit of all. In short, the major trade-off will be the benefit to Japan and South Korea of adherence by China, North Korea, and Russia to a predictable regime with common minimum standards of discharge, in exchange for training and technical assistance from Japan and South Korea. In addition, South Korea will enhance its regional status.

The necessary conditions for the regime to be successful can be summarized as follows:

- One nation, preferably South Korea, must be willing to lead overall, and the dominant nations, China and Japan, must be willing to play supportive roles, perhaps in exchange for acceptance of leadership in other sectors, for example, fisheries in the case of Japan.
- Sufficient incentives in the form of technology and training must be made available to North Korea and China.
- North Korea must not object to the arrangement—at least not strenuously; its active participation can come when it is ready.
- Participants must accept the superficial principle of equity in costs and benefits, although side-payments and barter exchange will supplement any financial exchanges.
- The initial regime must be consultative and 'self-help' in nature, but involve conferences of high-level environmental officials.
- The threshold of bureaucratic inertia in Japan must be breached to move the concept from theory to reality.

Notes

1. Douglas Johnston, 'Vulnerable coastal and marine areas: a framework for the planning of environmental security zones in the ocean', *Ocean Development and International Law Journal*, 24 (1993): 72.
2. 'The dying Yellow Sea', *The Korea Herald*, 10 January 1995, reprinted in *The Japan Times* (14 January 1995): 18.
3. Dong Soo Lee and Mark J. Valencia, 'Pollution', in Joseph R. Morgan and Mark J. Valencia (eds.), *Atlas for Marine Policy in East Asian Seas*, Berkeley: University of California Press, 1992, p. 29.
4. Lee Sanghoon, 'Population resources and environmental consideration in the Yellow Sea Rimland', *Report on Regional Development in the Yellow Sea Rim*, September 1991, Honolulu, Hawaii: East–West Center, 1991.

5. Ibid.
6. Ibid., p. 13.
7. Huang Yunlin, 'Sea Planning: The Case of the Yellow Sea', Ph.D. thesis, University of Hawaii, 1993.
8. *The Bo Hai–Yellow Sea Region Economic Development Plan*, Ministry of Communication, Beijing: People's Communication Press, 1989, pp. 6, 11.
9. Ibid., p. 7.
10. Melana Zyla, 'Deep trouble', *Far Eastern Economic Review*, 18 March 1993, p. 21; Sergei Shargorodsky, 'Thousands died producing Soviet chemical weapons', *Honolulu Advertiser* (24 December 1993): A-7.
11. Robert Scalapino, 'Northeast Asia—prospects for cooperation', *Pacific Review*, 5 (1992): 101–11; Jane Khanna, 'Economic interdependence and challenges to the nation state: the emergence of Natural Economic Territories in the Asia-Pacific', Concept paper for the Pacific Forum/CSIS, 1 July 1993.
12. Mark J. Valencia and John Klarquist, 'National marine environmental policies and transnational issues', in Joseph R. Morgan and Mark J. Valencia (eds.), *Atlas for Marine Policy in East Asian Seas*, Berkeley: University of California Press, 1992, p. 139.
13. Oran R. Young, *International Co-operation: Building Regimes for Natural Resources and the Environment*, Ithaca: Cornell University, 1989, pp. 37–44.
14. Valencia and Klarquist, 'National marine environmental policies and transnational issues', p. 139; For example, the northern and southern parts of the Sea of Japan differ sharply with respect to hydrography, circulation, continental shelf width, submarine topography, and coastal geomorphology. Indeed, hydrography and circulation in the Sea resemble that of a 'mini-ocean' with regard to temperature and salinity differentials between northern and southern zones, circulation patterns, and zonal mixing. There are also contrasts between the narrow continental shelves and the smoother coastline, with its complex links between coastal and submarine geomorphological processes in places like the Yamato Basin and Toyama Bay.
15. Wang Tieya, 'China and the Law of the Sea', in D. Johnston and N. Letalik (eds.), *The Law of the Sea and Ocean Industry: New Opportunities and Restraints*, Honolulu: The Law of the Sea Institute, 1984, pp. 581–9.
16. Lee Dong Soo and Mark J. Valencia, 'Pollution', in Joseph R. Morgan and Mark J. Valencia (eds.), *Atlas for Marine Policy in Northeast Asian Seas*, Berkeley: University of California Press, 1993, pp. 129, 131–4.
17. Peter Hayes and Lyuba Zarsky, 'Regional cooperation and environmental issues in Northeast Asia', *Nautilus Pacific Research*, 1 October 1993, p. 9.
18. Garrett Hardin, 'The tragedy of the commons', *Science*, 162 (1968): 1243–8.
19. Mark J. Valencia, 'East Sea: sea of co-operation?', *The SISA Journal*, 23 December 1993; 'DPRK wants joint nuclear waste probe with Russia', *FBIS-EAS*, 6 July 1993, p. 35; 'Russia's radioactive waste disposal: a matter of grave concern', *US Department of State Dispatch*, 4, 47, 22 November 1993, p. 807; 'Tokyo hosts GLOBE environmental conference', *FBIS-EAS*, 1 September 1993, pp. 1–3; *Nikkei Weekly*, 25 October 1993, pp. 1, 4; 'US to ask ocean ban on N-waste', *Los Angeles Times*, 4 November 1993, p. A-16; 'Russia to join survey on East Sea nuclear waste', *FBIS-EAS*, 12 April 1993, p. 29; *FBIS-EAS-93-063*, 5 April 1993, p. 2; 'Kono views nuclear waste impact on Russian aid', *FBIS-EAS*, 6 July 1993, p. 35 (*Chungang Ilbo*, 3 July 1993, p. 3).
20. International Maritime Organization, Marine Environmental Pollution Committee, *33/WP.7/Add.1*, para. 13.4, p. 21.
21. '*ROK, Japan, Russia to test for waste dumping*', FBIS-EAS, 14 March 1994, p. 38.

POLLUTION AND ENVIRONMENTAL PROTECTION   241

22. Shargorodsky, 'Thousands died producing Soviet chemical weapons', p. A-7.
23. 'Russia said to jettison ammunition off Japan', *Japan Times*, 10 February 1995, p. 4; Russia may resume waste dumping, *Japan Times*, 14 July 1995, p. 2.
24. David Pitt, 'Nuke dumping: Russia pressed on nuke waste accord', *New York Times*, 5 December 1993, p. 6.
25. Symposium on Sea of Japan International Cooperation, Russia in Asia Report No. 15, July 1993; 'China decides to sign the London Dumping Convention on waste disposal', *World Journal*, 19 February 1994, p. 9.
26. 'Proposed marine waste dumping rules viewed', *FBIS-EAS-93-043*, 8 March 1993, p. 10.
27. *United Nations Convention on the Law of the Sea*, New York: United Nations, 1982.
28. UNEP/WP, *Protection of the Marine Environment Against Pollution from Land–Based Sources* (Montreal Guidelines), 120/3 (Part IV), Ad hoc Working Group of Experts, Montreal, 11–19 April 1985.
29. United Nations Convention on the Law of the Sea, Article 194(1).
30. Peter M. Haas, 'Protecting the Baltic and North Seas', in Peter M. Haas, Robert O. Keohane, and Marc A. Levy (eds.), *Institutions for the Earth, Sources of Effective Environmental Protection*, Cambridge, Massachusetts: The MIT Press, 1993, pp. 135–6.
31. United Nations Convention on the Law of the Sea, Article 194(2).
32. Hayes and Zarsky, 'Regional cooperation and environmental issues in Northeast Asia'.
33. Ibid., p. 7.
34. Ibid., p. 14.
35. The Third Northeast Asian Conference on Environmental Cooperation, *EMECS Newsletter*, 6, 25 November 1994, p. 8.
36. Lyuba Zarsky, Peter Hayes and Keith Openshaw, 'Regional Environmental Cooperation in Northeast Asia', *Draft Report to Regional Bureau for Asia and the Pacific*, United Nations Development Programme, 1994.
37. Economic and Social Commission for Asia and the Pacific, *Report of Expert Group Meeting on Environmental Co-operation in North-East Asia, 24–6 November 1994*.
38. 'Northeast Asia drafting environmental plan', *FBIS-EAS-93-025*, 9 February 1993, p. 1; 'Talks open on Asia-Pacific environment', *Japan Times*, 24 June 1994.
39. NGOs are beginning to gain influence. In China, a committee of government and non-government persons has drafted a national 'Agenda 21', a comprehensive industrial and environmental program to clean up China's cities, waterways, and fields. It has also established the China Biosphere Nature Reserve Network to improve domestic and international environmental co-ordination. The network will co-operate with UNESCO's Man and the Biosphere Program. In South Korea, a coalition of seven environmentalist groups threatened to lead the nation's first boycott against products of twenty-four local companies if they do not end pollution violations by 1996. In Russia, environmental rules are becoming more stringent due to growing 'green' awareness among grassroots groups. P.T. Bangsberg, *Journal of Commerce*, 1 November 1993; *China Daily*, 13 July 1993; 'Business Briefing', *Far Eastern Economic Review*, 23 December 1993; 'C.I.S. environmental rules growing tougher', *Oil and Gas Journal*, 20 September 1993, p. 31.
40. 'Asia NGOs will gather in Kyoto', *Japan Times*, 23 November 1994, p. 2: 'Fears of Asian environmental crisis spur NGO networking', *Kyodo*, 22 November 1994.

41. 'Tokyo hosts GLOBE environmental conference', *FBIS-EAS*, 1 September 1993, p. 1.
42. APEC Environmental Vision Statement and Framework of Principles for Integrating Economy and Environment in APEC, mimeograph, 1994.
43. 'NE Asians to meet on reducing tension', *The Korea Times*, 29 September 1993, p. 1; 'Regional Briefing', *Far Eastern Economic Review*, 2 December 1993, p. 15.
44. Arthur A. Stein, 'Coordination and collaboration: regimes in an anarchic world', in Stephen D. Krasner (ed.), *International Regimes, International Organization* (Special Issue), 36, 2 (1982): 299–324.
45. Boleslaw A. Boczek, 'Concept of regime and the protection and conservation of the marine environment', in Elisabeth Mann Borgese and Norton Ginsburg (eds.), *Ocean Yearbook 6*, Chicago: University of Chicago Press, 1986, pp. 288–92; 294.
46. United Nations Environment Programme (UNEP), *Draft Action Plan for the Protection, Management and Development of the Marine and Coastal Environment of the North-West Pacific Region*, Nairobi, September 1993.
47. Ivan Zrajevskij, 'The North-West Pacific Region Action Plan: Progress problems and lessons learned', in Hyung Tack Huh, Chang IK Zhang, and Mark J. Valencia (eds.), *Proceedings of the International Conference on East Asian Seas: Cooperative Solutions to Transnational Issues*, Seoul: Korea Ocean Research and Development Institute and East–West Center, 1992.
48. *International Environment Reporter*, 4 December 1991, p. 657.
49. United Nations Environment Programme, Draft Action Plan for the Protection, Management and Development of the Marine and Coastal Environment of the North-West Pacific Region.
50. 'Environmental Conference drops "Sea of Japan",' *FBIS-EAS-94-180*, 16 September 1994, p. 4; United Nations Environment Programme, *Preparatory Meeting of Experts and National Focal Points for the Intergovernmental Meeting on the Northwest Pacific Action Plan*, Report of the Meeting, UNEP (OCA)/NOWPAP WG.4/5, 13 September 1994; United Nations Environment Programme, Draft Action Plan for the Protection, Management and Development of the Marine and Coastal Environment of the North-West Pacific Region.
51. Alicia Barcena, 'Cooperation for development in regional seas. General evaluation of the Regional Seas Programme', Paper presented at *Pacem in Maribus XXI*, Takaoka, Japan, September 1993, pp. 2–4; Peter M. Haas, 'Save the seas. UNEP's Regional Seas Programme and the coordination of regional pollution control efforts', in Elisabeth M. Borgese, Norton S. Ginsburg, and Joseph R. Morgan (eds.), *Ocean Yearbook 9*, Chicago: The University of Chicago Press, 1991, pp. 188–212.
52. Chua Thia Eng, Robert Cordover, Miles Hayes, Celso Roque, David Shirley, Gurpreet Singhota and Philip Tortell, 'Prevention and Management of Marine Pollution in East Asian Seas', Formulation Mission Report Prepared for the United Nations Development Programme Division, Regional Bureau for Asia and the Pacific, April 1993, pp. 2.11–2.27; *The Nation*, 16 February 1993.
53. Hayes and Zarsky, 'Regional cooperation and environmental issues in Northeast Asia' p. 13.
54. United Nations Environment Programme, *Third Meeting of Experts and National Focal Points on the Development of the North-West Pacific Action Plan*, Report of the Meeting, UNEP (OCA)/NOWP.WG3/6, 10–12 November 1993.
55. Ibid.
56. Boczek, 'Concept of regime and the protection and conservation of the marine environment', p. 285.
57. Han Taek-Whan, 'Northeast Asia environmental cooperation: progress

POLLUTION AND ENVIRONMENTAL PROTECTION

and prospects', Paper presented to the Workshop on Trade and Environment in Asia-Pacific: Prospects for Regional Cooperation, East–West Center, Honolulu, 23–5 September 1994.

58. Huang Yunlin, 'Sea Use Planning: The Case of the Yellow Sea', Ph.D. thesis, University of Hawaii, 1993.

59. Lee and Valencia, 'Pollution', pp. 132–3.

60. Jon Birger Skjaerseth, 'The "effectiveness" of the Mediterranean Action Plan', *International Environmental Affairs*, 5, 4 (Winter 1993), p. 323.

61. Mark J. Valencia, 'Economic cooperation in Northeast Asia: the proposed Tumen River scheme', *The Pacific Review*, 4, 3 (1991), p. 266; Mark J. Valencia, 'The proposed Tumen River scheme', in Hong Yung Lee and Chung Chongwook (eds.), *Korean Options in a Changing International Order*, Berkeley: Institute of East Asian Studies and Center for Korean Studies, University of California, 1993, pp. 193, 194, 198.

62. United Nations, Convention on the Law of the Sea, Article 83(3).

63. Haas, 'Protecting the Baltic and the North Sea', pp. 173–5.

64. Steiner Andresen, 'The effectiveness of regional environmental cooperation in the northern seas', Paper presented at the 36th Annual Conference of the International Studies Association, Chicago, February 1995; Haas, 'Protecting the Baltic and the North Sea'.

65. Mark J. Valencia, *International Conference on the Yellow Sea*, East–West Environment and Policy Institute Occasional Papers, No. 3, pp. 88–90.

66. Boczek, 'Concept of regime and the conservation and protection of the marine environment', p. 294.

67. Lee Sang Don, 'Remarks', in Hyung Tack Huh, Chung Ik Zhang and Mark J. Valencia (eds.), *Proceedings of the International Conference on East Asian Seas: Cooperative Solutions to Transnational Issues*, Korea Ocean Research and Development Institute Seoul, 21–3 September 1992.

68. 'ROK–China anti-pollution accord will beef up regional efforts', *Korea Times*, 29 October 1993, p. 3.

6 Fisheries

Introduction

This chapter defines the fisheries issues, describes existing international fisheries regimes, delineates their inadequacies, reviews alternative multilateral approaches and precedents, and then focuses on the Yellow Sea to illustrate the issues and the possible solutions. The chapter concludes with an analysis of the realistic possibilities and necessary next steps for regime formation, and summarizes the critical elements of a model multilateral fisheries regime.

The global fisheries situation establishes the general context for fisheries issues in North-East Asia. Globally, fisheries are in crisis.[1] The global catch has been decreasing since 1989, and thirteen of the seventeen major global fisheries are now depleted or in serious decline. The other four are overexploited or fully exploited. Before 1989, rising catches masked an ecological shift from valuable species to less desirable ones such as spiny dogfish, skate, and shark. The situation has become sufficiently dire for the United Nations to sponsor talks to protect the remaining fisheries both within and outside EEZs.

Although fish prices have been rising since the early 1980s, fish nonetheless remains the most important source of animal protein in many countries, especially poor ones. When rich countries fish out their stocks, they have alternatives. Euphemistically, they can switch from sole to steak. The developing world can't. Moreover, overfishing threatens jobs. Artisanal fisheries employ twenty times as many people as the industrial fisheries that are replacing them. And fishermen tend to live in places where few other jobs are available.[2]

It seems that only the dramatic collapse of local stocks can persuade individual groups of fishermen and individual governments of the need for radical, tough regulation. Too often, politicians have been reluctant to conserve stocks for fear of the political reaction from fishermen whose income is reduced and managers have been unwilling to follow scientists' advice. When countries have banned unregulated foreign fleets from their EEZs, domestic fleets have expanded to take their place. The size of the global fishing fleet has nearly doubled in the last 20 years. And in North-East Asia, the chief thrust of fisheries policy in the 1970s, particularly among countries bordering the Sea of Japan, was the development of distant-water fishing capacity. This reduced competition for increasingly scarce coastal resources and made regulation easier, but it also diverted attention away from management of coastal fisheries.

Ironically, the improved relations and lessening tension in North-East Asia have allowed North-East Asian fishermen to venture further from home. And now UNCLOS and the 200-nautical mile extended fisheries jurisdiction era is forcing a re-examination of domestic fisheries management policies and regulations throughout North-East Asia. Although North-East Asian countries are still reluctant to alter what has been a relatively stable fisheries regime for many years, pressure is mounting for all countries to fully extend fisheries jurisdiction and to control access by foreign vessels. In the absence of extended jurisdiction, North-East Asian seas are under an open-access regime. In most other regions of the world which have had open-access regimes, regional and international organizations provided a forum for distant-water fishing states and coastal states to consider both their effect on fishing and the necessary management recommendations.[3] These recommendations usually resulted in a single, overall annual quota for each stock, and a minimum size of fish which could be caught. But the quotas generally did not constrain the amount of fishing. And they promoted economic waste by encouraging each state to maximize its share of the catch by adding more boats to its fleet. And when the capacity of their fishing industries exceeded quotas, states agreed to raise the quotas, often beyond the maximum sustainable yield of the stocks, in order to satisfy that capacity.

When these states extended their jurisdiction, they ceased their participation in the relevant regional and international organizations. With extended jurisdiction, though, the need for co-operation has greatly increased. Organizations are needed to manage shared stocks, to allocate catch, and to obtain economies of scale in costly research, management, and training, and in the marketing and trade of fisheries products. When North-East Asian countries extend their fisheries jurisdiction to encompass the Yellow/East

China and Japan Seas, new regional organizations will be needed for effective fishery management.

In the late stages of negotiations on the UNCLOS, it was believed that a new international legal 'system' for ocean fisheries would emerge from the bilateral agreements negotiated between coastal states claiming EEZs and other states.[4] What could not be foreseen then, in the global arena, was the diversity of ocean development and management arrangements that would have to be negotiated at the regional level. And there is still no blueprint. Each marine region is virtually free to find its own way of building a bridge between the aspirations of national autonomy and the responsibilities of membership in the world community.

Extended jurisdiction would clearly force a change in the essential structure of the North-East Asian region's fisheries regime. Coastal states would acquire exclusive authority to manage stocks in their waters. In this respect, UNCLOS requires a system for licensing domestic and foreign fisheries, and for setting of catch quotas, season and gear restrictions, and age and size limitations, as well as terms for joint ventures and enforcement. For fish stocks within the EEZ of two or more coastal states, UNCLOS Article 63 stipulates that the states concerned 'shall seek, either directly or through appropriate subregional or regional organizations, to agree upon the measures necessary to co-ordinate and ensure the conservation and development of such stocks'.

The community of interest defined in UNCLOS Article 63(1) consists of coastal states with adjacent EEZs through which a stock migrates.[5] Based on its sovereign rights over the fisheries resources in its EEZ, each coastal state is entitled to part of the benefits to be obtained from the transboundary marine fisheries resources located on its side of the boundary. If the stock is exploited in the EEZ of any of the states involved, that state has a duty to co-operate with the other states through whose EEZ the stock migrates. Only if the other states involved are of the opinion that the duty of due diligence is not breached by the state in whose waters the stock is exploited, would co-operation not be required.

UNCLOS Article 63(2) also defines a community of interest consisting of the states actually fishing for the resource on the high seas and the coastal states through whose EEZ the stock migrates. Based on a coastal state's sovereign rights over the fisheries resources of its EEZ, it is entitled to participate in the co-operative arrangements concluded by states fishing on the high seas. Interest in conservation is thereby secured.

Moreover, the duty of due diligence requires that states fishing in adjacent high seas areas co-operate with the coastal state in the adoption of conservation measures. But it is at the discretion of the coastal state to determine whether such co-operation is required to satisfy the duty of due diligence owed it by the states fishing on the high seas. Despite these specific provisions, there will undoubtedly be differences and disagreements among the coastal states over their rights and duties with respect to fisheries. To prevent potential conflicts and solve existing problems, fisheries co-operation will require frequent and frank consultation and negotiation. The 1995 UN Conference on Straddling and Highly Migratory Fish Stocks provided a forum for such negotiations. However, China, Japan, and South Korea joined together with other distant-water fishing countries to oppose adoption of a legally binding and comprehensive document aimed at managing living resources of the high seas.[6]

The new architecture of fishery management will need to be based on a policy structure that considers the changes caused by extended jurisdiction and the requirements it imposes on management. It must take into account the successes and failures of the organizations that operated under open-access regimes, as well as the special needs of the coastal states. Specifically, this policy structure must address the following facts:

1. Fundamental factors that influence fisheries are economic, and management must address these factors as much as fishermen-fish interactions.
2. Although coastal states will have exclusive authority to control fishing in their own waters, their authority over the stocks is not exclusive; some stocks are shared, and others enter the High Seas.
3. The requirement for information and data under an active management system, particularly for predicting recruitment, will be far greater than it would be under a passive management, open-access regime.
4. Since fisheries constitute an important basis for economic and social development in the developing countries, and in economically disadvantaged portions of developed countries, fisheries policy must be viewed as a problem of rural development and reform rather than as a pure sectoral issue.

In North-East Asia, fisheries policy must address the common problems of overfishing, management and allocation of shared stocks, scattered and fragmentary data, unresolved jurisdictional boundaries, difficult political relationships, and the inadequacies of the current fisheries management regimes.

Multilateral fisheries regimes for North-East Asian seas are already being officially explored. In March 1993, Russia announced that it wanted to hold multinational negotiations with Japan, South Korea, China, Poland, and other countries on fishing in the Sea of Okhotsk. South Korea agreed to have unofficial discussions with the other countries first because Russia considers it to be the main offender in the area.[7]

In June 1993, China and Japan reached an agreement on cooperation between their respective law enforcement authorities to prevent piracy and poaching in areas of claim overlap in the East China Sea. And China and South Korea have agreed to co-operate in combating marine pollution in the Yellow Sea.[8] Such *ad hoc* solutions are one way of avoiding the consequences of formally declaring EEZs while actually excluding pirates and poachers from non-claimant countries.

In what is probably the most notable regional fishery initiative to date, participants from Japan, Russia, China, and South Korea agreed to establish an international committee in 1994 on protection of the fishery resources in the Sea of Japan.[9] The agreement came at a conference sponsored by the Hyogo prefectural government in western Japan. The meeting also agreed to urge North Korea to participate, and to attend the international forum planned for 1994. This forum could have served as the basis and catalyst for a multilateral regime. However there has been no report of any follow-up and it is not clear whether this was an NGO effort or if it has the support of central government.

The Problems and Issues Defined

Yellow Sea/East China Sea[10]

1. Overfishing: how can it best be regulated?

Although the total fish catch from the Yellow/East China Seas has been steady or increasing, the catch of particularly valued species has declined. Major target species are demersal (bottom-dwelling, relatively sedentary) such as small yellow croaker, hairtail, large yellow croaker, flatfish, and cod; and pelagic (free-swimming, highly mobile) such as Pacific herring, chub mackerel, Spanish mackerel, and butterfish. Almost all of these species are overfished and some larger, higher trophic level species have been replaced by smaller, lower trophic level species. Thus the catch-per-unit-effort has declined in quantity and quality. The stocks are fished by China (2.5 MMT), South Korea (1.32 MMT), Japan (0.474 MMT), and North Korea (amount unknown). While Japan's demersal catch has been decreasing, Korean and Chinese catches have been steady or increasing but with changing species composition.

2. Unresolved and overlapping jurisdictional claims: how can these impediments to fisheries management be overcome?

Multilateral fisheries management will be complicated by the multiplicity of actual claims or hypothetical boundaries. Large areas are not covered by any formal jurisdiction or agreement, while some areas are subject to overlapping jurisdiction and agreements.

3. Shared or migratory stocks: how should they be regulated and allocated?

Many species migrate freely across boundaries, while spawning and wintering grounds straddle various jurisdictional lines (see Figure 5.4). Fish migration patterns in the Bo Hai and the northern Yellow Sea are extremely complex. Although this area is almost completely within the jurisdiction of China and North Korea, it harbours many spawning grounds for the demersal and pelagic species that are distributed throughout the Yellow Sea. One spawning area off the Yalu River mouth is divided by these countries' respective claims. Moreover, demersal and pelagic species migrate to and from the northern Yellow Sea, and between Chinese and North Korean waters. The major stocks shared are hairtail, chub mackerel, croakers, Japanese Spanish mackerel, crabs, shrimps, and prawns.

The central and southern Yellow Sea is ringed with spawning grounds for both species groups, and pelagic spawning grounds and demersal wintering grounds occupy the central part of the Sea. Both species groups migrate between North and South Korean waters, and across a hypothetical equidistance line between China and South Korea. The lack of a defined fishing zone around the five islands near the North/South Korea Armistice Line may lead to conflict. Moreover, these species move back and forth across the declared Chinese fisheries zones and in and out of South Korea's fisheries zone (which coincides with its territorial seas), including that around Cheju Island. Most transnational migrations in the southern Yellow Sea are by demersal species.

Like the central Yellow Sea, pelagic spawning grounds and wintering grounds in the East China Sea encompass the entire offshore. Extensive spawning grounds for both commercial demersal and pelagic species occupy the eastern portion of the Senkaku/Tiao Yu Tai (Japan/China) overlap. A similar spawning area extends into the South Korea/Japan joint development area, as do wintering grounds for both types of fishes. Extensive movements of the two species occur throughout the Sea: in and out of the joint development zone, and across the hypothetical equidistance line and China's fisheries regulation zones. Some demersal species also migrate between waters claimed by China and Taiwan.

4. Scattered and fragmentary data: how to persuade countries co-operatively to integrate, analyse, and supplement it?

A single comprehensive view of the fisheries of the Yellow Sea, East China Sea, and the Bo Hai is needed.[11] The necessary data include catch statistics by species, effort, and country; yield-independent surveys of major resource populations on mesoscale spatial and temporal sampling frequencies; and process-oriented studies of ecosystem structure and function. Co-operative research is especially needed for the migratory species, concentrating on life history, stock assessment, and biological dynamics. Data standardization, collection, and exchange are necessary to provide a comparable statistical basis for decisions.

5. Tenuous political and economic relationships: how to balance competing national interests within the fishing sector

The issues include the appropriate role of North Korea in Yellow Sea fisheries and their management, and the possible declaration by China and South Korea of EEZs and the resultant exclusion of Japan from fishing in the Yellow Sea.[12]

Sea of Japan[13]

1. Data on and knowledge of the status of stocks are seriously incomplete

Because there is little publicly available information on the catch and status of stocks in the western part of the Sea, the exchange of information and co-operation in fisheries research and management are necessary and urgent. Scanty information indicates that North Korea's catch is quite large, almost as large as Japan's.

2. Some important species may be overfished

Production from the Sea of Japan, the Sea of Okhotsk, and part of the North-West Pacific, excluding that by North Korea, has increased from about 9 million tons in 1982 to 12 million tons in 1985. Most conventional species are fully exploited, but the total catch might be increased to about 13 million tons. The species composition of the catch has changed, implying changes in the ecosystems of both demersal and pelagic fish. In the coastal fisheries there is concern about the stocks of flying fish, Pacific herring, sandfish, halibut, Alaska pollack, and Japanese sardine.

3. Jurisdiction is uncertain and, in some areas, disputed. Many commercial stocks are shared or migratory. Who owns the stocks and how should they be allocated?

The UNCLOS proscription to ensure proper conservation and management measures is inoperative in the Sea of Japan. Each state bordering it has different conservation and management measures, and they conflict. All simply prefer to catch more fish rather than to conserve them. This is particularly true with respect to the shared stocks.

In the south-eastern Sea of Okhotsk, both demersal and pelagic stocks migrate within and across the territorial and EEZ waters attached to the disputed southern Kurile Islands/Northern Territories. A spawning ground for mixed demersal species lies within these territorial and EEZ waters, and another in the extreme southern Sea of Okhotsk extends across the equidistance line between Japan and Russia, as do migrating pelagic and demersal species.

In the southern Sea of Japan, the pattern is more complex. Pelagic spawning grounds occupy the southern part of the overlapping claims area around Tok Do/Takeshima (South Korea/Japan), and are also divided by the South Korea/Japan continental shelf boundary. Pelagic species migrate through the disputed area and across the boundary. An extensive demersal spawning area reaches north and south of North Korea's claimed EEZ, and both demersal and pelagic stocks migrate in and out of this zone as well as its Military Warning Zone. The Korea Strait is a confluence of demersal and pelagic spawning and wintering grounds and their migration routes, and the South Korea/Japan boundary artificially divides these natural fisheries distributions.

Existing International Regimes

There is presently a web of nine bilateral fisheries agreements in force in the region. Each regional government is involved in at least one.[14] These agreements provide a solid background of experience on which to build a multilateral regime.

Fisheries Agreements

1. Other than for salmon, the agreement between Japan and Russia establishes reciprocal fishing rights and conditions, including fishing grounds, number of vessels to be licensed, total allowable catch, and catch quotas for major species. This agreement provides access for Japanese fishermen to Russian fish—for a fee. Five fishing grounds in the Russian EEZ are stipulated in the agreement. Japanese fishing rights in Russian waters are subdivided into two categories: free fishing and fee fishing.

Bilateral consultation among scientists and fisheries experts has become a permanent feature of the Japan-Russia fisheries regime. It also includes several private-level arrangements: for Japanese crab fishing off Sakhalin, in the Sea of Okhotsk, and in the Sea of Japan in exchange for fisheries co-operation fees; for Japanese kelp and sea urchin fisheries around the Russian-controlled Kaigara Island, east of Hokkaido; for Japanese purchase at sea of Alaska pollack and Pacific herring; and for Pacific cod dragnet fishing in the Russian EEZ. During Soviet Premier Gorbachev's April 1991 visit to Japan, the two countries agreed to co-operate in privatising the Russian Far East fishing industry but their territorial dispute over the southern Kurile islands has prevented progress.[15]

2. Japan and the USSR have several agreements on high seas salmon fishing. In 1988 the two nations agreed to construct salmon hatcheries in Sakhalin, and to harvest the spawn of their joint efforts. To generate funds for this venture, Japan was allowed to catch up to 2,000 tons of salmon in the Soviet EEZ east of the Kurile Islands.

In early 1992, in the face of deteriorating stocks, Japan and Russia agreed to cease salmon fishing outside their respective EEZs, ending Japan's high-seas salmon fishing in the region. Japan was still allowed a catch of 4,000 tons per year in its own fisheries zone.[16] In 1995, this was increased to 5,123 tons reflecting an improvement in the condition of salmon resources due to stocking. As a consequence, Japan and Russia, plus Canada and the United States set up the North Pacific Anadromous Fish Commission, headquartered in Vancouver, Canada. It is designed to assess, 'promote,' and conserve anadromous fish stocks in the North Pacific region. The treaty also called on the signatories to co-operate in the conservation and management of all living resources of the North-West Pacific which lie outside their EEZs. This was an extension of a treaty signed in 1985 between Japan and Russia.[17]

3. The North Korea–Japan agreement was established on a non-governmental basis. It allows access of Japanese fishermen to North Korean fish—for a fee. It defines the conditions of Japanese fishing in the EEZ of North Korea, including squid jigging, salmon gillnet, salmon longline, and crab-pot fisheries. The DPR Korea–Japan Joint Fisheries Co-operative Committee oversees the agreement.

A new provisional agreement was concluded in December 1987. It allowed Japanese small-scale fishing in North Korea's provisional operation zone between its 50 nautical mile Military Warning Zone[18] and its 200 nautical mile EEZ boundary. The agreement also reaffirmed an earlier agreement on Japanese purchase at sea of 50,000 tons of Alaska pollack from North Korea. When Japan had earlier rejected the North Korean demand for a 300,000 ton Japanese purchase, Pyongyang banned all Japanese fishing activities in waters under its jurisdiction. Since 1988 Japan has been required to pay fishing fees in return for the resumption of fishing in the North Korean EEZ. Another provisional agreement was concluded in 1989, providing for continued Japanese fishing in the provisional operation zone in exchange for fees for squid angling, and 'co-operation fees' for driftnet and long-line fisheries.

4. The Japan–South Korea agreement was reached shortly after normalization of relations between the two countries in 1965. It attempts to avoid conflict by regulating the fishing activities of both nations in each other's waters. Its components are shown in Figure 6.1. Each nation has a 12 nautical mile wide exclusive fishing zone around its territories. The South Korean exclusive fishing zone is surrounded by a joint regulatory zone, inside which fishing conditions for both countries are controlled by agreement as to number and size of boats, type of gear, dates of operation, and catch to be shared equally—currently a maximum total of 150,000 tons. There are also two types of subdivision within

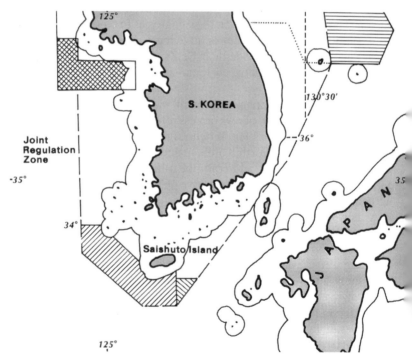

Figure 6.1 Japan–South Korea Fisheries Regulation Zones
Source: Ronald P. Weidenbach and John E. Bardach, 'Fisheries and Aquaculture' in Joseph R. Morgan and Mark J. Valencia (eds.), *Atlas for Marine Policy in Northeast Asian Seas*, Berkeley: University of California Press, 1993, p. 121.

the joint regulatory zone: trawlers of more than 50 tons are not allowed to operate east of meridian 128°E, and mackerel fishing from boats of less than 60 tons is restricted to a zone south of 35°309N, north-east of the peninsula, and south of 33°309N, north-west of Cheju Island. Only fifteen such mackerel boats are allowed to fish. Beyond the joint regulatory zone, the agreement provides for a joint fishery resources zone in which the two governments would conduct scientific surveys to assess the condition of the resources.

The 1965 agreement also established a binational fisheries commission to discuss and provide advice on scientific studies and regulatory measures based on such studies, as well as the delimitation of the joint resource survey zones, provisional fisheries regulatory measures, and matters concerning the safety of fishing and general procedures for handling accidents involving fishing boats. Final decisions on fisheries policies were delegated to the governments. The 1965 agreement also provided for dispute settlement procedures.

More recently, Japan and South Korea have concluded The Agreement on Voluntary Fishing Regulations, a separate domestic, self-regulatory scheme for the fishermen of one operating in the other's coastal waters beyond the joint regulatory zone. The agreement adopted the flag-state principle for enforcement and court jurisdiction outside the 12 nautical mile fishery zones. It was established because South Korea has not yet extended (and might not extend) its jurisdiction, while Japan's extended jurisdiction had not yet been applied to South Korean vessels. The scheme covers the Japanese trawl and purse seine fisheries operating along the South Korean coast, particularly south and west of Cheju Island, and trawl, squid jigging, and pot fishing by South Korea along the Japanese coast, particularly around western Hokkaido (Table 6.1). Although Japanese fishing off Cheju Island virtually stopped under this agreement, South Korea's unwanted fishing in western Japanese waters decreased but is still significant nonetheless (Table 6.2).

In October 1987, Seoul and Tokyo agreed that the South Korean trawl fisheries would gradually be phased out, and that by April 1991 they would be totally eliminated inside the otter trawl prohibition line in Japanese waters. In return Japanese bull trawlers operating in South Korean waters would be reduced by 50 per cent, and the fishing period for the remaining boats halved by April 1991.

Table 6.1 Regulations Inside the Joint Areas Set by the Japan-South Korea Fisheries Treaty

Type of fishery	Size of boat (tons)	Maximum effort (boats)	Mesh size	Light power	Catch (tons)	Remarks
Trawler <50 tons	Other trawlers 100–550	115	33 mm	—	10,000	Following domestic area restrictions are to be observed by the other party: (1) Closed area to trawl fishery set by Japan (2) Closed area to trawl fishery set by Korea (3) Closed area to purse seine fishery set by Japan
Trawler ≥50 tons	Other trawlers 30–170	Nov 1–Apr 30 270; May 1–Oct 31 100	54 mm	—	30,000	
Purse seine	40–100	Jan 16–May 15 60 nets; May 16–Jan 15 120 nets	30 mm	10.0 kw × 2; 7.5 kw × 1; Total 27.5 kw	110,000	
Mackerel angling—60 tons	60–100	Jan 1–May 31 0; Jun 15–Dec 31	—	10.0 kw		Area limited

Source: Mark J. Valencia, *International Conference on the Yellow Sea*, Occasional Paper No. 3, East-West Environment and Policy Institute,

Table 6.2 Incidents of South Korean Illegal Fishing in Japanese Waters

1988	2,117 cases
1989	2,136 cases
1990	1,430 cases
1991	936 cases
1992	1,055 cases
1993 (to October)	1,291 cases

Source: Young-Hoon Chung, 'South and North Korea Cooperation in Fisheries: Toward Establishing a Joint Fishing Zone', Master's Thesis, University of Delaware, 1993.

The two also agreed that South Korean dragnet squid angling, and conger eel pot fisheries in Japanese waters would observe strict prohibition of area and period, and that Japanese dragnet, squid angling, and coastal fishing would be subject to similar prohibitions. Under the current agreement, which became effective in March 1992, they further agreed to strengthen and expand their control of fishing operations by extending the joint control arrangement, including the assignment of officials from both governments on the same patrol boats.

5. In 1975 Japan and China signed an agreement that formalized and extended a succession of non-governmental fisheries agreements. The agreement regulates Japanese fishing in Chinese waters. The first of these non-governmental agreements had been signed in 1955 between the Japan-China Fishery Association of Japan and its Chinese counterpart, the Fishery Association of China. They agreed that the fishermen of both countries should comply with Chinese fishery regulations set forth in 1950 and amended in 1955 to protect the coastal fishery resources of the East China Sea and Yellow Sea from motorized trawl fishing. A Military Warning Zone in the northern part of the Yellow Sea was also delineated. A second non-governmental agreement was concluded between the fishery associations in 1963, and in 1970, they further agreed to regulate motorized purse seine fishing with attractant lights in the East China Sea.

The 1975 governmental agreement has been revised several times, producing a complex pattern of regulations (Table 6.3; Figure 6.2) which include or have included the establishment of a horsepower (hp) restriction line inside which trawlers and purse seiners of 660 hp or more are prohibited; closed areas or suspension areas which are completely closed during designated periods; conservation areas or protected areas in which the number of boats is restricted during designated periods; and fishing restrictions regarding minimum body length, minimum mesh size, light intensity fish-attracting devices, incidental catch limit, and others.

A published regulations map based on the Japan-China fisheries agreement as amended in 1985 indicates that all of the closed areas and trawl conservation areas designated in earlier agreements have been retained and several new areas have been added. The map also shows a westward shift in the eastern boundary of the horsepower restriction line and the purse seine conservation areas, or their elimination altogether. It does not list other previous restrictions or distinguish between different boat sizes and types.

6. A North Korea–Russia agreement establishes reciprocal fishing rights, with catch quotas for each in the other's waters to be decided at regular meetings of the Joint Fisheries Committee set up by the two countries. Scientific and technical co-operation has been discussed in detail at these committee meetings. Large North Korean trawlers have been reported fishing in both the Sea of Japan and the Sea of Okhotsk. The agreement also provides for Russian small and medium-size fishing within the triangular shaped EEZ of North Korea outside its Military Warning Zone. It has been renewed several times, most recently on 20 December 1992.

Table 6.3 History of the Japan–South Korea Agreement

The First:	3-year Agreement	16 November 1980–31 October 1983
Extended:	3-year Agreement	1 November 1983–31 October 1986
Revised:	4-year Agreement	January 1987–December 1991
Extended:	3-year Agreement	March 1992–31 December 1994

Source: Letter to author, 1995, from Douglas Ancona, Oceans and Natural Resources Regional Attache, United States Embassy, Tokyo, Japan.

FISHERIES

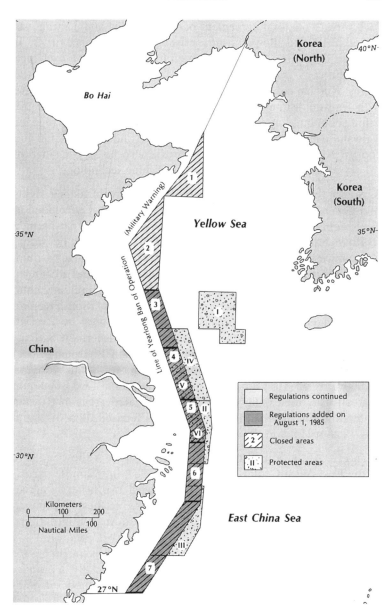

Figure 6.2 Japan–China Fishery Regulation Zones
Source: Ronald P. Weidenbach and John E. Bardach, 'Fisheries and Aquaculture' in Joseph R. Morgan and Mark J. Valencia (eds.), *Atlas for Marine Policy in Northeast Asian Seas*, Berkeley: University of California Press, 1993, p. 122.

7. A private bilateral agreement between Russia and Taiwan was signed in 1991 to allow Taiwanese fishing in the Russian EEZ.[19] It provides access for Taiwan fishermen to Russian fish in exchange for technology transfer and probably fees. Under its terms, the largest Russian conglomerate, Sovrybflot, will form a joint venture with Taiwan fishing companies to facilitate fisheries trade and fishing boat building and repairs. Sovrybflot will further arrange for Taiwanese and Russian firms jointly to catch and process squid and cod off Russia's Pacific coast. There will also be an exchange of fishery experts and researchers, and of fishing technology.

8. South Korea and Russia established reciprocal fishing rights in a government-to-government fisheries agreement concluded in 1991.[20] This agreement provides access for South Korean fishermen to Russian fish in return for a fee or fish in kind. Russia permitted Korean vessels to catch 30,000 tons of Alaskan pollack, squid, saury, and other species in Russia's fishery zone and in return, South Korea transferred 30,000 tons of saury, mackerel, and chub mackerel to Russia. Because there are about 100,000 South Korean fishing boats in South Korean waters, fishing there by Russian boats would create conflicts with South Korean fishermen. So Russia agreed to receive fish caught by South Korean fishermen instead of catching it themselves, and also to a quota for South Korea of 50,000 tons of Alaska pollack in exchange for a fee. But South Korean fishermen believed that the designated areas in Russian waters did not have sufficient stocks, and focused their fishing in the 'Peanut Hole' in the Okhotsk Sea instead.

9. and 10. At the International Conference on the Yellow Sea held at the East–West Center in Honolulu in 1987, the Chinese participants reported that there had been at least two fisheries agreements between North Korea and China. One of these pertains to fishing near the Yalu River mouth, North Korea's international border with China.

Joint ventures are also a form of fisheries relations. Japan has for centuries been a leader in their development, especially in primary industries like fishing. In 1980, Japan still maintained 276 joint ventures with investments totalling US$92 million, and numbers have increased slowly but steadily since.[21] At present, there are twenty Russo-Japanese joint ventures in fisheries, although many are in financial trouble.[22]

While it does not pertain directly to North-East Asian waters,

China, Japan, South Korea, Russia, Poland, and the United States agreed in Tokyo in July 1993 to suspend pollack fishing in the Bering Sea 'Doughnut Hole' for two years to give the resource a chance to recover. They also decided to negotiate an agreement on the management of pollack resources and fishing quotas after the suspension ends.

An analysis of this web of agreements and arrangements reveals certain commonalities. First of all, North-East Asian countries do appreciate the value of scientific information, and they can act in multilateral concert when stocks are threatened.[23] Except for the North Korea–Japan and Russia–Taiwan arrangements, all agreements established specific fisheries committees as advisory bodies. The committees hold joint annual meetings to discuss the state of the resources and fisheries, and to revise the fishing conditions for the next fishing season. Scientific meetings are usually organized as a scientific subcommittee under the parent bodies. These meetings are regularly held just before the annual meeting of the main committees, and occasionally, when required by the parent committee. Details of the discussions held or the documents used at such meetings are not made public.

Second, Japan is by tradition and mastery of techniques the most important fishing nation in the region, a fact further demonstrated by the extent of its deployment and size of catch, and its bilateral agreements with every country. But because of the tenuous relations between Japan and several of the region's nations, these agreements are often between fishermen's organizations rather than governments. And they apply more to bottom fisheries than to pelagic fisheries, which migrate and occupy different areas at different times for spawning and feeding.

Third, overall political relations provide the context for the agreements. For the Sea of Japan, both the Japan–South Korea and Japan–Russia agreements were reached within the context of the normalization of respective bilateral diplomatic relations. In both cases, the fisheries talks affected, and were affected by, negotiations concerning overall diplomatic relations. Despite their shortcomings, the Japan–Russia and Japan–South Korea fisheries regimes have provided a modicum of stability and predictability for fisheries relations in the postwar period. The annual fisheries committee meetings under the two regimes have allowed fisheries experts to exchange their respective assessments of the fisheries

Table 6.4 Incidents in the East China Sea, 1992–1993*

Year	Date	Ship Name (Type)	Nationality	Event
1992	Apr. 9	Luna (Cargo)	Panama	Δ
	Apr. 9	Ryoei-Maru Nr. 5 (Fishing)	Japan	○
	May 19	Takara-Maru Nr. 5 (Fishing)	Japan	○
	May 20	Vinh (Cargo)	Vietnam	Δ
	July 24	Dae Bong Nr. 1 (Type ?)	South Korea	○
	Oct. 2	Gyosei-Maru Nr. 56 (Fishing)	Japan	○
	Oct. 28	Shotoku-Maru Nr. 21 (Fishing)	Japan	○
	Oct. 29	Yahata-Maru (Fishing V)	Japan	○
	Oct. 29	Tosan-Maru Nr. 25 (Fishing)	Japan	←
	Nov. 3	Tosan-Maru Nr. 21 (Fishing)	Japan	Δ
	Nov. 4	Daikichi-Maru Nr. 16 (Fishing)	Japan	○
	Nov. 12	Kwaeki 616 (Fishing V)	Taiwan	○
	Nov. 12	Chosei-Maru (Fishing V)	Japan	←
	Nov. 14	Taiyo-Maru Nr. 8 (Fishing)	Japan	○
	Nov. 18	Eishin-Maru Nr. 36 (Fishing)	Japan	←
	Nov. 19	Shoei-Maru (Fishing V)	Japan	←
	Nov. 24	Meiriki-Maru (Fishing V)	Japan	○
	Dec. 4	Seiwa-Maru Nr. 18 (Fishing V)	Japan	←
	Dec. 7	Koyo-Maru (Fishing V)	Japan	←
	Dec. 8	Arman (Tanker)	Panama	χ
	Dec. 9	Kyokuto Maru (Container)	Singapore	◊
	Dec. 9	Miki-Maru Nr. 31 (Fishing)	Japan	○
	Dec. 9	Matsuyoshi Maru (Fishing)	Japan	←
	Dec. 10	Shoko-Maru Nr. 1 (Fishing)	Japan	←
	Dec. 11	Kakuei-Maru (Cargo)	Japan	○
	Dec. 12	Syukur (Cargo)	Indonesia	◊
	Dec. 18	Tanikawa-Maru (Fishing)	Japan	←
	Dec. 19	Manila Shine (Cargo)	Japan	Δ
	Dec. 25	Hose-Maru Nr. 21 (Fishing)	Japan	○
	Dec. 26	Ofunayama-Maru (Cargo)	Japan	Δ
	Dec. 27	Hosei-Maru (Fishing V)	Japan	○
	Dec. 28	Shinanogawa-Maru (Tanker)	Japan	Δ
	Dec. 30	Harimatun (Cargo)	Panama	◊
1993	Jan. 1	Bohai 2 (Cargo)	Panama	◊
	Jan. 7	Shinkai-Maru (Survey ship)	Japan	←
	Jan. 14	Pacific Queen (Cargo)	Honduras	Δ, ◊

Table 6.4 *Continued*

Year	Date	Ship Name (Type)	Nationality	Event
1993	Jan. 14	Genfuku-Maru Nr. 73 (Fishing)	Japan	◊
	Jan. 31	Uchimura-Maru Nr. 83 (Fishing)	Japan	Δ
	Feb. 2	Yushou (Cargo)	Japan	◊
	Feb. 2	Orange Ocean (Cargo)	Japan	◊
	Feb. 3	Tenryu-Maru Nr. 7 (Tanker)	Japan	◊
	Feb. 4	Beccles (Cargo)	Panama	←
	Feb. 4	Sukhnichi (Cargo)	Russia	◊
	Feb. 6	Ryuzan-Maru Nr. 15 (Cargo)	Japan	←
		Total:	45 cases	

* To 10 February 1993
Δ: Vessel inspected involuntarily
○: Vessel chased
←: Vessel approached
X: Vessel fired on without warning
◊: Vessel received warning shots across the bow

Source: CSCAP Maritime Cooperation Working Group.

situation had become particularly dangerous with the increase of Chinese fishery vessels in this militarily sensitive zone because South Korea's navy cannot always immediately distinguish a Chinese fishing vessel from a North Korean vessel.[48] And in late May 1995, a South Korean fishing boat was seized by North Korean gunboats 16 miles north of the North/South Korean western border, heightening already tense relations.[49]

Japanese fishing vessels poaching in the disputed waters around the Kuriles threaten to aggravate Tokyo's already tense relations with Russia. In August 1994 Moscow issued an order allowing Russian border guards to fire on foreign vessels violating Russian borders and disobeying lawful orders. In September 1994 the Russian Coast Guard shot and killed two Chinese fishermen and seized a South Korean boat; both vessels attacked were under charter to a Japanese company. And in October 1994, the Coast Guard sank a Japanese fishing boat.[50] In November 1994, Japan's Maritime Safety Agency announced that since the start of 1993, eighteen Japanese fishing boats had been seized and four Japanese citizens wounded.[51] In March 1995, Russia launched a new campaign against poaching.[52]

The situation is becoming increasingly dangerous. More than 5,600 Japanese fishing vessels operate around the frontier, most of them small and quick enough to dodge coastal patrol vessels. Between 5,000 and 9,000 violations are registered annually, and many more go undetected. The fishing grounds are irresistible to Japanese fishermen despairing of their own overfished waters. A confluence of warm and cold currents in the deep straits gives the area a yield of as much as 1.5 million tons of sea products annually—about 25 per cent of Russia's total annual seafood supply—with a value estimated at US$1 billion. With a quick haul of a few nets, Japanese fishermen can catch tens of thousands of dollars worth of crab, trout, and sea urchin which is far more than they could catch at home.

The August 1994 incident has prompted Japan to propose a private non-governmental fishery agreement with Russian firms to ensure safe fishing in the waters around the disputed islands. The agreement would guarantee that Japanese vessels would not be harassed, in exchange for Japan's provision of funds for joint management of the resources and for enhancing the social infrastructure and welfare of residents on the four disputed islands. But Japan has refused to pay fees to the Russian government for fishing in the area as Moscow proposes, because that might undermine its position in the dispute.[53] And Russia continues to insist that it has the authority and the will to prevent illegal fishing there.[54] One positive sign is the ongoing discussion between Japan's Maritime Safety Agency and the Russian border guard organization concerning co-operation on maritime safety and the prevention of drug smuggling by sea in the area.[55]

Alternative Regimes

Many models are available, ranging from the most formal organizations and commissions at one end of the spectrum, to a hybrid of non-governmental fora or networks at the other end that offer the advantages of informal back-channel diplomacy. In discussing alternative arrangements, it is useful to classify them according to the objectives that they might fulfill.[56]

Objectives

1. To exchange and disseminate scientific and technical information within the region

Data exchange is a task frequently assigned to regional organizations. The North Pacific Marine Science Organization (PICES) has a direct concern with research data, since this is the service it was designed to provide to other bodies charged with the responsibility for fishery policy, development, and management. Of the North-East Asian fishing states, only three, China, Japan, and Russia, were involved in the negotiation of the PICES treaty, along with Canada and the United States. One option for data-sharing arrangements in North-East Asia is to expand PICES, including in it all the countries that fish in the North Pacific, and perhaps even extraregional fishing states such as Poland. Still, arguments can be made against creating an unwieldy, pan-North Pacific scientific organization of this kind. Interpersonal trust and professional credibility may be the most important ingredients in a successful data-sharing arrangement, and the non-governmental network approach may be more beneficial than an all-inclusive PICES forum.[57]

2. To promote, co-ordinate and conduct co-operative research within North-East Asia and to facilitate research with adjacent and other relevant regions

One of the features of the UNCLOS is the regulatory system for marine scientific research contained in Part XIII. After much controversy, the delegations agreed that research in the EEZ and on the continental shelf may be conducted only with the consent of the coastal state.[58] In these areas adjacent to its coastline, a coastal state may withhold consent for the conduct of a research project of another state or 'competent international organization', if that project 'is of direct significance for the exploration and exploitation of natural resources . . .'.[59] This language reflects, and indeed reinforces the reluctance of many countries, especially developing coastal states, to permit foreign research in their offshore waters that might put them at a knowledge disadvantage in the development and management of fishery resources there.

Clearly, then, one incentive for regional co-operative research arrangements is the existence of current 'commons' areas of unknown resource potential. And there are still considerable high seas areas in North-East Asia beyond the limits of national jurisdiction, although this may not remain the case for long. Nevertheless, most states in the region have an interest in co-operative research as a basis for effective high seas fishery management, particularly of straddling stocks. It is expected that the present limited-membership version of PICES will facilitate experiments in co-operative research within the region. It might also be expected that the non-governmental Pacific Economic Co-operation Council (PECC) Fisheries Task Force and/or an APEC subcommittee will create research-related linkages between North-East Asia (or the North Pacific) and adjacent South-East Asia.[60]

Fishery resource management requires reliable information on the status of the resources in demand. But parties to bilateral regimes in the region often come up with conflicting resource assessments. This fact alone argues strongly for co-ordinated efforts in resource studies. Yet in each country of the region, there seems to be a dichotomy between government scientists and academic scientists. This means that those with the best access to the fishery resource data are least able to divorce themselves from the politics of their government's position, while those able to maintain an independent posture have less access to the data, or are unable or unwilling to share it. In addition, the motivation of representatives of the fishing industry may be difficult to anticipate or control. In South Korea, for example, whenever a species becomes the focus of popular concern, the source of information on it tends to dry up.[61] The problem of inadequate data is exacerbated by the inability of governments to support coastal fishery resource conservation policy. Fishery biologists and economists have emphasized the necessity of an integrated data base for management purposes but government response has been inadequate.

However, one role of the suggested organization could be to co-ordinate interregional co-operation among scientists. National researchers do not seem to have adequate access to the published literature of colleagues in other nations in the region. It should be feasible to organize at least an exchange of publications among the principal researchers around a given sea. FAO regional fisheries bodies have done extensive scientific work in areas where international co-operation is difficult and could be of assistance. Another approach could be the establishment of a Yellow Sea or Sea of Japan working group through existing international organizations like the Intergovernmental Oceanographic Commission's WESTPAC. Yellow Sea fishing nations are already members. One advantage of such an arrangement would be the augmentation of existing expertise by collaboration and capabilities from out-side the region. However, it is doubtful that national decisions would be taken based solely on research by an outside scientific organization.

3. To exchange views on current fishery policy and management issues and developments within North-East Asia, and to promote and conduct such consultations with adjacent and other relevant regions

Most major marine fisheries are part of a larger marine ecosystem, and thus international in scope, and almost all are international in significance from legal, political, diplomatic, and commercial perspectives. Few of the fundamental issues in fishery policy and management can be confined to the national level of analysis without the risk of serious lacunae and distortions. Even in the age of extended coastal state jurisdiction, most questions of fishery policy and management have an international dimension. And the regional level of consultation is the most convenient level for the exchange of policy and management ideas and proposals.

Since the end of the UNCLOS negotiations, most opportunities for relevant 'back-channel diplomacy' have been provided by nongovernmental organizations and networks. Given the difficulties of promoting intergovernmental co-operation in North-East Asia for the fisheries sector, it might be wise for the governments of this region to follow the same course, at least initially, and encourage institutions such as the East–West Center, the United Nations University, and the PECC Fisheries Task Force to organize regular consultative meetings and co-ordinate linkages with counterpart officials and specialists in adjacent and other relevant regions, perhaps in co-operation with extra-regional networks.[62]

An agenda might include such topics as the conditions of access to the EEZ of another state, high seas fishery management issues (such as those affecting straddling and highly migratory stocks), problems associated with joint ventures, opportunities for co-operative management, allegations of illegal fishing and enforcement issues, relevant GATT issues and rulings, and the extraterritorial application of national environmental standards to fishing practices.

4. To provide fishery development and management assistance within North-East Asia and to co-ordinate such assistance to adjacent and other relevant regions

In the context of fishery development and management assistance, the North-East Asian region consists of both contributor and recipient states. Training programs are expensive undertakings, and although private foundations might be expected to provide some support, most funding for this purpose in North-East Asia will have to come, directly or indirectly, from regional governments. Only a small fraction of regional fishery training requirements can be met through long-term academic degree and diploma programs conducted by universities and major oceanographic institutions. Most training of this kind will have to depend on short-term intensive certificate courses.

North-East Asian trainees currently participate with trainees from other regions in fishery-related training programs offered by institutions as diverse as ESCAP, FAO, International Center for Living Aquatic Resources Management (ICLARM), the International Oceans Institute (IOI), PECC, and JICA. Interregional training courses might be developed further in coming years through co-operation among other institutions such as the Asian Institute of Technology (AIT), East–West Center, the South-East Asia Fisheries Development Council (SEAFDEC), the South-East Asia Program on Ocean Law, Policy and Management (SEAPOL), and the United Nations University.

A case could, of course, be made for creating a new, wholly North-East Asian organization for regional training in the fisheries sector, given the insufficiency of existing training courses, both short and long term, elsewhere in the world. But training requirements are a 'bottomless pit' and training resources always scarce. It might be more cost-effective for the governments and other institutions of North-East Asia to avoid the costs of establishing their own training facility and invest instead in the strengthening of existing arrangements.

5. To harmonize national fishery development and management policies and practices within North-East Asia

Policy harmonization is a principal function of an organization committed to regional integration such as the European Community. In rhetoric, if not in substance, it is also an objective of continental organizations such as the Organization of American States (OAS) and the Organization of African Unity (OAU). No such regional organization is immediately in prospect for North-East Asia. Nevertheless, intraregional harmonization of national policy and legislation in selected sectors can be undertaken by less ambitious consultative networks such as the Association of Southeast Asian Nations (ASEAN) for certain states in South-East Asia, and the Nordic Council for Scandinavia and neighbouring countries (Denmark, Finland, Iceland, Norway, and Sweden). A similar consultative mechanism for fisheries in North-East Asian (or North Pacific) countries is a possibility in the 1990s, given the conflict potential of fishery issues in the post-UNCLOS era.[63] The initiative would presumably have to be taken by the governments themselves, not least to ensure the effective participation of all relevant ministries and agencies.

6. To negotiate and conduct co-operative fishery management activities within North-East Asia (or the North Pacific)

'Co-operative fishery management' is a concept that could include virtually every facet of fishery development, management, and conservation: exchange of data, research, surveillance, enforcement, entry limitation policy-making, quota allocation, gear regulation, vessel inspection, joint ventures, and marketing. Intergovernmental co-operation is, of course, more difficult to promote in some of these areas, such as surveillance and enforcement, than in others.

7. To facilitate measures designed to avoid or control fishery-related disputes and conflicts in the region, and to co-ordinate efforts to deal effectively with similar disputes and conflicts with fishing states in adjacent and other relevant regions

An ideal fisheries management system would have several components.[64] The fishery would be managed as a unit, which means that its boundaries should not exclude any large portions of individual stocks. For the purpose of fishing, it would be agreed that no boundaries would intersect the fishing unit thereby eliminating the access problem and giving a single entity authority to manage the stocks. The unit would be leased to a single industry entity chosen on the basis of its capability, qualifications, and potential performance. It would be accountable to government(s) to maintain production above a certain level (to avoid a monopoly) and to prevent the stock from declining below a certain level (to conserve the stock). The selection of an entity would not mean that it had a perpetual right to fish the unit. Its performance would be continually reviewed by the concerned governments and, if found lacking, it would be replaced. There would be no need for enforcement in the usual sense because it would be in the entity's self-interest to abide by its own rules in order to maximize its profits.

Under the idealized system, government would not be a regulator but a monitor of the performance of the industry, ensuring that it is working within acceptable bounds. It would set standards of performance, both economic and regarding conservation. Government would also provide information to keep the market working efficiently, and conduct long-term research that individual entities could not afford to undertake and which would, when accomplished, improve the fishery management process.

If the stocks in a management unit are unfished, then the establishment of a fishery entity should be relatively simple. If the stocks are already fished, then implementation will be a much more difficult process. If it is to be at all successful, the existing participants must be compensated for their capital and livelihood. Financing the compensation might be done through the sale of bonds, using the potential rent from the fishery as collateral as well as public-sector funds. The existing participants could be compensated either directly with cash payments or with shares of stock in the fishery entity. Initially, fishermen who participated in the fishery and wanted to go on fishing could do so. While they would not own capital, they would own shares of stock and continue to fish on their former vessels.

At some point, the management entity would make the business decision either to increase or decrease its capital stock. A strategy for increase would follow standard capital development procedures. A decision to decrease capital, however, would dislocate fishermen. Nevertheless, the problem would not be as difficult as in the traditional system because fishermen would be given incentives to leave an overcapitalized system. That is to say, if the fishery was overcapitalized, the management entity would decide on the optimal amount of capital, and sell or otherwise salvage the inefficient units. Inefficient fishermen would have an incentive to leave the system because the value of their shares of stock would increase.

The Example of the Yellow Sea

For the Yellow Sea, there are three general alternatives: maintenance of the status quo, implementing international trends, or a *de nouveau* Yellow Sea approach.

Maintenance of the Status Quo

The existing mix of national regulations and international agreements has survived with minor changes for more than two decades. The lack of agreed maritime boundaries between China and South Korea, and of diplomatic relations between South and North Korea makes regulations in the central portions of the Sea ambiguous but coastal regulations relatively certain. Although this has been a positive circumstance for international relations rather than a negative one, the arrangement requires considerable voluntary restraint by all nations fishing in the Sea. And these voluntary restraints are an increasing irritant to all parties.

Several factors are causing regime instability: continued deterioration of the resource, better fish detection and capture technology, expansion of fishing capabilities and grounds by all three coastal states, and impending declaration of EEZs by China and South Korea. The declarations will affect their economic and political relations with Japan. They will not be taken lightly. Furthermore, a coastal state's demand for extended jurisdiction necessitates either direct negotiations to establish fishing zone boundaries and agreement on regulations, or indirect recognition of respective claims. A further destabilizing factor could be the entry of another fishing nation such as Russia, or a markedly increased take by Hong Kong or Taiwan in Yellow Sea fisheries. Without a more precise definition of national jurisdictions in the area, new entry is theoretically possible outside, or even within the boundaries of existing bilateral agreements.

Implementing the International Trend

Implementing the international trend can be defined as implementing the provisions relevant to fisheries in the UNCLOS. They establish a 12-nautical-mile territorial sea, and a 200-nautical-mile EEZ, within which coastal states regulate fisheries subject to a variety of responsibilities. The provisions also state that coastal states shall seek to agree with neighbouring states on the conservation and management of shared stocks, and that, together with fishing states, they shall seek to agree on conservation measures for straddling stocks in adjacent high seas areas.

The four major fishing nations in the Yellow Sea (China, North Korea, South Korea, and Japan) are signatories to the Convention. Although none has yet ratified it, their signatures are substantial evidence that there is agreement, in principle, with the object and purpose of the Convention. All four have been observed to respect the claims of other nations outside the region in implementing their EEZ regulations. And all have joint venture or fishing access agreements with other states that attest to this fact.

Why then have the provisions of the Convention not been implemented in the Yellow Sea? There are several reasons, beginning with the absence of agreement on a boundary, the reluctance to raise the boundary issue, and the fact that the Convention does not *require* states to claim EEZs; it only establishes some parameters for those who wish to do so. Second, declaration of EEZs would disrupt a sensitively balanced fishery regime and replace it with a fragile and possibly unstable arrangement. Third, as long as diplomatic relations do not exist between North and South Korea, it would be difficult to implement a more satisfactory fishery resource regulatory regime. Fourth, declaration of EEZs, even if limited to the Yellow Sea, would pose substantial problems of precedent for other issues in fisheries and for the broader economic and political arenas.

What would coastal states stand to gain from establishment of EEZs? Extended jurisdiction holds the promise of marginally more catch for China and South Korea at the expense of the current small harvests taken by Japan. Japan might respond by instituting restrictions on imports, or engage in a 'fish war' by increasing harvests in the East China Sea that would affect harvests in the Yellow Sea. China and South Korea might retaliate by forming a united front against Japan regarding fish marketing. In sum, the benefits to be gained by extended jurisdiction come only through the ability successfully to improve regulatory measures over a much wider area than previously administered. Extended jurisdiction is no panacea for improved management.[65]

A *De Nouveau* Yellow Sea Approach

The Yellow Sea is a unique area, and the political, economic, and social relationships within it are also unique. Possible management responses that move incrementally from the *status quo* and do not require extension of jurisdiction include: modification of the existing arrangement; creation of a quadrapartite non-governmental arrangement; and the establishment of a scientific organization.[66]

The present bilateral agreements could be used as a basis for discussing coastal state/distant-water fishing concerns as well as for developing a co-ordinated approach to improving the scientific basis for regulation. The key element in this fresh arrangement would be the extent to which Japan is willing or able to be a 'hinge' state on what may be a 'closing door' for its fisheries access. In this scenario, Japan could assume the role of mediator, information broker, and analyst in co-ordinating international scientific studies, in exchange for continued access to the fisheries. But there are impediments relating to lack of leadership and political contacts, obstinate bureaucracy, and mutual suspicion.

If Japan and the coastal states with which it has fisheries agreements in the Yellow Sea could agree that it should play this role, there could be relatively direct communication concerning all aspects of fisheries science, conservation, and eventually regulation. By providing the hinge-state function, Japan might ensure itself continued access to Yellow Sea fisheries as compensation for its invaluable co-ordinating services. The arrangement could stave off coastal state declarations of EEZs. And it would permit communication on fisheries regulation matters to be systematic and predictable regarding standardization of statistics, co-ordination of scientific research, delineation of shared stocks, or evaluation of overfishing.

Japan could perform a similar hinge-state function for non-governmental arrangements with a willing North Korea. In the event that Japan is unwilling to perform this role, or the arrangement is unacceptable to either China or South Korea, another third-party state or entity, for example, the Indo-Pacific Fisheries Commission or the FAO, could be asked to play it. Non-governmental fisheries arrangements were instrumental in the development of the China-Japan fisheries agreements, and Japan has used them with some success with North and South Korea. They have the advantage of avoiding the problems of direct government-to-government negotiation. But their ability to foster serious discussions of fishery regulation problems is likely to be limited.

Suggestions and Next Steps

Allocation of Shared Stocks

Scientific information and input is critical to the design of a fisheries regime, particularly one concerning the sharing of joint stocks.[67] The catch of most fish stocks is controlled by setting an upper limit to the amount that can be caught—the total allowable catch (TAC). TACs are set on the basis of advice from international teams of fisheries biologists, who usually pool their data and assess the stocks in question. Once the biological advice is available, the process of setting TACs begins. For exclusively owned stocks, it is done by the coastal states. For shared stocks, bi- or multinational negotiations are necessary. Since most stocks in North-East Asia are overexploited, fisheries biologists will probably recommend considerable—even drastic—cuts in catches. These recommendations would be unpopular among fishermen, and the industry throughout the region would lobby for higher TACs. If the lobbying is successful, the recovery of the stocks will be delayed or even prevented.

Once the TACs have been agreed, the share for each nation must be allocated. UNCLOS obliges states to use scientific evidence to allocate shares of stocks,[68] and there are two bases on which to do so: historical catch and zonal attachment, which is the share of the stock present in a country's EEZ (or what will be its EEZ). The latter principle has gained currency with the advent of EEZs, but in North-East Asia, China and South Korea have not declared EEZs and, moreover, the extent of EEZs has not been agreed. In the Yellow and East China Seas, then, historical catch may be the main guide to allocation, which would clearly favour Japan.

Zonal attachment may be used in the Sea of Japan. It can be based on spawning areas; distribution of regional larvae; the occurrence of juvenile fish; the occurrence and migrations of the fishable part of the stock; the history of the fishery, including the distribution of catch, rate of exploitation, and fishery regulations; and the state of exploitation of the stock. But available data on North-East Asian pelagic stocks does not permit the calculation of zonal attachment. Moreover, there is no objective scientific method for weighting the different criteria.

The distribution of the commercial catch can be used as a basis for the calculation but this approach excludes eggs, larvae, juvenile fish, and spawning grounds from consideration. The major criteria are the distribution of adult fish and the availability of particular species, calculated on the basis of average allocation of catches in the various zones over a given period. The calculation is also influenced by fishing effort (number of vessels and fishing time), distance to home port, and the gear employed. The zonal attachment method cannot be used at all if there are insufficient data, the EEZ boundaries are not precisely known, or the countries cannot agree on the size of the stocks. For North-East Asia, then, the allocation of at least shared pelagic stocks might be based on historical performance and political realities; that is, Japan would get the larger share, moderated by the necessity to make sidepayments to neighbours who might otherwise defect from the regime.

One of the most elaborate schemes to regulate fisheries is that in the Australian state of New South Wales.[69] The state gives fishermen shares, which are registered, like land titles, in each fishery, and distributed as a proportion of their past catch. Each year owners give back 2.5 per cent of their shares to the government, which sells them and keeps the proceeds. This reinforces the concept that the fisheries are a communal resource even though the right to fish can be owned by individual fishermen. Fishermen can sell all or part of a year's fishing rights, which allows them to buy quota from other boats to cover fish in their catch of the wrong quality or species, rather than throw them away.

One of the terms of the shares is the acceptance of restrictions, set for ten years, on fishing inputs, such as the kinds of nets and boats fishermen can use. Penalties for cheating are steep. If the rules are changed early, fishermen can fish under the old rules up to the end of the ten-year period already underway, or accept the new rules for a new ten-year share. The aim is to ensure that the scheme evolves in tandem with scientific understanding of different fisheries, but also to make changes in the rules gradual and predictable, so that they do not undermine confidence in the system. Simpler but more controversial is a system which determines how many boats can be licensed, and then allocates the licenses on some equitable basis. If a multilateral fisheries organization could be established, one of these schemes could be applied in North-East Asia. Surplus boats can be weaned from the industry through cash incentives.

Definition of the 'Region' for Fisheries Management

For the purposes of fisheries management, there are several overlapping international marine regions in North-East Asia. Although the Sea of Japan and the Yellow/East China Sea are semi-enclosed seas and can be 'core areas' for multilateral fisheries management, many species of interest to one or more of the coastal nations are distributed between and beyond these discrete geographic units. And the North-West Pacific Ocean is a major fishery for several of the littoral nations, as well as for extra-regional states. For statistical reporting purposes, FAO treats the broad north-west Pacific as a single area[70] (Figure 6.3). But significant parts of this area fall outside the jurisdictional claims of the Pacific-facing entities: Russia, Japan, and Taiwan. Given these incongruent natural, institutional, and political fisheries regions, *ad hoc* agreement on a core area or areas within the potential jurisdictional purview of the littoral nations seems the wisest approach. So North-East Asian fisheries regimes could focus, at least initially, on the Sea of Japan (Japan, North Korea, and Russia) and on the Yellow/East China Seas (China, Japan, South Korea, and Taiwan), perhaps with consultations between the two groups regarding management of shared stocks.

For maximization of profit, one might envision an ideal combination of (surplus) Japanese vessels, equipment and technology, manned by low-labour cost Chinese or North Koreans, fishing rich Chinese and Russian fisheries, processing at sea or in Russia, marketing the product in Japan, and sharing the profits. But this system would not necessarily ensure the sustainability of the stocks, nor be the most advantageous to each of the players. Furthermore, it is probably not politically feasible on a formal, regional scale. For these reasons, bilateral fisheries agreements will be the norm for the foreseeable future. But management regimes must be multilateral if they are to be successful.

Clearly the developing countries—China and North Korea—and Russia wish to enhance both their ability to harvest fish from their own waters and control over their resources. China recognises that regional co-operation on fisheries is necessary and inevitable.[71] Because of the history of conflict in North-East Asia, the first stage of regional fishery co-operation might not be expected to emerge directly from collective governmental initiatives, such as treaty negotiations. Rather it might evolve from a common willingness to

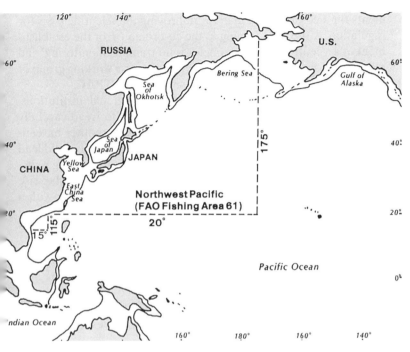

Figure 6.3 FAO Reporting Area for the North-West Pacific
Source: Ronald P. Weidenbach and John E. Bardach, 'Fisheries and Aquaculture' in Joseph R. Morgan and Mark J. Valencia (eds.), *Atlas for Marine Policy in Northeast Asian Seas*, Berkeley: University of California Press, 1993, p. 100.

participate in regular informal meetings and training programs, which would be, at least initially, the responsibility of a coalition of respected non-governmental institutions. Regular activities of this kind could facilitate the establishment of a network of government officials, scientists, and other experts—an espistemic community—and eventually provide the basis and stimulus for the governments of the region to proceed to the negotiation of more formal co-operative arrangements.

Given the socio-economic dimensions of fishery policy and management,[72] it might be wise to extend the suggested regional network to representatives of the various sectors of the fishing industry, the relevant trade unions, and the fishing communities, as well as to academic specialists in the field of ocean development and management, including the Law of the Sea. Possible specific co-operative approaches include modification of existing arrangements for a co-ordinated plan for improving the scientific basis for regulation, with Japan serving as the co-ordinator; or the establishment of a scientific organization, for joint training, monitoring, and research on stock status, either *de nouveau* or as a working group of existing international organizations such as WESTPAC.

Because of their lack of experience in regional institutions, North-East Asian countries might be expected to tread carefully before committing themselves to the building of another bureaucratic organization. Opening this topic at the wrong time could lead to chaos in the existing fisheries arrangements, or to a strong redistribution of current allocations away from Japan. It might be wise for the governments of the region to begin experimentally with a variety of relatively low-risk initiatives involving decentralized power and authority. The initiatives might include some of the following options:

1. government participation in a SEAPOL-type network for the North Pacific, which might be developed under the auspices of PECC or APEC, and through co-operation among the East–West Center, relevant universities, and other non-governmental institutions with a special interest in the ocean affairs of that region;
2. expansion of PICES to become a fully representative forum for the ocean scientists of the North Pacific;
3. organization of a research project to evaluate the effectiveness of existing regional fishery commissions in various parts of the world, with special reference to their roles, structure, and financial arrangements in light of new conditions of the Law of the Sea and of the recommendations of the United Nations Conference on Environment and Development;
4. establishment of intergovernmental task forces to study the case for and against the establishment of formal fishery management or consultation mechanisms for North-East Asia;
5. organization of a workshop to discuss the design of a proposed fishery conflict/dispute settlement system for North-East Asia;
6. an informal intergovernmental meeting to compare public participation policies and practices, and to review alternative modes of consultation with non-governmental bodies in the context of fishery policy and management and related sectors; and
7. establishment of an informal intergovernmental forum de-

signed to facilitate the harmonization of national fishery development and management policies and practices within North-East Asia and eventually, the North Pacific.

Joint Development

Joint development is a concept that has been used primarily in situations where nations have overlapping claims to areas with petroleum potential and negotiations are stalled. It involves putting aside of the sovereignty issue for future generations to resolve, and the agreement jointly to manage the exploration and exploitation of resources in the area of overlap. There are numerous precedents throughout the world and in Asia, including an agreement between Japan and South Korea in the northern East China Sea.[73] It provides for continued fishing in the area by both parties, and for safeguards to protect the fisheries from possible petroleum pollution. Malaysia and the Philippines have a similar agreement.[74]

The joint development concept has been extended to fisheries, usually in the form of separate but equal access to an area. Relevant precedents creating common fisheries zones include the 3 November 1976 Common Fisheries Policy of the European Economic Community (Common Fisheries Policy) as related to the North-East Atlantic Ocean and the North Sea;[75] the 19 November 1973 agreement between Argentina and Uruguay for purposes of regulating jurisdiction in the Plate River and the ocean areas adjacent and beyond this river (Plate River Agreement);[76] the 18 November 1978 agreement between Australia and Papua New Guinea concerning sovereignty and maritime boundaries in the area between the two countries, including the area known as the Torres Strait, and related matters (Torres Strait Agreement);[77] the 2 March 1953 Convention for the Preservation of the Halibut Fishery of the North Pacific Ocean and Bering Sea (Pacific Halibut Convention) concluded between Canada and the United States and its Protocol of 29 March 1979;[78] the 13 January 1978 agreement between Colombia and the Dominican Republic on the delimitation of marine and subsoil areas and maritime co-operation;[79] the 27 February 1980 fisheries agreement between the European Economic Community and Norway;[80] the 28 May 1980 agreement between Iceland and Norway on the fisheries resources of the Jan Mayen area (Jan Mayen Agreement);[81] the 11 January 1978 agreement between Norway and the Soviet Union to establish a provisional joint fishing zone in the Barents Sea (Grey Zone Agreement);[82] and the 13 September 1973 Convention on Fishing and Conservation of the Living Resources in the Baltic Sea and Belts (Gdansk Convention) and its Protocol of 11 November 1982.[83]

The Common Fisheries Policy of the European Economic Community is different from the other eight arrangements in that it is not limited to implementing the provisions of UNCLOS. States seeking to implement article 63(1) are supposed to jointly adopt or co-ordinate the conservation measures applicable to the shared stock; jointly determine the TAC for the shared stock; and to allocate the stock among themselves. Regarding geographical scope, such arrangements should cover the full migratory range of the stock within the exclusive economic zone or fisheries zone of the parties.

The Common Fisheries Policy, though, seeks to integrate the fisheries policies of member states to attain a common market for fish and fish products as part of the objective of attaining a common market for all products. The member states have transferred their authority to the Community in the regulation of exploitation, conservation, and access. As a result, it is the Community which is responsible for implementing the general provisions related to the exploitation of marine fisheries resources in the exclusive economic/fishery zone—the internal fisheries policy—and the Community which is responsible for implementing article 63(1) with states whose waters border those in which the Community regulates fishing activities—the external fisheries policy. The agreements between the European Economic Community and Norway and the participation of the Community in the International Baltic Sea Fisheries Commission are examples of the Community exercising these responsibilities. For purposes of the fisheries provisions of the 1982 UNCLOS, the Community is a state, and the obligations resting upon states with respect to transboundary marine fisheries resources rest with it, instead of with each member state. This harmonization of policy and practice may be a model for a multilateral regime in North-East Asia.

The other eight co-operative arrangements illustrate different ways in which states regulate exploitation of, access to, and conservation of transboundary stocks between exclusive economic zones/ exclusive fishery zones. Although the arrangements concluded between Canada and the United States, and between Norway and the Soviet Union apply to areas where no boundaries have been determined, the issues dealt with are identical: how to regulate exploitation to ensure conservation, and how to ensure that each state obtains access to a resource that migrates through areas in which it exercises jurisdiction. UNCLOS Article 74(3) provides that, pending agreement on the delimitation of neighbouring exclusive economic zones, 'the States concerned, in a spirit of understanding and co-operation, shall make every effort to enter into provisional arrangements of a practical nature'. The agreements between Canada and the United States, and between Norway and the Soviet Union could be regarded as implementing article 74(3).

The Gdansk Convention is also different from the other arrangements in the sense that it is a regional arrangement. Insofar as transboundary and associated species are concerned, however, the relevant issues are the same. Article 63(1) of the 1982 Convention refers to co-operation between 'two or more states', and so would not prevent several states through whose waters a stock migrates from jointly implementing it. Conservation requirements, in fact, would seem to favour such a choice. In addition, the Gdansk Convention relates to UNCLOS Article 123 on co-operation between states bordering enclosed and semi-enclosed seas.

Allocation of jurisdiction varies among the agreements. Flag state jurisdiction may be applied in disputed boundary areas, or flag state and coastal state enforcement may both apply outside boundary areas. In most cases coastal state jurisdiction applies. The Convention on the Special Maritime Frontier Zone,[84] concluded between Chile, Ecuador, and Peru within the framework of the Permanent Commission of the South Pacific is also of interest in this respect, although it is not concerned with transboundary stocks. The Convention establishes a 10 nautical mile zone on either side of the maritime boundaries in the areas outside the 12 nautical mile zone. In the ten mile zone the coastal states have agreed not to consider fishing by vessels of the neighbouring state an infraction of the fisheries regulations in force. Although this provision does not grant those vessels a right to exploit the resources, it does enable them to operate more freely in boundary areas and in relation to transboundary stocks.

All arrangements require co-ordinated or joint adoption of conservation measures by the parties. The agreements concluded between Colombia and the Dominican Republic, and between the European Economic Community and Norway require co-ordination. In all the other cases, conservation measures are to be adopted jointly by the parties to the agreement, although some of the agreements provide for a state to 'opt out' of the measures. In addition, the agreement between Canada and the United States gives each party the right to adopt more restrictive measures for its own vessels and vessels licensed by it. And the agreement between Colombia and the Dominican Republic lets each party retain the right to adopt within its area of jurisdiction the measures which it deems pertinent.

Most regional fisheries commissions alter the contractual environment by creating iterated negotiating processes that move from principles to rules, and then change the rules as better knowledge becames available.[85] When the first generation of rules, closed seasons, closed areas, minimum landing size, and minimum net mesh proved unsatisfactory, the commissions moved to total allowable catches (TAC) or quotas, eventually supplemented by national allocations. Most of the agreements require co-operation between the states concerned with regard to the adoption of TACs and their allocation. Except for the arrangements between Colombia and the Dominican Republic, and between the European Economic Community and Norway, all require the joint adoption of TACs either directly by the parties or by a fisheries commission. The allocation of TACs, except in the agreement between Colombia and the Dominican Republic, is determined by co-operation between the parties. In several agreements, the issue of who gets what is regulated in the agreement itself. Annual negotiations on the most sensitive issue of fisheries management are thereby avoided. As a result, the parties can concentrate on the more technical aspects of fisheries management. Other agreements have a provision allowing a party to 'opt out' of the allocation.

Turning to access, all arrangements provide that vessels of all parties are entitled to exploit the resource in question. None of the arrangements provides for the possibility of exploitation in one of the zones by or under the auspices of one of the parties not taking place, even though benefits nonetheless accrue to the party concerned. But actual participation in the fishery by the coastal state is not required in order for it to obtain benefits. Once the TACs have been allocated, third states, subject to the provisions of the arrangement in question, may be given access to the resource. Fishing by third states may only take place within the area where the licensing state exercises jurisdiction, and may require the consultation or permission of all states. Third-party vessels may be prevented from fishing near disputed boundaries, or banned altogether. This means that once allocated, the resources, for purposes of regulating access, are not treated as the exclusive economic zone stocks of a single state. The other party retains a say over access. And the responsibility to ensure the optimum utilization of the stocks rests on both parties.

Precedents regarding co-operative management of transboundary stocks between exclusive economic zones/fishery zones and adjacent high seas are also relevant. These include the Convention on Future Multilateral Co-operation in the North-East Atlantic Fisheries (the North-East Atlantic Fisheries Convention)[86] and the Convention on Future Co-operation in the North-West Atlantic Fisheries (the North-West Atlantic Fisheries Convention).[87] Although the main object of both conventions is to regulate the exploitation of high seas stocks, each contains special provisions referring to transboundary stocks between exclusive economic zones/fishery zones and adjacent high seas areas. Under these agreements, coastal states, regardless of whether or not they exploit the stocks in their exclusive economic zone, are entitled to participate in the co-operative arrangements concluded. States implementing UNCLOS Article 63(2) are to implement the high seas regime, and to that effect jointly adopt conservation measures and TACs.

These measures should be consistent with those applied to relevant stocks in the high seas areas and those for stocks in the areas in which the coastal states exercise fisheries jurisdiction. The objective of the regime is to maintain or restore populations of harvested species at levels which can produce the maximum sustainable yield, as qualified by relevant environmental and economic factors. But the objective may also be optimum utilization, in which case the TACs may be integrated into national quotas.

The concept of joint development could be applied to the overlapping maritime claims around the southern Kuriles/Northern Territories, around Tok Do/Takeshima in the Sea of Japan, in the central Yellow Sea, and in the central and southern East China Sea. A Russian/Japanese joint venture for fishing in the Kuriles could use North Korean or South Korean labour, and jointly process and market the catch. As a possible first step in this direction, the Hokkaido and Sakhalin governments have agreed to establish a joint co-operative body to oversee fishing in the southern Kuriles and the processing of products therefrom. But Japanese fishing boats would be required to pay fees and would remain under Russian control.[88] And Japan is taking the position that joint development at least on the islands will be possible only after its sovereignty over then is confirmed.[89]

As another example, South Korea and North Korea could consider establishing a joint fishing zone in the Sea of Japan around their troublesome border.[90] Their current fisheries issues include shared stocks, depletion of stocks, and lack of an agreed boundary, all of which lead to inefficient use of the resource and potential conflict. During the 1980s, ten South Korean fishing boats and 161 fishermen were captured by North Korean military forces. Target species in the joint fishing zone would be, in order of descending amount and value: Alaska pollack, squid, saury, crab, shrimp, mackerel, and sardine. If a joint fishing zone were agreed, North Korea could catch squid, sardine, saury, and mackerel on the South Korean side, while South Korea could fish the highly desirable and increasingly scarce Alaska pollack on the North Korean side. The wider advantages of a joint fishing zone between the two Koreas would be the rational use of the resources, increased benefits to fishermen, a stable fishing environment, and strengthened relations between the two Koreas.

North and South Korea should agree on a single merged zone with a single license providing equal access to all waters of the joint zone. The Sea of Japan should be the initial target because of its richer fish stocks and less complex boundary problems. The zone could extend 30 miles north and south from an agreed extension of the Armistice Line and between 12 and 200 nautical miles out to sea. South Korea already has a Special Maritime Zone in the area approximating its half of such a joint zone (Figure 6.4). Territorial waters would be excluded to decrease security concerns. North Korea would have to make an exception to its 50 nautical mile Military Warning Zone, which would overlap the area, and both South and North would have to exclude Japan and Russia from fishing there. A scientific body would assess fish stocks and report to a Joint Commission, which would promulgate regulations. The Commission would also set quotas for each species for each party, as well as the number and size of vessels authorized to fish in the zone. Before entering and leaving the zone all fishing vessels would be checked by both parties for compliance with the regulations. Each state would enforce violations of the regulations by vessels of

FISHERIES 291

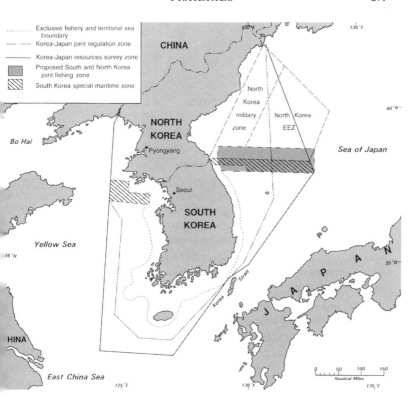

Figure 6.4 Proposed Joint Fishing Zone for North and South Korea
Source: Chung Young-Hoon, 'South and North Korea Cooperation in Fisheries: Toward Establishing a Joint Fishing Zone', Master's thesis, College of Marine Studies, University of Delaware, 1993, p. 69.

the other party in its half of the zone, but owing to differences in legal systems, the flag state would have court jurisdiction. But each state would have to report to the other the result of such process.

Elements of a Model Regime

Any fisheries regime in North-East Asian seas must consider the following factors.

The Scientific Basis for Regulation

Regulation of fishing requires an understanding of the hydrography, primary productivity, and fisheries ecology of a region.[91] It is also necessary to establish an adequate system of data collection and analysis to provide a comparable statistical base for decisions.[92] The efforts in this region to study most of the important features of the marine circulation and biological systems have been inadequate. No one nation has access to all the existing relevant materials. Co-operative studies have not taken place. National statistics are not standardized or intercalibrated. For the Yellow Sea, national catches are almost always reported jointly with those of the East China Sea, creating a statistical nightmare. And the two international commissions established by bilateral agreements do not publish reports documenting the scientific basis for management.

Specific studies must be undertaken on the hydrography of the Chinese and Korean coasts in the Yellow Sea, plankton and primary productivity, overall assessments of the demersal fish stocks with emphasis on coastal catches, further research on the biology of pelagic stocks, assessment of the cephalopod resources, assessment of the fleshy prawn stock, and identification and research on coastal pelagic shrimps.[93] The project to investigate the Yellow Sea as one large marine ecosystem is a step in the right direction.[94]

Regulation of Shared Stocks

Virtually all the major species of fish caught commercially are subject to seasonal movements from one jurisdiction to another. Even with the limited claims to jurisdiction of coastal nations and the bilateral agreements, numerous seasonal changes of location bring species back and forth across different regulatory regimes. Furthermore, migration patterns in the Yellow Sea and East

China Sea are particularly complicated, and the two areas can probably not be separated with respect to regulations for many species.

Regulation of Overfishing

Probably all the major fisheries resources of the region are fully exploited, with the possible exception of filefish and mackerel.[95] It is unlikely that new fisheries resources will be discovered in commercial quantities. Therefore it is important to conserve the existing stocks.

Regulation of a Mixed Stock Fishery

Total catch has been steady or increasing, while that of particularly valued species has declined. An appropriate regulatory goal must be decided and pursued, perhaps regulation to maintain or improve stocks of single species, or to maintain maximum productivity from the area.

Role of Allocation in Regulation

There is no forum wherein all fishing nations meet to discuss the distribution of catches. Coastal states have made some attempts to reserve coastal fisheries for their nationals. International agreements make some modest attempts to indicate sharing of catches within jointly regulated areas or limiting effort to stay within catch quotas for designated areas. There must be a comprehensive approach to allocation for the entire region. Discussion of this issue would entail a systematic review of the existing *de facto* allocations and re-examination of the legal regime of fishing rights.

Structure

Because of the migratory nature of many of the commercial species, shared stocks, uncertain jurisdiction, and distant-water fishing within the region by the region's nations, an overarching consultative body which includes all the littoral parties seems necessary. It should initially be consultative rather than a formal intergovernmental body because of the need to include the divided states and to avoid issues of diplomatic recognition. Working groups would focus on each Sea separately, that is, on the Sea of Japan, and the Yellow/East China Seas. These working groups would be the core of the regime and governments could sign on only for them and not the plenary consultative body.

Current National Interests

The approach must satisfy the fundamental national interests of all parties and emphasize complementarities. In summary, the respective national interests appear to be as follows:

- China: reservation of its mid-distance fisheries for its nationals; acquisition of foreign exchange; technology transfer; training.
- Japan: access to desired species and the reservation of its northern coastal fisheries for its nationals.
- North Korea: acquisition of foreign exchange through fees and market access; technology transfer; training; information.
- South Korea: access to desired species, technology, and information.
- Russia: acquisition of foreign exchange through fees and market access; technology transfer; control of foreign fishermen.
- Taiwan: access to desired species.

On the face of it, Russia, and to a lesser extent North Korea and China, could trade *responsible* access by Japan, South Korea, and Taiwan for fees, technology, training, and information. For domestic political reasons, each needs to reserve its nearshore fisheries for its own nationals.

Functions

These might include: scientific and technological co-operation building on the existing agreements between Japan and Russia and Japan and South Korea; monitoring of stocks which show erratic fluctuations or deterioration; comparisons and integration of resource assessments; the determination of quotas; establishment of joint conservation zones; and compensation schemes for displaced fishermen and boats.

Specific Issues

The regime must address the questions of fishing rights and allocation in disputed areas like Tok Do/Takeshima and the Northern

Territories/Southern Kuriles, as well as in areas with uncertain jurisdiction like the East China Sea. This would involve trade-offs patterned, for example, on the Japan/Russia salmon agreement in the Northern Territories, in which Japan builds salmon hatcheries in Sakhalin and harvests salmon jointly with Russia. Japan's financing of the hatchery comes from its salmon catch in the Russian zone, for which it pays a fee. Similar agreements could be struck for highly desired species in Japan like scallops, shrimp, kelp, sea urchin, and crab.

Leadership

South Korea is a middle power situated on all the Seas and has a dual role as a coastal and distant-water fishing nation. But Japan is the region's 'hinge' state for fisheries. Perhaps South Korea should take the lead with Japan ostensibly following, but really co-ordinating the regime in a *de facto* sense through its bilateral agreements. As the regime evolves, Japan's bilateral agreements would be phased out or integrated into the regime, and information on the status of the stocks and allocation would become more public.

The Policy Environment

Although a widespread positive perception of the benefits and necessity of co-operation on fisheries management does not yet exist, an epistemic community of fisheries scientists and policy-makers is beginning to form. Media and advocacy groups have already begun to bring the issues to public attention and will increasingly do so. The exogenous pressure necessary to push the issues to the forefront is at hand—the expected extension of jurisdiction by China and South Korea, the global and regional deterioration of fish stocks, the increasing frequency of poaching and violent incidents connected therewith, and for the public, the increasing price of fish and their decreasing availability. Although the specific long-term benefits to each participant are unclear, maintenance of the stocks at optimum sustainable yield will benefit all.

Negotiations

The negotiations must be informed by the best scientific knowledge and that knowledge must be equally available to all. Issues and objectives and alternative regimes to meet them, including those outlined herein, must be clearly defined and frankly discussed. Japan and South Korea should finance the initial meeting, but Japan must not attempt to use its hegemonic position in fisheries to its sole advantage. Hosting of meetings would rotate annually among the members. Negotiations must be aimed at identifying and reaching explicit agreement on a common concern, such as regulation and the method thereof for overfished shared or migratory stocks. The resulting agreements may then be extended to other species.

Regime Formation

The regime would be negotiated by skilled diplomats and be built in stages, beginning with the international committee on protection of fishery resources in the Sea of Japan. As the regime evolves, it may incorporate elements of the existing bilateral agreements and address more contentious issues such as allocation, regulation, including conservation areas, monitoring, and enforcement. An *ad hoc* agreement between China, Japan, and South Korea on maritime security in the East China Sea may be the germ of a regime there, and one will almost certainly form in the Yellow Sea. These will eventually become working groups under the umbrella of a fisheries commission involving all relevant governments. But the agreed rules and regulations should be enforced by the respective governments in their own waters, or jointly, where more than one claim an area.

Notes

1. Jessica Matthews, 'Depleted fisheries: world crisis is now', *Honolulu Advertiser*, 20 March 1994, p. B-1; 'Fish: The tragedy of the oceans', *The Economist*, 19 March 1994, p. 21; 'The catch about fish', *The Economist*, 19 March 1994, pp. 13–14; Anne Swardson, 'World's fishermen hit bottom in pursuit of ocean's bounty', *Honolulu Advertiser*, 14 August 1994, p. A-21; William Montalbano, 'Overfishing is emptying the oceans of food', *Los Angeles Times*, 12 March 1995.
2. Matthews, 'Depleted fisheries: world crisis is now'.
3. Brian J. Rothschild, 'Achievement of fisheries management goals in the 1980s', in Brian J. Rothschild (ed.), *Global Fisheries Perspectives for the 1980s*, New York: Springer-Verlag, 1983, p. 165.

4. Douglas M. Johnston and Mark J. Valencia, 'Fisheries', in Mark J. Valencia, (ed.), *The Russian Far East and the North Pacific Region*, Boulder: Westview Press, 1995, p. 149.
5. Ellen Hey, *The Regime for the Exploitation of Transboundary Marine Fisheries Resources*, Dordrecht: Martinus Nijhoff, 1989, pp. 81–2.
6. William Branigan, 'World's fishing nations net pact to end conflicts', *International Herald Tribune*, 5–6 August 1995, p. 1.
7. 'Russia seeks Sea of Okhotsk fishing talks', *FBIS-EAS*, 9 March 1993, p. 24.
8. *Xinhua News Agency*, 3 July 1993; 'China, Seoul agree to involve DPRK in Project', *FBIS-EAS-94-067*, 7 April 1994, p. 33.
9. *FBIS-EAS*, 9 November 1992, p. 10.
10. This section is derived from Mark J. Valencia, *International Conference on the Yellow Sea*. East–West Environment and Policy Institute Occasional Papers No. 3, 1987, pp. 61–5.
11. Valencia, *International Conference on the Yellow Sea*.
12. Valencia, *International Conference on the Yellow Sea*.
13. This section is derived from Mark J. Valencia, *International Conference on the Sea of Japan*, East–West Environment and Policy Institute Occasional Paper No. 10, 1989, pp. 75–6, 85–94.
14. Ibid., pp. 77–85.
15. Johnston and Valencia, 'Fisheries'.
16. 'Agreement reached with Russia on salmon hauls', *FBIS-EAS*, 23 March 1993, p. 4.
17. Valencia, *International Conference on the Sea of Japan*.
18. Chung Young-Hoon, 'South and North Korea cooperation in fisheries: toward establishing a joint fishing zone', Master's thesis, University of Delaware, 1993 pp. 32–3.
19. *EBIS-CHI*, 4 March 1992.
20. Chung, 'South and North Korean cooperation in fisheries: towards establishing a joint fishing zone'.
21. Tsuneo Akaha, 'Bilateral fisheries relations in the Seas of Japan and Okhotsk: a catalyst for cooperation or seed of conflict', in Elisabeth Mann Borgese, Norton Ginsburg and Joseph R. Morgan, eds., *Ocean Yearbook* 11, Chicago: The University of Chicago Press, pp. 389–92.
22. Ibid., p. 391.
23. 'Nations suspend pollack fishing in Bering Sea', *FBIS-EAS*, 7 July 1993, p. 3.
24. Tsuneo Akaha, *Japan in Global Ocean Politics*, Honolulu: Law of the Sea Institute and the University of Hawaii Press, 1985, pp. 18–23.
25. Valencia, *International Conference on the Sea of Japan*, pp. 68–75.
26. M. J. Peterson, 'International fisheries management', in Peter M. Haas, Robert O. Keohane, and Marc A. Levy (eds.), *Institutions for the Earth: Sources of Effective Environmental Protection*, Massachusetts: The MIT Press, 1993, p. 250.
27. *Yonhap*, 22 December 1993.
28. Chung, 'South and North Korea cooperation in fisheries: toward establishing a joint fishing zone', pp. 26–30.
29. Ibid., p. 75.
30. Ibid.
31. Paik Jin-Hyun, 'Exploitation of natural resources: potential for conflicts'. Paper presented to the Maritime Cooperation Working Group, First Meeting, Kuala Lumpur, 2–3 June 1995.
32. Paik Jin-Hyun, 'Strengthening maritime security in Northeast Asia', Paper presented at the 8th Asia Pacific Roundtable on Confidence Building and Conflict Reduction in the Pacific, Institute for Strategic and International Studies, Malaysia,

5–8 June 1994; 'Seoul eyes expansion of its territorial waters', *Japan Times*, 4 March 1995, p. 4.

33. Liu Rongzi, 'Exploitation and management of the fishery resources and regional cooperation in the Yellow Sea and the East China Sea', Paper presented to the China-Korea Seminar on Ocean Affairs, Seoul, 15–16 September 1993.

34. Paik, 'Exploration of natural resources: potential for conflicts'.

35. 'Seoul eyes expansion of its territorial waters', p. 4.

36. 'Talks held with Japan on fishing dispute', *FBIS-EAS*, 23 February 1994, p. 24.

37. Akaha, 1994, 'Bilateral fisheries relations in the Seas of Japan and Okhotsk: a catalyst for cooperation or seed of conflict?' p. 394.

38. Ibid., pp. 395–6.

39. Mark J. Valencia, 'For peace in the Seas of China', *International Herald Tribune*, 3 February, p. 9.

40. *Kyodo News Service*, 19 August 1993.

41. Radio-1, 16 July 1993.

42. *FBIS-USR*, 11 November 1993, pp. 86–8.

43. *FBIS-EAS*, 14 June 1993, p. 8.

44. *Xinhua News Agency*, 3 July 1993; *IBRU Boundary and Security Bulletin*, October 1993.

45. Sumihiko Kawamura, 'Maritime transport and communications—including marine safety and SLOC security'. Paper presented to CSCAP Working Group on Maritime Cooperation, First Meeting, Kuala Lumpur, 2–3 June 1995; Stanley Weeks, 'Law and order at sea: Pacific co-operation with piracy, drugs and illegal migration.' Paper presented to CSCAP Working Group on Maritime Co-operation, First Meeting, Kuala Lumpur, 2–3 June 1995.

46. 'DPRK gunboats fired on PRC fishing boats', *FBIS-EAS*, 27 January 1993, p. 18.

47. 'Seoul to protect PRC on illegal fishing', *FBIS-EAS*, 24 August 1993, p. 25.

48. *Korea Herald*, 24 August 1993.

49. S. Korean fishing boat seized by N. Korea, *Honolulu Advertiser*, 30 May 1995, p. A-5.

50. *ITAR-TASS News Agency*, 22 January 1994; 'Crab boat skipper shot by Russians', *Japan Times*, 18 August 1994, p. 2; 'Russia kills two Chinese fishermen', *Japan Times*, 14 September 1994, p. 2; 'Russians sink Japanese boat, *Japan Times*, 7 October 1994, p. 4.

51. 'Russia says fishing deal seems difficult in '95', *Kyodo*, 28 November 1994.

52. *The Japan Times*, 12 March 1995, p. 2.

53. *Honolulu Advertiser*, 7 February 1994, p. A8; 'Safe fishing sought near disputed isles', *Japan Times*, 23 August 1994, p. 3.

54. 'World', *Japan Times Weekly*, International Edition, 3–9 October, 1994, p. 2.

55. 'Japan, Russia sound out ways to fight drug trafficking', *Kyodo*, 17 November 1994; 'Russia starts poaching campaign', *The Japan Times*, 12 March 1995, p. 2.

56. Johnston and Valencia, 'Fisheries', pp. 169–72.

57. The PICES treaty, negotiated in 1991, creates a forum for scientists and technical experts to exchange and evaluate fisheries data, and promote and co-ordinate co-operative research. It is expected that PICES experts will participate in a technical rather than representative capacity.

58. *United Nations Convention on the Law of the Sea*, New York: United Nations, 1982, Article 246(2).

59. Ibid., Article 246(5a).

60. PECC has concluded a preliminary co-operative research project between

South-East Asia and the South Pacific, which seems to establish that the tuna resources of these two regions belong to the same biological population.

61. Park Seong Kwae, 'The status of fisheries in Korea with emphasis on distant-water operations', Paper presented at the 27th Annual Conference of the Law of the Sea Institute, Korea Ocean Research and Development Institute, 13–16 July 1993, pp. 6–12.

62. Interregional networking has been the purpose of several programs, such as the PECC Fisheries Task Force which provides a linkage between fisheries specialists in North-East Asia, South-East Asia, the South Pacific, and the Central Eastern Pacific.

63. Douglas M. Johnston, *Fishery Politics and Diplomacy in the North Pacific: New Sources of Instability*, Working Paper No. 7, Center for International and Strategic Studies, York University, 1992.

64. Rothschild, 'Achievement of fisheries management goals in the 1980s', pp. 172–6.

65. Ibid., pp. 153–4; B. Cleaveland, 'National adjustments to changes in fisheries law and economic conditions: a synopsis of 26 case studies', FAO Fisheries Circular No. 783, Rome, 1985.

66. Valencia, *International Conference on the Yellow Sea*, pp. 68–9.

67. Sigmund Engesaeter, 'Scientific input to international fishery agreement', *International Challenges*, 13, 2 (1993): 85–100.

68. United Nations Convention on the Law of the Sea, Article 61.

69. 'Fish: The tragedy of the oceans', *The Economist*, 19 March 1994, p. 24.

70. Ronald P. Weidenbach and John E. Bardach, 'Fisheries and aquaculture', in Joseph R. Morgan and Mark J. Valencia (eds.), *Atlas for Marine Policy in East Asian Seas*, Berkeley: University of California Press, 1992, p. 100.

71. Liu Rongzi, 'Exploration and management of the fishery resources and regional co-operation in the Yellow Sea and the East China Sea'.

72. Michael H. Glantz, 'Man, state, and fisheries: an inquiry into some societal constraints that affect fisheries management', *Ocean Development and International Law Journal*, 17, 1, 2 and 3 (1986): 239–44.

73. Masahiro Miyoshi, 'The Japan-South Korea Agreement on Joint Development of the Continental Shelf', in Mark J. Valencia (ed.), *Geology and Hydrocarbon Potential of the South China Sea and Possibilities of Joint Development*, Oxford: Pergamon Press, 1985, pp. 545–53; Choon-Ho Park, 'Joint development of mineral resources in disputed waters: the case of Japan and South Korea in the East China Sea', in Mark J. Valencia (ed.), *The South China Sea: Hydrocarbon Potential and Possibilities of Joint Development*, Oxford: Pergamon Press, 1981, pp. 1335–54; M. Takeyama, 'Japan's foreign negotiations over offshore petroleum development: an analysis of decision-making in the Japan-Korea continental shelf joint development program', in Robert Freidheim (ed.), *Japan and the New Ocean Regime*, Boulder: Westview Press, 1984, p. 276; Masahiro Miyoshi, 'Licensing in Japan-South Korea joint development arrangement', Paper presented at the Third East–West Center Workshop on the Hydrocarbon Potential of the South China Sea and Possibilities of Joint Development, Bangkok, Thailand, February 1985.

74. *Indochina Digest*, 24 December 1993, p. 4.

75. For relevant provisions of the agreement and comparative analysis see Ellen Hey, *The Regime for the Exploitation of Transboundary Marine Fisheries Resources*, Dordrecht: Maritinus Nijhoff, 1989, pp. 133–41.

76. Ibid., pp. 143–5.
77. Ibid., pp. 147–53.
78. Ibid., pp. 155–8.
79. Ibid., pp. 159–60.

80. Ibid., pp. 161–3.
81. Ibid., pp. 165–8.
82. Ibid., pp. 169–71.
83. Ibid., pp. 173–7.
84. Santiago Declaration Regarding Maritime Zones, 18 August 1952, *1006 United Nations Treaty Series*, 324.
85. Peterson, 'International Fisheries Management', p. 302.
86. Hey, *The Regime for the Exploitation of Transboundary Marine Fisheries Resources*, pp. 181–5.
87. Douglas M. Johnston, 'Legal and diplomatic developments in the Northwest Atlantic fisheries', *Dalhousie Law Journal*, 4 (1977): 37–61.
88. 'Fishing north of Hokkaido on line', *Japan Times*, 9 September 1994, p. 10; 'Russia claims right to police fishing', *Japan Times*, 16 September 1994, p. 2.
89. 'Russia asks Japan to join economic zone on island', *Japan Times*, 13 October 1994, p. 1.
90. Chung, 'South and North Korea cooperation in fisheries', pp. 66–84.
91. J. A. Gulland, *The Management of Marine Fisheries*. Seattle: University of Washington Press, 1974, pp. 156–165.
92. 'Interim Report of the ACMRR Working Party on the scientific basis of determining management measures', *FAO Fisheries Circular No. 718*, Rome, 1979.
93. Valencia, *International Conference on the Sea of Japan*, pp. 64–6; Shiro Chikuni, 'The fish resources of the northwest Pacific', FAO Fisheries Technical Paper No. 266, Rome, 1985.
94. *FBIS-EAS*, 12 January 1994, p. 25.
95. D. Zhu, 'A brief introduction to the fisheries of China', FAO Fisheries Circular No. 726, Rome, 1980.

7 Conclusions

Although North-East Asia has long been a locus of tension and conflict among the world's major powers, the regions's political system is now in the midst of a transformation—one in which economic interdependence, multipolarity and multilateralism are increasingly important factors. Although traditional security concerns are likely to persist, confrontation is being replaced by dialogue, and tension is relaxing. Indeed, regionalism is becoming more attractive as the influence of the United States declines, and developing North-East Asian states promote their own economic and political interests, as well as a sense of international community.

Within this milieu, maritime issues are rising to the surface of current regional security concerns. And the end of the Cold War, the warming of relations in the region, the extension of maritime jurisdiction, and the coming into force of the United Nations Convention on the Law of the Sea (UNCLOS) provide opportunities to build confidence and cultivate a habit of dialogue and working together through the establishment of maritime regimes. Moreover, maritime regime building is favoured by multilateralism and regionalism, combined with the growing acceptance of the concept of comprehensive security.

Regimes are a special form of co-operation in which state action is regulated. They are created to resolve problems for the collective or common good. A major function of international regimes is to avoid structural anarchy. In particular, regimes establish a clear legal framework with liability for actions; improve the quality, quantity, and availability of information; reduce transaction costs; and offer a mechanism for presenting a united stand against outside actors.

Despite regime rules, unilateral actions, bargaining, and coercion remain central to the processes of arriving at social choice. Regimes are also valuable for their potential contribution to the solution of problems yet to be defined, thus creating the promise of future gains by meeting present difficult commitments. Epistemic communities, such as those forming in the marine policy sphere in North-East Asia, enhance co-operation in general, and regime formation in particular. Regimes are supplied when there is sufficient demand for the functions they perform. Such demand is currently developing in North-East Asia.

Institutions arise from the intersection of expectations and patterns of behaviour or practice. Successful institutions seize opportunities, expand their scope, multiply roles, emphasize integrative bargaining, simplify implementation, and mobilize leadership. Both regimes and institutions for resource and environmental management are favoured by, and promote high levels of, government concern, a hospitable contractual environment, and sufficient national political and administrative capacity. A resource or environmental management regime should be described in terms of its institutional character, jurisdictional boundaries, and the conditions or consequences of its operation. Not all regimes require organizations: many are anarchical, employing self-help arrangements and mutually agreed rules, enforced by each individual state within its own jurisdiction. But in the marine environment and fisheries sectors there is a need for organizations to address international commons or transboundary problems such as migratory fish or pollution, and to allocate the benefits and costs of their management. Criteria for evaluation of maritime regimes include their political feasibility, distributive implications, and social consequences. Regimes may be spontaneous, negotiated, imposed, or 'led.' Once formed, though resistant to change, regimes do evolve in response to shifting power structure, technological change, and national interests.

A marine policy regime is a system of governance for a particular maritime sector or region. As such it has structure, objectives, functions, powers, processes, and programmes. Co-operation in the maritime sphere includes joint activities, regional organizations, treaty arrangements, harmonization of laws and policies, and informal contacts. Objectives or priorities usually differ between developed and developing states, the former preferring an enhancement of the robustness and stability of the system, and benefits for their nationals, while the latter are more interested in influencing management decisions and enhancing their economic development through training, education, and technical assistance. Thus trade-offs are possible and frequent.

CONCLUSIONS 303

The Law of the Sea Conference and publicization of the importance of marine uses and resources, as well as limits, have led to an increased marine awareness and widespread claims to maritime space. Marine regionalism is the natural result of recognition that global standards or regimes may not adequately address the special needs of a region. The Law of the Sea Convention puts special emphasis on co-operation in semi-enclosed seas, recognizing the mobility of water, fish, pollutants, and ships within a region. Full or partial marine regionalism has been attempted in the Baltic, the Mediterranean, the North Sea, and the Arctic. The results are mixed; the approaches in the Baltic and the North Sea being the most successful.

Factors retarding success include inadequate leadership, limited management authority, lack of effective enforcement powers, effects on non-parties, disagreement among member states, shortage of funds, trained personnel, and equipment, limited information or the ability to use it, jurisdictional limitations, differing time horizons, and different perceptions of the costs of surrendering national autonomy.

The formation of a model regional maritime regime would be enhanced by the following: a common positive perception of co-operation that the regime is equitable, and that the benefits of participation outweigh the costs; a negotiating process spurred by an exogenous shock or crisis and participants willing and able to seize the opportunity; well-defined issues and sector specific, clear objectives and functions; reasonable expectations of progress; formation in stages, from limited and temporary objectives to a broader convention; an uncomplicated but reasonably well-defined structure; a match between the regime and the natural system; decentralized decision-making; a process which concentrates on policy questions, is negotiated by skilled diplomats and results in explicit agreements; available alternate regime designs and a willingness to adjust to new ideas; avoidance of explicit use of power in institutional bargaining; willing, able, and widely acceptable leadership; and a strong individual leader.

The geography, circulation pattern, and ecology of the semi-enclosed seas of North-East Asia—the Sea of Japan and the Yellow/East China Sea—necessitate that they be managed as complete systems. The littoral states are North Korea, South Korea, Japan, and Russia for the Sea of Japan, and China, North Korea, South

Korea, and Japan for the Yellow/East China Sea. However, Russia fishes in the East China Sea and China pollutes and desires access to the Sea of Japan through the Tumen River. All these nations have made various maritime jurisdictional claims; some claims do or will overlap and many boundaries remain unresolved.

Integrative forces in the region include the transnationality of living resources, pollution, and shipping; advances in technology and marine use patterns and concepts; growing conflicts among uses and users; the coming into force of UNCLOS and its emphasis on regional management of semi-enclosed seas; the newfound US support for multilateral initiatives; and the growth of an epistemic community of maritime affairs specialists. Dis-integrative forces include the remaining isolation of North Korea and the tense relations between North and South Korea, between China and Taiwan, and to a lesser degree between Japan and Russia. A destabilized Russia, disputes over islands and maritime boundaries, and the conceptual dichotomy in UNCLOS between co-operation and sovereignty would also have dis-integrative influence. Given this set of circumstances, an *ad hoc,* issue-specific, evolutionary process toward multilateral maritime regime building appears necessary.

Before the recent displays of Chinese anger at Taiwan's campaign for greater recognition, the maritime interests of China and Taiwan were converging in such areas as management of cross-Strait trade, piracy, drug trafficking, and illegal immigrants, together with fisheries management, representational issues regarding distant-water fishing; common claims to the Senkakus and continental shelf of the East China Sea (including agreement to jointly explore for oil and gas in the East China Sea), and a common stand on the South China Sea.

For China, regime participation could lead to a fairer international legal environment, technology transfer from Japan, South Korea and Taiwan, and confidence-building in itself as a member of the international community. But China would have to limit its flexibility in the marine sphere and commit scarce resources to fulfill its regime responsibilities. Furthermore Beijing prefers bilateral relationships which it can dominate. For Taiwan, regime participation would expand its channels for discussions with China and enable maritime issues to be addressed; and it would enhance its status *vis-à-vis* China. On the other hand, Taiwan might have to share its technological know-how and possibly sensitive data with the other participants, including China. And it would probably have to pay more than its share for the implementation of the regime. If China and Taiwan are to be included in the regime, it must be non-governmental—and the East China Sea must initially be excluded.

For Japan, participation in a maritime regime is favoured by its economic and technological dominance, its knowledge and experience, and its web of bilateral maritime agreements. Benefits include the possible delay of implementation of EEZs by China and South Korea; protection of regional fisheries resources and the environment; elimination of the transaction costs of annual bilateral fisheries negotiations; and enhancement of its status in the region. But a prominent role for Japan in a multilateral regime is inhibited by its preference for bilateral relationships which it can dominate; the Kuriles dispute with Russia; the memories of Japan's wartime behaviour; Japan's priority on immediate national economic gain; and its bureaucratic conservativeness and generally reactive posture regarding international affairs.

North Korea could use participation in a regime as a 'coming out' into the international community; to feel out potential regional partners; to increase its financial, technical, and knowledge capacity; and to gain a cleaner environment. But for North Korea, opening of its society to foreigners and their cultures and practices could help undermine control and expose serious pollution problems. Moreover, scarce resources would have to be diverted to fulfill its obligations to the regime.

For South Korea, participation in a multilateral maritime regime could help avoid costly fisheries disputes; eliminate fishing by extra-regional states; enhance conservation and management of fishery resources; enhance co-ordination with neighbouring countries regarding the management of transnational fish stocks; possibly improve its access to neighbours' stocks; increase the efficiency of its fishing effort; improve its international stature; and expand its points of diplomatic contact with North Korea.

South Korea recognizes that environmental degradation, depletion of fishery resources, maritime anarchy, and political conflict are not in its own, or any other nation's, interest. Indeed, South Korea supports a multilateral marine environmental protection regime and might even be willing to exercise leadership thereof. Although South Korea is the only state to border all three Seas, it puts more emphasis on the Yellow Sea because it is more polluted, and because Korea's western coast is more developed than its eastern shore. Moreover, a regime for the Yellow Sea could employ China as a go-between to help broaden its contacts with North Korea.

Russia has much to gain and little to lose politically and economically by participating in a regional maritime regime, especially for the Sea of Japan. It could gain technological and financial assistance as well as diminished poaching by foreigners. But fisheries co-operation as well as co-operation in environmental protection is hampered by bureaucratic confusion, ineffectiveness, lack of infrastructure, economic malaise, and a relative lack of interest in maritime affairs, particularly in North-East Asian seas. Nevertheless, Russia does have considerable experience in fisheries arrangements and international co-operation in marine environmental protection, a legislative base upon which to build, and considerable fisheries and oceanographic expertise, and it is interested in fulfilling its proper role as a member of the international community. The rise of a domestic Green movement and pressure from aid organizations could promote its participation in a regional marine environmental protection regime. However, Russia's participation in regional maritime regimes will likely be determined by the progress of Russian market reform, political relations between Russia and North-East Asia, particularly Japan, and the impact of incipient non-governmental organizations.

Based on theory and lessons learnt, this book develops separate model maritime regimes for both environmental and fisheries management. The Yellow Sea is one of the most polluted in the world and projected development around its rim will surely make it worse. It also contains valuable fisheries and ecological resources. The Sea of Japan is relatively free from pollution, and has abundant fisheries and ecological resources. However the revelation of former Soviet dumping of nuclear waste has raised concern.

Regarding multilateral marine environmental management, although the transnationality of North-East Asian seas and of their pollution sources is becoming obvious, multilateral measures or infrastructure for marine pollution control are virtually absent and there is a general dearth of capacity and will to co-operatively monitor marine pollution. Indeed relatively little is known about the state of marine pollution in this region. The degree of concern and practice are quite varied, with Japan possessing the most advanced information. There has been little regional marine environmental law drafting and policy development, and what does exist is a varied response to broader global initiatives emanating from the IMO and UNCLOS. Theoretically the littoral nations must now adjust their entire national framework to be compatible with UNCLOS.

There is now a plethora of fora and proposals for both general and specific marine environmental protection in the region, including most prominently UNEP's NOWPAP. But each of these initiatives has problems, particularly NOWPAP. These include differences over the definition of the region for co-operation, the countries to be included in a convention, the priorities for projects and approaches as well as discharge and water quality standards, and allocation of costs. Regionally there is considerable redundancy among activities; a near-sighted preoccupation with coastal pollution; and a lack of understanding of the causes and consequences of marine pollution.

Taking the Yellow Sea as an example, China, South Korea, and North Korea could set unified seawater quality and effluent standards; co-operate on protection of the Yalu River mouth and the northern Yellow Sea; co-ordinate marine environmental monitoring; co-operate in modelling and contingency planning for transnational oil spills; establish a unified regional marine dumping policy; co-operate on protection of valuable fish spawning, breeding, and wintering grounds, and on research and information exchange. These co-operative efforts might culminate in a comprehensive sea-use plan for the Yellow Sea.

An ideal regional marine environmental protection regime must rationalize the redundancy of existing and proposed international programmes; and provide the infrastructure for co-operation on a broad front, including co-ordination of policies, regulations, and research in national zones. The regime should have a mutually acceptable geographic scope and also approximately fit natural features and processes; educate the public and policymakers; and, most important, even up knowledge, concern, and capacity among the participating countries.

For the Sea of Japan, there is a convergence of factors which makes formation of a marine environmental protection regime likely—the low number of states involved; the growing recognition that such a regime could have positive effects in other sectors; the interconnectivity of the waters; the exogenous shock of the nuclear waste dumping controversy; and the diversity of financial and technical capacity, thus facilitating trade-offs. In the Yellow Sea, there is already incipient co-operation between South Korea and China motivated by the growing recognition of its severe pollution.

The regimes should initially be consultative and of the self-help genre. A capable medium power—perhaps centrally located South Korea—should lead, and China and Japan must be supportive. Areas of overlapping claims should be excluded. The regime should begin with a limited and temporary focus on monitoring and clean-up of radioactive waste in the Sea of Japan and evolve into a broader and more co-ordinated regime. Over time, a higher functional level may be reached at which laws and policies are harmonized, and common standards and regulations are set, monitored, and enforced—albeit by national teams in their own waters. Although the scope should include all major sources of pollutants, the regime should not be multisectoral. Further, it should be simple yet well-defined. Decision-making would be by consensus, implementation would be voluntary, and compliance would be enhanced by detection, publicization, and persuasion. Eventually an organization with a secretariat might develop acceptable recommendations for regional management. Most important is that the level of marine environmental technology and expertise would be evened up throughout the region. Indeed the major trade-off would be the benefit to Japan and South Korea of the adherence by China, North Korea, and Russia to a predictable regime in exchange for training and technical assistance from Japan and South Korea.

Regarding multilateral fisheries management, although North-East Asian states have been reluctant to alter what has been a stable fisheries regime, the decline in fish production combined with expanding Chinese fishing, the coming into force of UNCLOS and extended jurisdiction are forcing a re-examination of fisheries policies and regulations. The present web of bilateral regimes is inadequate, involves serious compromises of national interests, and is unlikely to remain stable for long. Most fisheries are overfished, stocks are deteriorating, and conflicts between fishermen are rapidly increasing. None of the regimes include all coastal or fishing nations and thus there is no multilateral forum to discuss the condition of transnational or regionally shared stocks, or their allocation. There are no clearly separate national fisheries areas. Although bilateral commissions are established under the bilateral agreements, they do not publish their decisions or data. New regimes will have to address difficult political relationships;

inequities; underlying economic and social factors influencing fisheries; overfishing; increasing technological and fishing capacity in the region; regulation of shared and mixed stocks; unresolved jurisdictional boundaries; and increased information requirements; allocation; and the structure and functions of the regime.

Objectives of fisheries regimes can include exchange of information; co-operative research; exchange of views on policy and management issues; provision of assistance in fisheries development and management; harmonization of national policies and practices; negotiation and conduct of co-operative fisheries management; and management of disputes and conflicts.

Multilateral regimes for the Sea of Okhotsk and the Sea of Japan are already being explored. And there is presently a web of nine bilateral fisheries agreements in force in the region. They indicate that North-East Asian countries do appreciate the value of scientific information and can act in concert when stocks are threatened. Japan is the common factor in most of the bilateral agreements and is clearly the dominant fishing nation in the region in terms of capacity, technology, and information. As such it is the key to any regional fisheries management regime.

For the Yellow Sea there are three general fisheries management alternatives—maintenance of the *status quo*; accepting and implementing international trends as defined in UNCLOS; or designing a *de nouveau* Yellow Sea approach, including creation of a quadripartite non-governmental arrangement supported by a scientific organization. The regime should include allocation of shared stocks or fishing effort perhaps based on historical catch distribution and zonal attachment.

Ad hoc agreement on a 'core' area for a regional regime is necessary, perhaps beginning with two separate regimes for the Yellow Sea and the Sea of Japan. Trade-offs will also be necessary, such as transfer of technology, training, information, and fees from Japan and perhaps South Korea to China, North Korea, and Russia in exchange for responsible access to their fish stocks. Governments might begin experimentally with a variety of relatively low-risk initiatives with decentralized power and authority such as academic networks and meetings; co-operative research on the effectiveness of existing regional fisheries commissions; intergovernmental task forces; discussions of possible regional fisheries regimes for North-East Asia; and establishment of an informal intergovernmental forum to facilitate harmonization of national policies and practices. Meanwhile joint development of fisheries in overlapping claim areas may ameliorate disputes such as those over the southern Kuriles/Northern Territories, Tok Do/Takeshima, Senkaku/Diaoyutai, and the North Korea/South Korea maritime frontier. There are numerous relevant precedents—both bilateral and multilateral.

As a middle power situated on all the Seas, South Korea may play a leadership role in fisheries regime formation—but Japan must be supportive. Negotiations must be informed by the best knowledge made equally available. Issues, objectives, and alternative regimes must be clearly defined. The regime should be negotiated by skilled diplomats and form in stages perhaps beginning with an international committee for the protection of fishery resources in the Sea of Japan. As it evolves, it can build on and incorporate elements of existing bilateral agreements.

The prospects for regime formation in the marine environmental protection sector appear better than those for fisheries management since the former involves fewer competitive and conflictual issues. Initiatives by NGOs, governments, and international organizations are numerous, and clear trade-offs are possible. However, one nation, acceptable to most, must be willing to take the lead.

Formation of a fisheries management regime is more problematic, primarily because of the reluctance of Japan to alter what it perceives as a system generally in its interest. However, the expansion of China's fleet, the increasing competition for fish, deteriorating fish stocks, and the expected extension of jurisdiction by China and South Korea may combine to force the system towards multilateral regime formation.

Because of the continuing tension involving North Korea and because the Korean Peninsula borders all three Seas, the involvement of North Korea in these regimes at some stage is quite important. As North Korea ever so slowly improves its relations in the region, the chances for comprehensive maritime regime formation will improve. And with such a regime will come the building of confidence—so critical, yet still absent in the troubled waters of North-East Asia.

Appendix I
An Ocean Zoning System for the Yellow Sea

Ocean zoning is a method for delineating marine regions for appropriate sea uses. In general, ocean zones can either be natural regions or functional regions. A 'natural region' is 'a portion of the Earth's surface whose physical conditions are homogeneous.'[1] Natural regions can be defined using uniformity of sea surface temperature, salinity, oxygen concentration, nutrient concentrations, circulation patterns, and the presence of certain flora and fauna uniquely adapted to the environment. Sea use zones can also be functional marine regions, defined by the types and intensity of existing or potential sea uses. Such functional marine regions may include: shipping regions, fishery regions, mineral exploitation regions, defence regions, marine dumping regions, and marine scientific research regions. In this formulation, the establishment of a zoning system for sea use planning is based on an understanding of how and to what degree humans use and protect the ocean.

In the Yellow Sea, such a system should consider: (1) all possible sea uses in the Yellow Sea; (2) existing and/or potential areas for marine resource conservation; (3) special areas for marine environment and ecosystem protection; and (4) the interaction among different sea uses (see Figure 5.3).

The following ocean zones have been defined based on the present state of knowledge of the marine environment, resources, and economic development in the Yellow Sea.[2]

Fishery Zones

1. Protected spawning grounds for fully-exploited fish species (March–July);
2. Protected wintering ground for fishes (32°00′–34°00′N, and 124°00′–126°′00′E);
3. Protected zones for juvenile fishes:
   - Yanwei fishery ground juvenile fish protection zone (37°30′–38°30′N and 121°00′–122°00′–122°30′E);
   - Shidao and Lianqinshi fishery ground juvenile fish protection zone (34°00′–37°00′N and 122°00′–123°00′E);

4. Protected wintering ground for Chinese shrimp (33°00'–36°00'N and 122°00'–125°00'E);
5. Protected zones for juvenile Chinese shrimp (36°00'–37°30'N and 122°00–123°00'E);
6. Protected spawning zones for Southern Rough Shrimp (37°30'–37°50'N and 12°10'–122°30'E); and
7. Coastal and nearshore aquaculture zones:
   - aquaculture zone to the south of the Yalu River mouth (15 and 40 metres);
   - aquaculture zone to the east of the Changshan islands (20 metres);
   - Yanwei prawn propagation zone;
   - Jiangsu nearshore aquaculture zone (15 and 40 metres).

Zones for Marine Transportation

1. Traffic separation schemes (TSS):
   - suggested Qingdao TSS;
   - suggested Dalian TSS;
2. Oil tanker routes:
   Shipping routes of oil tankers from or to Dalian, Qinhuangdao, Huangdao (Qingdao), Lianyungang in China, and Inchon, Kunsan, and Pusan in South Korea.

Zones for Offshore Petroleum Resources

1. Suggested oil exploration zones:
   - the northeastern segment of NSB;
   - the No. 8 and No. 2 segments of NB;
   - the No. 4 segment of SSB;
   - the deposition centre of the 'Fu Nan Sha uplift';
2. Zones most prospective for petroleum resources:
   These include the north-eastern and northern segments of the NSB, segments 4, 7, and 5 of the SSB, and segments 8 and 2 of the NB.

Zones for Marine Dumping and Marine Environmental Protection

Dumping

As long as dumping does not conflict with other sea uses and causes no significant marine environmental impacts, it should be permitted, especially in areas with a high capacity for self-purification.

1. Existing zones for marine dumping:
   - Dalian Zhoushueizhi Airport aviation fuel release zone
   - nearshore Jiaozhou Bay marine dumping zone
   - Yantai port marine dumping zone for dredged material (38°15′N, 122°00′E; 38°40′N, 122°00′E; 38°15′N, 123°30′E; 38°40′N, 123°30′E).
2. Planned zone for marine dumping:
   A marine dumping area needs to be defined offshore Lianyungang;
3. Ecologically sensitive zones

Annexes I, II, and V of MARPOL 73/78 encourage countries to identify 'special areas' and/or 'ecologically-sensitive zones' for the protection of marine living resources and marine ecosystems. The Mediterranean, Baltic, Black, and Red Seas, and the Persian, Oman, and Aden Gulf areas have been designated as 'special areas' for this purpose. The North Sea countries and the United States have recently taken actions to designate portions of the North Sea and the Gulf of Mexico respectively, as Annex V special areas.[3] In the Yellow Sea, candidate areas are nearshore areas with a water depth of between 15 to 40–50 metres. In general, these are fishery spawning and breeding grounds and, as aquaculture sites, are very sensitive to marine pollutants. The prawn spawning and breeding grounds may also be defined as ecologically sensitive zones.

Marine Environmental Protection

Regions of Existing and Potential Conflicts

- Sea use conflicts may occur in the Northern Yellow Sea (38°00′–39°00′N and 124°00′–125°00′E) between offshore oil exploitation and nearshore aquaculture. Military uses may also complicate the issue. Since North Korea has few petroleum resources, offshore oil exploitation is of special concern to it. Therefore the north-eastern Yellow Sea should be reserved as a potential oil exploitation zone.
- The major conflict in the nearshore of Yantai and Chengshantou is between coastal and nearshore fishery resources/aquaculture and maritime transportation. The area is at the entrance to the Bo Hai, where marine traffic density is relatively high. It is also conducive to coastal and nearshore aquaculture, and marine tourism. Some parts are in a Chinese military warning zone. China has also specified emergency dumping areas there for aviation fuel from planes using

Dalian airport. There are also potential offshore oil resources in the region. Furthermore, the area lies in the path of surface oil slicks moving from the Bo Hai to the Yellow Sea, and there is already a small dumping area off Yantai. Juvenile fish and shrimp are threatened and need protection. Traffic separation schemes and deep water routes are possibly needed for oil tankers and other vessels departing or entering the Bo Hai.

- The area between Chengshantou and Qingdao has traditionally been an important fishery ground and is now an important aquaculture area. Many fish species spawn, breed, and migrate there. But surface oil pollution is already detectable.
- Qingdao and Shiqiusuo are developing rapidly and will experience more traffic in the near future. The area is frequently shrouded in fog, making it dangerous for mariners. Bottom sediments have high lead and zinc content. And there may be offshore petroleum.
- Aquaculture is projected in the nearshore areas to a depth of 40 metres. Potential conflicts exist with offshore petroleum exploitation, and maritime transportation. Two areas are defined as juvenile fish and hairtail protection zones, and one area is a nearshore aquaculture and juvenile fish protection zone. Two other areas are reserved for high density shipping.
- The southeastern part of the Yellow Sea is an important fishery and fish wintering ground. Winter fish catches should be limited or even suspended. The marine traffic density in the region is heavy. Because of its submarine topography, and bad weather, particularly typhoons, the potential for maritime casualties is high. Regulating the shipping routes between Cheju island and the south-west coast of the Korean peninsula could reduce the risk. Also South Korea is exploring for offshore oil and natural gas in the area.

Notes

1. Joseph Morgan, 'Marine regions and the Law of the Sea', *Ocean and Shoreline Management*, 15 (1991): 262–3.
2. Huang Yunlin, 'Sea Use Planning for the Yellow Sea', Ph.D. thesis, University of Hawaii, 1994, pp. 252–68.
3. Ibid., p. 243.

Appendix II

Specific Measures for Protection of Fishery Resources in the Yellow Sea[1]

The existing fishery treaties have three basic objectives: to limit the amount of catch, to protect certain fish species (shrimp, seabream, hairtail, yellow croakers, and mackerel), and to reduce the incidental catch of certain species. Though the existing treaties do not directly set quotas for the Yellow Sea, they do regulate overfishing of particular species in specific areas by restricting access, the number of fishing boats, the fish and mesh size, the percentage of incidental catch, and the amount of catch by size of vessel.

But the specific regulations were based primarily on the condition of the stocks in the 1950s and 1960s, and so protect species that were abundant then but are not the major species caught now. Moreover, since the 1970s, fishing in spawning, breeding, and wintering grounds has increased rapidly and overfishing has become a serious problem. Species which have good commercial value and have not yet been seriously stressed should be the most important targets of protection. They include Chinese shrimp, southern rough shrimp, Spanish mackerel, and silver pomfret. Provisions for their protection should include regulation of net type and mesh size, fishing periods, and protected zones.

Regulation of Fishing Nets

Trawlnets are commonly used in the Yellow Sea for prawn but they also take an incidental catch of juvenile fish. In the 1960s, the Chinese trawlnet fishery contributed about 60 per cent of total fishery catch in the Yellow Sea, and 40 per cent in the 1980s.[2] The use of trawlnets causes overfishing of demersal fishes but underfishing of upper and middle layer species. The breeding communities of hairtail, for example, are often damaged by trawlnets. Perhaps purse seiners should be used instead to protect juvenile fish, especially little yellow croaker, Chinese herring, hairtail, and Spanish mackerel, as well as demersal fish species. And fishing should only be allowed after the spawning of shrimp and major commercial fish.[3] Fishing nets fixed in shallow waters where most spawning and breeding grounds are located also catch juvenile fish and should be restricted. The most seriously-damaged fishing grounds in Chinese waters are Lusi, Yanwei, and Shidao.

Purse seines and drift nets are used mainly for catching fish such as chub mackerel, Spanish mackerel, and Pacific herring in the upper and middle layers. In recent years, mesh size has become smaller as the quantity and density of fishery resources decline. The mesh size of drift nets to catch Spanish mackerel in the 1960s was 10.0 to 10.67 centimetres. In the 1970s, it was reduced to 9.0 to 9.33 centimetres, and in recent years, it has decreased to 7.67 to 8 centimetres. Consequently, sexually-immature Spanish mackerel and one-year-old chub mackerel are being caught.

Due to the United Nations ban on driftnets, only purse seines should be used and their mesh size regulated, thereby establishing the minimum width of fish which can be caught (width is related to length). According to the China/Japan fisheries treaty, the minimum mesh size (inside measure) of trawlnets should be 54 millimetres for yellow croaker and hairtail. For purse seines, the minimum mesh size is 35 millimetres. The minimum permitted lengths of yellow croaker and hairtail are 19 centimetres and 23 centimetres respectively. And the size limits for fork lengths of mackerel, jack mackerel, and marjali are 22 centimetres, 20 centimetres, and 18 centimetres respectively.

Regulation of Fishing Periods

To protect fishery resources, it is also necessary to establish fishing periods for major fishery grounds, especially during spring when many species spawn. The suggested fishing periods for drift net catch in some fishing grounds in Chinese waters are:[4] Dasha, and Lianqingshi fishery grounds, April 20–May 10; Shidao fishery ground, May 1–20; and Yanwei fishery ground, May 10–25. Although China closed the Lusi fishery ground for six years in 1981 to protect the little yellow croaker, South Korean fishermen negated China's conservation efforts by continuing to target this species in its wintering ground.

Suggested Protected/Prohibited Fishery Zones

These include protected spawning zones, wintering zones, and juvenile fish and shrimp zones (see Figure 5.3 and Table II.1).

Table II.1 The Proposed Fisheries Protection Zones

Name of Zone	Latitudes and Longitudes	Times	Policies for Protection
Protected Spawning Grounds for Fully Exploited Species	Chinese and Korean coasts	March–July	1. Set quotas 2. Close March–July 3. Control pollution 4. Delay the start of the fishing season
Protected Zones for Juvenile Chinese Shrimp	36°00'–37°30'N × 122°–123°E	July–September	1. Control fishing activities 2. Control pollution
Protected Wintering Zone for Fish	32°00'–34°00'N × 124°–126°E	November–March	Prohibit winter catch
Protected Wintering Zone of Chinese Shrimp	33°00'–36°00'N × 122°–125°E	December–March	Prohibit winter catch
Protected Zones for Juvenile Fish	37°30'–38°30'N × 121°–122°30'E; 34°00'–37°00'N × 122°–123°E	August–November	1. Catch of Spanish mackerel 2. Incidental catch <25%
Protected Zone for Juvenile Southern Rough Shrimp	37°30'–37°50'N × 121°10'–122°30'E	June–September	1. Prohibit fishing 2. Maximum 500 two-boat trawlers in spring

Protected Spawning Grounds for Fully Exploited Species

The spawning grounds of fully-exploited species along the coastline of China and in the Korean coastal and nearshore areas (offshore Jianghua island and offshore the south-west coast of the Korean Peninsula) should be protected immediately. Generally, there are three ways to protect spawning grounds: setting quotas or requiring certificates for catches in seriously-stressed spawning grounds; closing spawning grounds during certain periods; and reducing and/or controlling marine pollution in spawning grounds.

Protected Zones for Juvenile Chinese Shrimp

Because the eggs and juveniles of Chinese shrimp are vulnerable to pollution, regulations to control pollution in Chinese shrimp spawning grounds are necessary. A protected zone for juvenile Chinese shrimp is suggested for the area 36°00′–37°00′N and 122°00′–125°00′E. Catch of juvenile Chinese shrimp in this area should be prohibited from July to September, and marine pollution from ships prevented.

Protected Wintering Grounds for Fish

Overfishing on the wintering grounds causes serious problems for certain species, especially little yellow croaker and large yellow croaker. There are four major fish wintering areas in the Yellow Sea: between 35°00′–37°30′N and 122°00′–124°30′E (flounders, perch, Pacific cod, and red seabream); between 34°00′–36°00′N and 122°30′N–124°30′E (sillago, eel-pout, jewfish, and Pacific herring); between 33°00′–35°00′N and 122°00′–125°00′E (flounders and red seabream); and between 32°00′–34°00′N and 124°00′–126°00′E (hairtail and other fishery species). The last area is the most important wintering ground in the Yellow Sea. It should be defined as a protected fish wintering zone from November to March.

Protected Wintering Grounds for Chinese Shrimp

The present Chinese shrimp catch in winter of 2,500 tons is too high and leads to decreased recruitment.[5] A protected wintering ground for Chinese shrimp is suggested for the area 33°00′–36°00′N and 122°00′–125°00′E from December to March.

Protected Zone for Juvenile Fish

A juvenile fish protection zone for hairtail was promulgated by the Chinese government on 28 April 1981. It covers the area bounded by: 34°N, 121°23′E; 34°N, 121°53′E; 31°30′N, 123°27′E; and 37°30′N, 122°57′E. Trawlers are prohibited from entering this area from August to October.[6] The prohibition should be extended to include all motorized fishing boats in the area between 33°N to 37°N up to a depth of 40 metres. Motorized fishing boats should be prohibited from catching hairtail between July and November.[7]

Other protected areas should be offshore Yantai between 37°30′–38°30′N and 121°00′–122°30′E, and Shidao and Lianqinshi fishery grounds between 34°00′–37°00′N and 122°00′–123°00′E. The whole Bo Hai also needs to be considered a protected juvenile fish zone.[8] These zones will protect not only seriously-damaged stocks of hairtail, red seabream, and yellow croaker, but also Spanish mackerel, which is one of the major present resources in the Yellow Sea. The catch of juvenile Spanish mackerel in these protection areas should be prohibited from August to November. And the incidental catch of juvenile fish should not exceed 25 per cent of the total catch.

Protected Zones for Juvenile Southern Rough Shrimp

A maximum of 2,000 tons should be set for southern rough shrimp in winter, including the catch on the wintering ground. And a protected area for the species should be established in the area 37°30′–37°50′N and 121°10′–122°30′E. From June 1 to September 10, fishery activities should be prohibited in this area, and in spring only 500 two-boat trawlers should be allowed.[9]

New Approaches to Increasing Fishery Production

There are two possible approaches to increasing fishery production. One is development of mariculture, including 'ocean ranching' through the construction of artificial reefs and artificial propagation. The other is the development of new catch technology for species such as anchovy, which is still available and abundant in certain areas of the Yellow Sea.

Exploitation of New Fishery Resources

The anchovy is broadly distributed in the Yellow and East China Seas and found throughout the year from spring to winter. Total resources of anchovy in the two seas are estimated at about 3 million tons, of which 500,000 tons can be caught on a sustainable basis.[10] Because of technical obstacles in both catch and fish preservation, anchovy production is still low.

Nor has the sardine fishery in the Yellow Sea been fully developed. It is estimated that offshore Shandong province there are 7,290 tons of sardine, but only half this amount is caught. Although the sardine has long been an object of the Japanese fishery, potential to increase its catch still exists.[11]

To utilize these resources fully, it is necessary to know more about their spatial distribution, and migratory period and routes, and to develop new fishing gear.

Notes

1. This section is based on Huang Yunlin, 'Sea Use Planning: The Case of the Yellow Sea', Ph.D. thesis, University of Hawaii, 1993.

2. Editorial Group for Marine Fishery Resources of China, *Marine Fishery Resources of China 1990*, Hongzhou Zhejiang Science and Technology Press, p. 173.

3. Tang Qisheng and Yi Maozhong, 'Development and Protection of the Nearshore Fishery Resources of Shandong Province', Beijing: Agricultural Press, 1990, p. 150; Bureau of Aquaculture, Ministry of Agriculture, *Investigation and Regionalization for the Fishery Resources in the Bo Hai Sea and the Yellow Sea*, Beijing: Ocean Press, 1990, pp. 277–8.

4. Tang and Yi, *Development and Protection of the Nearshore Fishery Resources of Shandong Province*, p. 166.

5. Ibid., p. 176.

6. Jiang Tieming, *A study on China's marine regional economy*, Beijing: Ocean Press, 1990, p. 141.

7. Tang and Yi, *Development and Protection of the Nearshore Fishery Resources of Shandong Province*, p. 150.

8. Ibid., p. 166.

9. Bureau of Aquaculture, Ministry of Agriculture, *Investigation and Regionalization for the Fishery Resources in the Bo Hai Sea and the Yellow Sea*, Beijing: Ocean Press, 1990, p. 124.

10. Tang and Yi, *Development and Protection of the Nearshore Fishery Resources of Shandong Province*, p. 206.

11. Bureau of Aquaculture, *Investigation and Regionalization for the Fishery Resources in the Bo Hai Sea and the Yellow Sea*, p. 285.

Glossary

ACOPS	Advisory Commission on the Protection of the Sea
AIT	Asian Institute of Technology
APEC	Asia-Pacific Economic Co-operation Forum
ARF	ASEAN Regional Forum
ASEAN	Association of South-East Asian Nations
CFM	Committee on Fishery Management
CITES	Convention on International Trade in Endangered Species of Wild Fauna and Flora
COBSEA	Co-ordinating Body on the Seas of East Asia
COD	Chemical Oxygen Demand
CSCAP	Council for Security Co-operation in the Asia-Pacific
CSIS	Center for Strategic and International Studies
EEZ	Exclusive Economic Zone
EPB	Economic Planning Board
ESCAP	Economic and Social Commission for Asia and the Pacific
FAO	Food and Agriculture Organization of the United Nations
GEMS	Global Environmental Monitoring System
GEMSI	Group of Experts on Methods, Standards, and Intercalibration
GIPME	Global Investigation of Pollution in the Marine Environment
GNP	Gross National Product
IAEA	International Atomic Energy Agency
ICLARM	International Centre for Living Aquatic Resources Management
IIASA	International Insititute of Applied Systems Analysis
IMO	International Maritime Organization
IOC	Intergovernmental Oceanographic Commission
IOC-WESTPAC	IOC Subcommission for the Western Pacific
IOI	International Ocean Institute
ITQ	Individual Transfer Quota
IUCN	International Union for the Conservation of Nature
IWC	International Whaling Commission
JECSS	Japan-East China Sea Surveys
JFA	Japan Fisheries Association
JICA	Japan International Co-operation Agency

JV	Joint Venture
LBMP	Land Based Marine Pollution
LCD	Lesser Developed Countries
LDC	London Dumping Convention
MAB	Man and the Biosphere
MAFF	Ministry of Agriculture, Fisheries and Food
MAP	Mediterranean Action Plan
MARPOL	Maritime Pollution Convention (1978)
MARPOLMON	Marine Pollution Monitoring Programme
MMT	Million Metric Tons
MOE	Ministry of the Environment
MOF	Ministry of Finance
MOFA	Ministry of Foreign Affairs
MOHA	Ministry of Home Affairs
MOTIE	Ministry of Trade, Industry, and Energy
MPPA	Marine Pollution Prevention Act
SSB	Fleet Ballistic Submarine (Diesel)
NET	Natural Economic Territories
NGO	Non-government Organization
NOWPAP	North-West Pacific Region Action Plan
NSB	National Science Board
OAS	Organization of American States
OAU	Organization of African Unity
OCA/PAC	Oceans and Coastal Areas Programme/Pacific
ODA	Overseas Development Assistance
OECD	Organization for Economic Co-operation and Development
PAP	Priority Action Programme
PCB	Polychlorinated Biphenyl
PECC	Pacific Economic Co-operation Council
PICES	North Pacific Marine Science Organization
RCU	Regional Co-ordinating Unit
RFE	Russian Far East
SEAFDEC	South-East Asia Fisheries Development Council
SEAPOL	South-East Asia Program on Ocean Law, Policy, and Management
TAC	Total Allowable Catch
TBT	Tributyl Tin
TIAS	Treaties and Other International Acts Series
TINRO	Pacific Ocean Fishery and Oceanography Research Institute
TRADP	Tumen River Area Development Project
TSS	Traffic Separation Scheme
UN	United Nations

GLOSSARY

UNCED	UN Conference on the Environment and Development
UNCLOS	United Nations Convention on the Law of the Sea
UNDP/GEF	UN Development Programme/Global Environment Facility
UNEP	UN Environment Programme
UNEP/NOWPAP	UNEP North-West Pacific Region Action Plan
UNESCO	UN Education, Scientific, and Cultural Organization
UNITAR	United Nations Institute for Training and Research
WGIMP	Working Group for an Integrated Marine Policy
WHO	World Health Organization
WWF	World Wildlife Fund
YSLME	Yellow Sea Large Marine Ecosystems

Index

ACTION PLANS, 207–9
Advisory Committee on Protection of the Sea (ACOPS), 218–19
Antarctica, 32
Arctic, 55, 156
Asia Foundation, 25, 204
Asian Development Bank, 81, 216
Asia-Pacific Economic Co-operation Forum (APEC), 8–9, 205–6, 272
Association of South-East Asian Nations (ASEAN), 9, 56
Australia, 5, 216, 281, 285

BALTIC SEA, 51–2, 158–9, 286
Barcelona Convention for the Protection of the Mediterranean Sea Against Pollution (1976), 52
Bilateralism, 6, 108, 116, 118
Bo Hai, 75, 176, 177, 249, 250
Boundary agreements, 85–6

CANADA, 285, 287
Caribbean Sea, 32
Chile, 287
China: and environmental protection, 103, 219, 221; and Exclusive Economic Zones, 47; fisheries, 97–100, 294; fisheries agreements, 257–8, 259, 260; jurisdictional claims, 83; and Law of the Sea Convention, 89, 278; and marine policy regimes, 90, 92; and marine pollution, 180, 188; and maritime-regime participation, 102–4, 105, 304–5; and North-West Pacific Region Action Plan, 210, 211, 213, 214; relations with Japan, 3, 5, 25, 85, 115, 257–8; relations with North Korea, 25; relations with Russia, 4–5; relations with South Korea, 25, 139–40, 143–4, 265–6; relations with Taiwan, 90, 96–105, 304; reunification of, 101; territorial disputes, 90
China National Offshore Oil Corp., 100
Chung Won Shik, 142, 237–8
Clinton, William J., 10

Colombia, 285, 288
Compliance, 29–30, 32
Convention on the Continental Shelf (1958), 83
Convention on International Trade in Endangered Species of Wild Fauna and Flora (CITES), 32
Council for Security Cooperation in the Asia Pacific (CSCAP), 9

DAEWOO, 122–3
Democratic People's Republic of Korea, see North Korea
Diaoyutai, 85, 91
Dominican Republic, 285, 288

EAST CHINA SEA: environmental protection regimes in, 230–2; fisheries problems in, 248–50; natural characteristics of, 74–80; petroleum in, 90; pollution in, 175–8
East Korean Current, 72
East Sea, see Japan, Sea of
ECO ASIA 94, 203–4
Economic relations, 2
Economic and Social Commission for Asia and the Pacific (ESCAP), 106, 201, 202–3
Ecuador, 287
Effluent standards, 195–6
Enclosed seas, 88–9
Environmental Program for the Mediterranean, 52
Environmental protection: in China, 221; and economic development, 197–8; ideal regimes for, 229–39, 307–8; initiatives for, 206–7; institutional obstacles to, 197; in Japan, 111–15, 221–2; in North-East Asia, 178–83; in North Korea, 127–32; programmes for, 25, 219–23; regional initiatives for, 198–206; in Russia, 153–61, 162–3, 222; in South Korea, 134, 141–4, 177, 222; transnational approaches to, 223–9; and Tumen River Area

Development Programme, 204–5; in Yellow Sea, 223–9
Environmental regimes, 28–36; *see also* Regimes
European Economic Community, 285, 286, 288
Exclusive Economic Zones (EEZ): and China, 47, 86; created, 46; and Japan, 47, 107, 109; and Law of the Sea Convention, 91; and North Korea, 83, 85, 125; and Russia, 83, 85; and South Korea, 47, 86, 139; and Taiwan, 83; in Yellow Sea, 278–9

FISHERIES: agreements on, 25, 248, 252–62, 285–92, 289; allocation of stocks, 280–2, 288–9; alternative regimes for, 270–7; in Bo Hai, 249, 250; in East China Sea, 248–50; global conditions of, 244–5; ideal management of, 275–7; in Japan, 107–11; joint development of, 285–92; and Law of the Sea Convention, 246–7; management problems of, 262–70; model regimes for, 292–6; in North-East Asia, 245–8; in North Korea, 124–5; regional management of, 282–5, 308–9; research requirements of, 271–3; in Russia, 145–53, 161–2; in Sea of Japan, 73–4, 251–2; in Sea of Okhotsk, 251; in South Korea, 137–42; in Yellow Sea, 79–80, 226–7, *231*, 248–50
Food and Agriculture Organization (FAO), 50, 81, *283*
France, 54
Fujian province (China), 98, 100

GAS, 85
Gdansk Convention, 286, 287
Global Investigation of Pollution in the Marine Environment (GIPME), 189
Global Legislators' Organization for a Balanced Environment, 205
Gorbachev, Mikhail, 158

HAN SUNG-JO, 142
Harmonization of policies, 190–1, 275
Hata Tsutomu, 114, 184
Hironaka Wakako, 113–4
Hosokawa Morihiro, 106, 113, 116, 266

ICELAND, 286
Incidents at sea, 4, 267–70
Indonesia, 46

Indo-Pacific Fisheries Council (IPFC), 50
Intergovernmental Oceanographic Commission (IOC): created, 217; Group of Experts on Methods, Standards, and Intercalibration (GEMSI), 189; and regional marine issues, 50; Subcommission for the Western Pacific, 25, 81, 131, 217–18, 219
International Atomic Energy Agency (IAEA), 183–4
International Convention for the Prevention of Pollution from Ships, 188
International Maritime Organization, 81, 186–8
International Union for the Conservation of Nature (IUCN), 32
Italy, 54

JAPAN: environmental protection in, 111–15, 221–2; and Exclusive Economic Zones, 47, 107, 109; fisheries in, 107–11, 294; fisheries agreements, 25, 252–8, 258, *259*, 261–2, 285; and industrial-waste disposal, 188–9; and joint ventures, 260–1; jurisdictional claims, 83; and Law of the Sea Convention, 46, 89, 278; and marine policy regimes, 90, 92; and marine pollution, 180; and maritime-regime participation, 105–21, 305; non–governmental organizations in, 119; and North-West Pacific Region Action Plan, 209, 210, 211, 213, 214, 215; and nuclear-waste dumping, 130, 183–6; and pollution treaties, 186–8; relations with China, 3, 5, 25, 85, 115, 257–8; relations with North Korea, 25, 85, 253–4; relations with Russia, 4, 5, 25, 85, 115, 116, 183–6, 252–3, 260–1, 261–2, 290; relations with South Korea, 4, 5, 25, 85, 106, 114–15, 238, 254–7, 261–2, 265–7, 285; relations with Taiwan, 106; relations with United States, 115, 119
Japan, Sea of: environmental protection regimes in, 232–3, 234, 307–8; fisheries in, 73–4, 248; fisheries problems in, 251–2; natural characteristics of, 68–74; nuclear-waste dumping in, 111, 115, 130, 142–3; ocean currents in, 68–72; pollution in, 68, 72, 73, 111, 178, 306
Japan–East China Sea Surveys (JECSS), 81

Japan Fisheries Association (JFA), 118
Joint development, 285–92
Joint ventures, 147, 150–1, 260–1
Jurisdictional claims, 83–86, *231*

KIM WOO CHOONG, 122–3
Kim Young Sam, 106, 135–6, 144, 266
Korea, *see* North Korea, South Korea
Korea Straits, 80
Kurile Islands, 85, 91, 115, 116, 290
Kuroshio, 68, 72

LAGGARDS, 30
Law of the Sea, United Nations Convention on (1982), 11, 12, 32, 46; and China, 89; effects of, 86; and enclosed seas, 88–9; and environmental-protection regimes, 191; and fisheries, 139, 246–7; and harmonization of policies, 190–1; International Seabed Authority, 40; and Japan, 89; and marine regimes, 51; and marine research, 271–2; obligations under, 189–90, 191; and ocean management, 88–9, 91; and pollution control, 226; and regionalism, 89–90; and Russia, 89; and South Korea, 89, 139; and United States, 89; and Yellow Sea, 278–9
Lee Teng-hui, 96
London Dumping Convention, 127, 186, 188–9

MALAYSIA, 285
Management regions, 48
Marine regimes: advantages of participation in, 44–5; and Chinese reunification, 96; defined, 39, 302; and developing countries, 42–4; development of, 51; disadvantages of participation in, 44–5; and dis-integrative forces, 90–2; functions of, 40; and harmonization of laws and policies, 42; and informal contacts, 42; and integrative forces, 86–90; joint activities in, 41; model, 303; and North Korea, 90; objectives of, 39–40, 42–4; and regional organizations, 41–2; structure of, 39, 40–1; and treaties, 42
Mediterranean Centre for Research and Development in Marine Industrial Technology, 52
Mediterranean Sea, 32, 50, 52–4
Meeting of Senior Officials on Environment Co-operation in North-East Asia, 201–3
Military zones, 86
Mining of seabed, 46
Miyazawa Kiichi, 106
Model regimes, 229–39, 292–6, 303
Montreal Guidelines, 190
Montreal Protocol on Substances that Deplete the Ozone Layer, 156
Multilateralism: advantages of, 7–8; and big powers, 90; and Japan, 116, 117; and maritime regimes, 92, 306–7; and US policy, 89; versus bilateralism, 19
Murayama Tomoiichi, 114
Muto Kabun, 184

NATIONAL FEDERATION OF FISHERIES CO-OPERATIVES, 118
Nationalism, 3, 97
Natural economic territories (NET), 2, 144
Neorealism, 6–7
Nogami Yoshiji, 116
Non-governmental organizations (NGOs), 32, 115, 119, 160–1, 204
North-East Asia Economic Forum, 123
North-East Asian Conference on the Environment, 198–201
North-East Asian Cooperation Dialogue, 5
North-East Asian Environment Programme, 25, 131
North-East Asia and North Pacific Environmental Forum, 204
North-East Asia Security Dialogue, 9
North-East Atlantic Fisheries Convention, 289
Northern Territories, 85, 91, 92, 116
North Korea (Democratic People's Republic of Korea): and East Asian Seas Marine Pollution Program, 217; economic development of, 122–3; environmental protection in, 127–32; and environmental treaties, 128–9; and Exclusive Economic Zones, 125; fisheries in, 124–5, 290–2, 294; and fisheries agreements, 253–4, 258, 260, 310; and jurisdictional claims, 83; and Law of the Sea Convention, 278; and marine policy regimes, 90, 92; and maritime-regime participation, 305; and North-West Pacific Region Action Plan, 210, 211, 213, 214; pollution in, 126–32; and pollution treaties, 188;

and regional-regime participation, 121–32; and regional relations, 4; relations with China, 25; relations with Japan, 25, 85, 122; relations with Russia, 25, 85, 258; relations with South Korea, 122, 290–2; relations with United States, 122; and UNCED, 127
North Korean Cold Current, 73
North Pacific Anadromous Fish Commission, 253
North Pacific Cooperative Security Dialogue, 5
North Pacific Marine Science Organization (PICES), 81, 218, 271, 272
North Sea, 54–5, 56
North-West Atlantic Fisheries Convention, 289
North-West Pacific Region Action Plan, 25, 81, 106, 130, 144, 207–16; and China, 210, 211; geographic definition of, 210, 213; goals of, 211–12; and Japan, 209, 210, 211; and North Korea, 210, 211; and Russia, 210, 211; and South Korea, 210, 211
Norway, 286, 287, 288
Nuclear waste: in Arctic Ocean, 156; Russian dumping of, 156–7; in Sea of Japan, 111, 130, 142–3, 183–6

OCEAN CURRENTS: East Korean, 72; Kuroshio, 68, 72; North Korean Cold Current, 73; Primorye, 72, 74; Southern Liaoning Coastal Current, 78; Tsushima, 72, 73, 74; in Yellow Sea, 76, 77
Ocean management, 12
Ocean zones, 311–14
Oil spills, 223, *224–5*
Okhotsk, Sea of, 251
Organizations, 30–6
Overseas Petroleum Investment Co., 100

PACIFIC ECONOMIC CO-OPERATION (PECC), 8–9, 272
Pacific Ocean Fishery and Oceanography Research Institute (TINRO), 146, 149–50
Papua New Guinea, 285
Permanent Commission of the South Pacific, 287
Persian Gulf; pollution-control regimes in, 32

Peru, 287
Petroleum exploration, 85, 90, 100, 114
Philippines, 46, 285
Pollution: in Bo Hai, 176, 177; in China, 104; in East China Sea, 175–8; havens, 181–2; in North Korea, 126–32; in Sea of Japan, 68, 72, 73, 111, 178, 306; in South Korea, 141–3, 176; in Yangtze River, 176; in Yellow River, 176; in Yellow Sea, 76–8, 143, 175–8, 306
Pollution-control conventions, 186–8; *see also* London Dumping Convention
Primorye, 73
Primorye Current, 72, 74
Program on Prevention and Management of Marine Pollution in East Asian Seas, 25

RAMSAR CONVENTION, 112
Regimes: anarchical, 31–3; cognitive theories of, 23–4; and compliance, 29–30, 32; defined, 17–18, 301; demand for, 24–6; designing, 26–8, 38; development of, 19–24, 38–9; formation of, 37–8; functional theory of, 20–3; functions of, 18–19, 301–2; game theory of, 23; model fisheries, 292–6; model maritime, 58–61, 303; and organizations, 30–6; *see also* Marine regimes; social consequences of, 36–7; structural theory of, 20
Regionalism: in Arctic, 55; in Baltic Sea, 51–2; defined, 47–8; development of, 6–12; dis–integrative forces in, 44, 304; integrative forces in, 44, 304; and Law of the Sea Convention, 89–90; marine, 48–9, 51–8, 55–8, 86–92, 303; and maritime issues, 1; in Mediterranean Sea, 52–4; in North Sea, 54–5, 56; soft, 2
Republic of Korea, *see* South Korea
Resource extraction havens, 181–2
Resource regimes, 28–36; *see also* Regimes
Rhee Line, 266, 267
Roh Tae Woo, 106, 132
Russia: and Baltic Sea, 158–9; and environmental agreements, 158–9; environmental protection in, 153–61, 162–3, 222; fisheries in, 145–53, 294; and fisheries agreements, 151–2, 252–3, 258, 260, 261–2, 286, 287; and joint ventures, 260–1; and jurisdictional

claims, 83; and Law of the Sea Convention, 89; and marine pollution, 180; and maritime-regime participation, 306; non-governmental organizations in, 160–1; and North-West Pacific Region Action Plan, 210, 211, 213, 214; and nuclear-waste dumping, 115, 130, 156–7, 183–6; and pollution treaties, 186–8; relations with China, 4–5; relations with Japan, 4, 5, 25, 85, 115, 116, 183–6, 252–3, 260–1, 261–2, 290; relations with North Korea, 25, 85, 258; relations with South Korea, 4, 159, 260; relations with Taiwan, 25, 260; relations with United States, 5; resource management in, 144–5
Russian Far East, 90, 144–5, 264
Russian Federation Committee on Fishery Management, 146
Ryukyu archipelago, 80

SAKASHIMA GUNTO, 80
Sea-use planning, 227–9
Second International Symposium on Environmental Co-operation (1993), 204
Security: comprehensive, 9–10; co-operative, 10–11; discussions of, 5–6, 9; and maritime issues, 11–12
Senkaku Islands, 85, 91
South China Sea, 90, 92
Southern Liaoning Coastal Current, 78
South Korea (Republic of Korea): environmental protection in, 134, 141–4, 177, 219, 222, 238; and Exclusive Economic Zones, 47; fisheries in, 137–42, 290–2, 294; and fisheries agreements, 254–7, 258, 260, 261–2, 265–7, 285; and jurisdictional claims, 83; and Law of the Sea Convention, 46, 89, 139, 278; marine policy development in, 132–7; and marine pollution, 180; and maritime-regime participation, 305–6; and North-West Pacific Region Action Plan, 210, 211, 214, 215; and nuclear–waste dumping, 142–3; pollution in, 141–3, 176; and pollution treaties, 188; relations with China, 25, 143–4, 265–6; relations with Japan, 4, 5, 25, 85, 114–15, 139–40, 238, 254–7, 261–2, 265–7, 285; relations with North Korea, 290–2; relations with Russia, 159, 260; and Yellow Sea, 143

South Pacific Forum Fisheries Agency, 40
Soviet Union, see Russia

TAIWAN: fisheries in, 98–100, 294; and fisheries agreements, 260; and jurisdictional claims, 83–85; and Law of the Sea Convention, 104; and marine pollution, 180; and marine resources, 80; and maritime-regime participation, 104–5, 304–5; relations with China, 90, 96–105, 304; relations with Russia, 25, 260; and territorial disputes, 90; trade with China, 97
Takeshima, 85, 91
Territorial disputes, 85, 310; China, 90; Diaoyutai, 91, 102; Kurile Islands, 91, 115, 116, 152–3; Northern Territories, 91, 116, 152–3; Senkaku Islands, 91, 102; Taiwan, 90; Takeshima, 91; Tok Do, 91
Tok Do, 85, 91
Tokyo Electric Power Co., 183
Torres Strait, 285
Trade, 97
Training, 274
Tsushima Current, 72, 73, 74
Tumen River Area Development Program, 123, 204–5

UNITED NATIONS CONFERENCE ON ENVIRONMENT AND DEVELOPMENT (UNCED), 47, 127
United Nations Development Programme/Global Environmental Facility (UNDP/GEF), 25, 130–1, 216–17, 219
United Nations Environment Programme (UNEP), 32, 50, 201; Mediterranean Action Plan, 50, 52–4; Oceans and Coastal Areas Programme, 209–10; Regional Seas Programme, 207, 215–16; see also North-West Pacific Region Action Plan
United States: and fisheries agreements, 285, 287, 288; and Law of the Sea Convention, 46, 89; and North-East Asia, 7; relations with Japan, 115, 119; relations with Russia, 5; security policies of, 7, 10, 89
USSR, see Russia

VIETNAM, 46

Water-quality standards, 191, 192–4, 223
Western Pacific Naval Symposium, 5
WESTPAC, *see* Intergovernmental Oceanographic Commission, Subcommission for the Western Pacific
World Wide Fund for Nature(WWF), 32, 160

YAMATO BANK, 85
Yangtze River, 176
Yellow River, 176
Yellow Sea: alternative management approaches for, 279–80; destabilizing factors in, 277; environmental protection in, 223–9, 230–2, 233–4; fauna in, 79–80; fisheries in, 79–80, 226–7, *231*, 315–20; fisheries problems in, 248–50; jurisdictional claims in, *231*; and Law of the Sea Convention, 278–9; natural characteristics of, 74–80; ocean currents in, *77*; and ocean zoning, 311–14; oil spills in, 223, *224–5*; pollution in, 76–8, 143, 175–8, 223, 306; sea-use planning for, 227–9
Yellow Sea Cold Current, 76
Yellow Sea Large Marine Ecosystem, 219
Yellow Sea Warm Current, 76
Yeltsin, Boris, 146, 156, 184
Young, Oran R., 26

ZHIRINOVSKY, VLADIMIR, 156